Advances in Experimental Medicine and Biology

Volume 939

More information about this series at http://www.springer.com/series/5584

Bairong Shen • Haixu Tang • Xiaoqian Jiang
Editors

Translational Biomedical Informatics

A Precision Medicine Perspective

Springer

Editors
Bairong Shen
Center for Systems Biology
Soochow University
Jiangsu, China

Haixu Tang
School of Informatics and Computing
Indiana University
Bloomington, IA, USA

Xiaoqian Jiang
Department of Biomedical Informatics
University of California San Diego
La Jolla, CA, USA

ISSN 0065-2598 ISSN 2214-8019 (electronic)
Advances in Experimental Medicine and Biology
ISBN 978-981-10-1502-1 ISBN 978-981-10-1503-8 (eBook)
DOI 10.1007/978-981-10-1503-8

Library of Congress Control Number: 2016957788

Printed on acid-free paper

This Springer imprint is published by Springer Nature
The registered company is Springer Science+Business Media Singapore Pte Ltd.

Contents

Chapter 1
NGS for Sequence Variants

Shaolei Teng

Abstract Recent technological advances in next-generation sequencing (NGS) provide unprecedented power to sequence personal genomes, characterize genomic landscapes, and detect a large number of sequence variants. The discovery of disease-causing variants in patients' genomes has dramatically changed our perspective on precision medicine. This chapter provides an overview of sequence variant detection and analysis in NGS study. We outline the general methods for identifying different types of sequence variants from NGS data. We summarize the common approaches for analyzing and visualizing casual variants associated with complex diseases on precision medicine informatics.

Keywords Sequence variants • Next-generation sequencing • Sequence alignment • Variant calling • Association testing • Visualization • Precision medicine informatics

1.1 Introduction

Over the last decade, next-generation sequencing (NGS) has dramatically changed the precision medicine field by characterizing patients' genomic landscapes and identifying the casual variants associated with human diseases. The Sanger-based sequencing [48] ("first-generation sequencing") was used to sequence the first human reference genome for the Human Genome Project [3], which took 13 years to finish the draft genome at a total cost of $3 billion. NGS technologies make the sequencing at remarkable price and unprecedented speed by carrying out hundreds of millions of sequencing reactions at once [52, 57]. With the revolutionary technology, we can sequence thousands of genomes in just 1 month, address the biological questions at a large scale, identify the genetic risk factors for human diseases, and provide a more precise way to health care [24]. In particular, NGS can be used to detect a large number of sequence variants in the patients' genomes and identify the casual variants associated with human diseases, which has dramatically

S. Teng (✉)
Department of Biology, Howard University, Washington, DC 20059, USA
e-mail: shaolei.teng@howard.edu

B. Shen et al. (eds.), *Translational Biomedical Informatics*, Advances in Experimental Medicine and Biology 939, DOI 10.1007/978-981-10-1503-8_1

changed our perspective on genetic variants, human diseases, and precision medicine.

Discovery of casual sequence variants associated with certain traits or diseases has become a fundamental aim of genetics and biomedical research. The sequence variants can be classified to single nucleotide variants (SNVs), small insertions and deletions (INDELs), and large structural variants (SVs) based on their sequences in length. SNVs, the most common type of sequence variants, are single DNA base-pair differences in individuals. INDELs are defined as small DNA polymorphisms including both insertions and deletions ranging from 1 to 50 bp in length. SVs are large genomic alterations (>50 bp) including unbalanced variants (deletions, insertions, or duplications) and balanced changes (translocations and inversions). Copy number variants (CNVs), a large category of unbalanced SVs, are DNA alterations that result in the abnormal number of copies of particular DNA segments. Somatic mutations are tumor-specific variants in cancer–normal sample pairs. The different types of sequence variants play important roles in the development of human complex diseases. For example, the SNVs associated with major depression were found in the genes encoding serotonin transporter, serotonin receptor, catechol-o-methyltransferase, tryptophan hydroxylase, and tyrosine hydroxylase [29]. These sequence variants can influence the neurotransmitter functions in multiple ways including changing gene expression level, altering substrate binding affinity, or affecting transport kinetics [19]. A balanced t(1;11) (q42.1;q14.3) translocation in disrupted in schizophrenia 1 (*DISC1*) gene was discovered in a large Scottish family highly burdened for severe mental illnesses, and the family members with the translocation showed a reduced P300 event-related potential associated with schizophrenia [9]. Identifying the casual variants and their clinical effects provides important insight to understand the roles of sequence variants in the causation of human diseases.

Discovery of disease-causing variants from a large number of sequence polymorphisms detected from NGS data is a major challenge in precision medicine. Bioinformatics and statistical methods have been developed for detecting sequence variants and identifying disease-related casual variants. The schematic diagram of NGS variant analysis on precision medicine informatics is shown in Fig. 1.1. The DNA samples are extracted from patients (or normal individuals) and sequenced on NGS platforms. The billions of short sequence reads are produced by the sequencers, and sequence information is stored in FASTQ format files. From here, NGS variant analysis falls into two major frameworks. The first framework is the variant detection. The high-quality sequence reads passed quality control (QC) filters are aligned to a reference genome, and the sequence alignment data is deposited in SAM/BAM format files. Several variant detection tools are used to call small variants including SNVs and INDELs. The somatic mutation callers are applied to tumor–normal patient samples. Multiple SV callers are developed to detect large structural variants. The variants called from these tools can be stored in Variant Call Format (VCF) files or BED format files. The next framework is the variant analysis. The annotation tools are used to predict the functional effects of coding and regulatory variants. The association analysis can identify the common

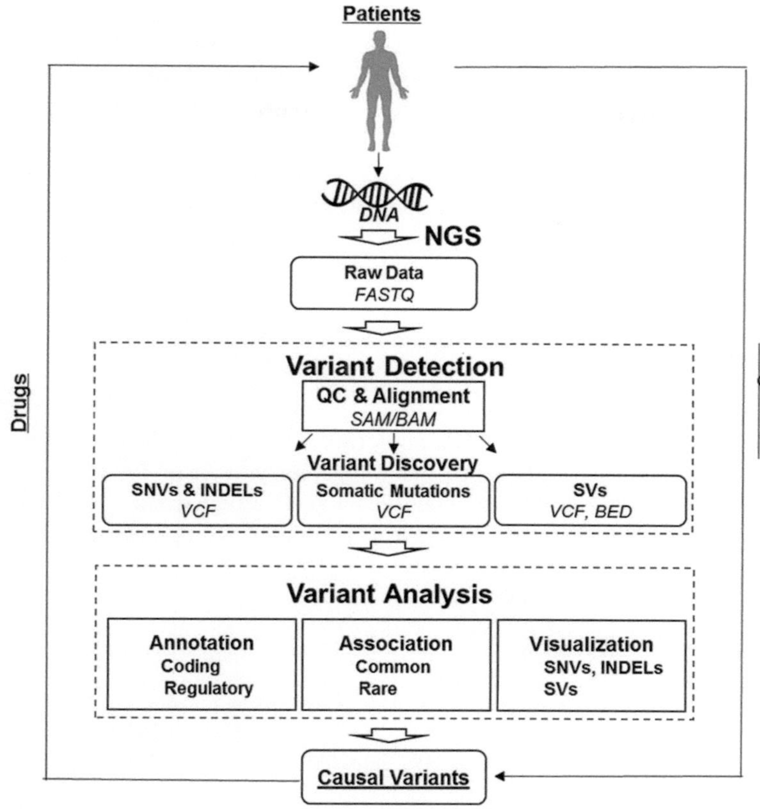

Fig. 1.1 A flowchart of NGS variant analysis in precision medicine

and rare variants associated with certain diseases or traits. Visualization tools are used to view the small and large candidate sequence variants. By combining numerous analyzing tools, the causal variants can be identified and connected with clinical information for precision medicine research. On the one hand, disease-related causal variants provide the genetic biomarkers for diagnostics of complex diseases. On the other hand, the candidate variants offer the targets for developing more precise treatments and drugs for patients. In the following sections, we will review the bioinformatics approaches and provide a guide for detecting and analyzing the sequence variants from NGS data.

1.2 Variant Detection

Variant detection consists of quality control (QC), sequence alignment, and variant calling. The raw data contains a large number of short reads generated by NGS sequencers. Preprocessing and post-processing QC are carried out to remove the potential artifacts and bias from data. The high-quality reads are mapped to positions on a reference genome. The variant calling is performed by comparing the aligned reads with known reference sequences to find which segments are different with the reference genomes. Multiple variant callers have been developed to detect different types of genetic variants including SNVs, INDELs, somatic mutations, and SVs. This section provides an overview on QC and alignment methods, SNV and INDEL callers, somatic mutation tools, and SV detection approaches.

1.2.1 QC and Alignment

The standard outputs of most NGS platforms are files in FASTQ format. The FASTQ files include raw sequence reads together with their Phred-scaled base quality scores. Several tools have been developed to perform preprocessing QC based on FASTQ files (Table 1.1). FastQC [7] provides a comprehensive QC report

Table 1.1 Variant quality control (QC) and alignment tools

Tool	Description	URL	Reference
Preprocessing QC			
FastQC	Tool can provide statistical QC summary report	http://www.bioinformatics.babraham.ac.uk/projects/fastqc/	[7]
Sickle	QC tool can trim low-quality bases	https://github.com/najoshi/sickle	[21]
Trimmomatic	QC tool can remove adaptor and bases	http://www.usadellab.org/cms/?page=trimmomatic	[10]
Hash table alignment			
MAQ	Hashing read aligner that allows two mismatches	http://maq.sourceforge.net/	[32]
SeqMap	Hashing read aligner that allows five mismatches	http://www-personal.umich.edu/~jianghui/seqmap/	[20]
SOAP	Hashing reference aligner	http://soap.genomics.org.cn/	[33]
Suffix tree alignment using Burrows–Wheeler transformation (BWT)			
BWA	BWT aligner using a backward search	http://bio-bwa.sourceforge.net/bwa.shtml	[30]
Bowtie	BWT aligner using a backtracking system	http://bowtie-bio.sourceforge.net	[26]
SOAP2	BWT and hash aligner	http://soap.genomics.org.cn/	[35]
Post-processing QC			
SAMtools	Tool can convert, sort, and index SAM/BAM files	http://samtools.sourceforge.net/	[34]
BamTools	Tool can manage BAM files and filter reads	https://github.com/pezmaster31/bamtools	[8]
Picard	QC tool can remove PCR duplicates	http://broadinstitute.github.io/picard/	

with tables and plots for basic statistics, quality score distribution, read length distribution, sequence duplication levels, and GC content distribution. With the FastQC summary, other QC tools, such as Sickle [21] and Trimmomatic [10], can be used to filter low-quality reads, remove undesired adaptors, and trim incorrectly called bases at the ends of the reads.

The sequence alignment is an essential step for identifying the sequence variants in patients' genomes. Since any errors in alignment will be carried through to the downstream analysis, each of the high-quality sequence reads must be accurately aligned to a reference genome. With the rapid development of NGS technologies, a wide variety of alignment tools have been developed to align the short sequence reads with high efficiency and accuracy (Table 1.1). Most of NGS aligners build indices for reads or references to quickly search potential alignment positions of reads in the reference sequences. Based on the property of the index, these tools can be briefly classified into two groups: hash table approaches or suffix tree approaches [31].

Hash Table Approaches These use a hash-based index to scan either read set or reference genome for rapid searching. Each position of reference is cut into equal-sized fragments and stored into a hash table. The species cut from the read with the same scheme are used as the keys to search the table. The approaches use a seed-and-extend paradigm to identify the matching positions in the reference for the reads. Here, we describe three common hash table tools: MAQ [32], SeqMap [20], and SOAP [33]. MAQ [32] can rapidly align a large number of short reads to the reference sequence and accurately detect small sequence variants including SNVs and INDELs. For sequence alignment, MAQ indexes and hashes the short reads before scanning reference sequence, which allows two mismatches in the first 28 bp of each read. It then searches ungapped match and extends the partial match when a seed match is identified. MAQ utilizes a Phred-scaled mapping quality score to evaluate the reliability of alignments, and the score can measure the probability that a true positive read is not the one found by the mapping algorithm. SeqMap [20] also applied an index filtering algorithm to create index tables for sequence reads. It allows up to five mismatches including substitutions and insertions/deletions. Instead of the construction of hash tables for reads that used in MAQ and SeqMap, SOAP [33] loads the reference genome into memory and constructs index tables for all references sequences. It utilizes a seed strategy for both ungapped and gapped alignments of either single read or paired-end reads.

Suffix Tree Approaches These use Burrows–Wheeler transformation (BWT) [11] to store all suffixes of a string. The reference genome can be converted to a transformed memory-efficient sequence using BWT. Reads are aligned base by base against the transformed reference sequence. The strategy can reduce memory footprint and increase mapping speed. Examples of BWT-based tools include BWA [30], Bowtie [26], and SOAP2 [35]. Burrows–Wheeler Alignment (BWA) tool [30] is the most commonly used NGS aligner. It uses backward search with BWT for exact matching and constructs inexact alignments supported by the exact matches. Bowtie [26] utilizes a novel backtracking system to account mismatches and allows up to two mismatches in the first 28 bp of sequence read. BWA and Bowtie compare the query reads and store the reference to short substrings. The tools compute all combinations of possible mismatches to align the entire reads to reference exactly. SOAP2 [35], an updated version of SOAP, uses BWT to index the reference genome in memory and constructs a hash table to search the location of a read in the reference index. The suffix tree methods run faster than hash table approaches due to the memory efficiency of BWT sequence. The indices of the entire human genome generated by BWT approaches are usually less than 2 GB, whereas the hash table approaches require more than 50 GB.

The sequence alignments are stored in SAM/BAM files [34]. Sequence Alignment/Map (SAM) file contains the read alignment data, and BAM file is the binary version of SAM file. SAMtools [34] can be used to convert SAM/BAM format and sort, index, and merge the alignment files. BamTools [8] can manage the BAM files and filter properly reads with high mapping quality. Picard can be used to remove PCR duplicates caused by the sorting from merged alignment files. These post-processing QC tools generate clean aligned sequencing files suitable for further variant detection.

1.2.2 SNV and INDEL Discovery

After mapping the short reads to a reference sequence, the variants can be discovered by comparing the sample genome to the reference genome. Many variant callers have been developed to detect small variants including SNVs and INDELs (Table 1.2). These computational tools use either heuristic or probabilistic approaches. Since probabilistic approaches can estimate sequencing error and monitor the accuracy of calling, they are more generally used for variant calling [40]. We introduce three probabilistic callers MAQ [32], SAMtools [34], and GATK [38] below.

MAQ [32] is the first widely used tool for variant calling in NGS data. It uses a Bayesian statistical model to generate consensus genotype sequence from the alignments. MAQ compares the consensus sequence to the reference genome to identify potential SNVs and filtered them using some predefined rules. SAMtools [34] uses a revised MAQ model to measure statistical uncertainty of called genotypes and applies given likelihood for each possible genotype. It uses a subset of commands, called BCFtools, to call SNVs and INDELs. The true small variants can be filtered by base alignment quality scores computed from the depth of coverage, numbers of reads in alternate and reference alleles, average quality scores, and mapping quality of reads.

Genome Analysis Toolkit (GATK) [38] is the most frequently used toolkit for small variant calling. It provides a structured Java programming MapReduce framework for NGS analysis. The GATK package includes coverage analyzer, local realigner, quality score recalibrator, and variant caller. It inputs SAM/BAM

Table 1.2 Variant discovery tools

Tool	Description	URL	Reference
SNV and INDEL discovery			
MAQ	Tool can detect small variants using a Bayesian statistical model	http://maq.sourceforge.net/	[32]
SAMtools	Tool can detect genotypes and small variants using a revised MAQ model	http://samtools.sourceforge.net/	[34]
GATK	Package including coverage analyzer, local realigner, quality score recalibrator, and variant caller	https://www.broadinstitute.org/gatk/	[38]
Somatic mutation discovery			
VarScan2	Caller can detect somatic mutations using Fisher's exact test	http://varscan.sourceforge.net/	[23]
Strelka	Caller can detect somatic mutations using Bayesian probability model	https://sites.google.com/site/strelkasomaticvariantcaller/	[49]
SomaticSniper	Caller can compute Phred-scaled scores to detect somatic mutations using Bayesian probability model	http://gmt.genome.wustl.edu/packages/somatic-sniper/	[27]
JointSNVMix	Caller can detect somatic mutations using two Bayesian probability-based models	http://compbio.bccrc.ca/software/jointsnvmix/	[47]
Structural variant discovery			
CNVnator	Read-depth caller can detect deletions and duplications	http://sv.gersteinlab.org/cnvnator/	[2]
BreakDancer	Read-pair caller can detect insertion, deletions, inversions, and translocations	http://breakdancer.sourceforge.net/	[12]
Pindel	Split-read caller can detect large deletions and medium insertions	https://github.com/genome/pindel	[61]
CONTRA	Read-depth caller can detect CNVs from exome sequencing data	https://sourceforge.net/projects/contra-cnv/	[36]
XHMM	Read-depth caller can detect CNVs using hidden Markov model from exome sequencing data	https://atgu.mgh.harvard.edu/xhmm/	[18]

files from initial read mapping. Then, the tool carries out a local INDEL realignment and computes base quality scores for recalibration. A program, called UnifiedGenotyper, is used to identify all potential SNVs and INDELs. GATK applies machine learning approaches to filter true variants from machine artifacts in NGS technologies [56]. GATK recently developed a HaplotypeCaller [16] program which performs a local de novo assembly of aligned reads and calls SNVs and INDELs simultaneously. It provides a greater quality for INDELs calling than UnifiedGenotyper program [42]. In addition, HaplotypeCaller can handle the non-diploid samples and work well for the region including different types of sequence variants close to each other. The outputs of the most variant callers are

Variant Call Format (VCF) files. The VCF is used for storing sequence variants such as SNVs, INDELs, as well as SVs. The genetic variant information in VCF files includes variant positions, unique identifiers, reference and alternate alleles, quality scores, filters, annotations, and genotypes.

1.2.3 Somatic Mutation Discovery

Discovery of somatic mutations associated with oncogenesis is essential for identifying appropriate treatments for cancer patients. Several callers have been developed for detecting the somatic mutations that present in tumor cells but not in normal tissue (Table 1.2). VarScan2 [23] screens the genotypes that are above certain coverage and quality thresholds from the cancer and normal samples, respectively. The variant calls with minimum variant frequency of all reads greater than 20 % are classified as either heterozygous calls or homozygous calls. For each position with genotypes that do not match in tumor and normal, VarScan2 uses a one-tailed Fisher's exact test to check significant difference of allele frequency across samples. The somatic mutations are called if the normal samples are homozygous reference or heterozygous as loss of heterozygosity, but the calls in tumor samples do not match.

Several tools based on Bayesian probability model have been developed for the discovery of somatic mutations in matched cancer–normal pairs. Strelka [49] carries out a realignment around INDELs in the tumor and normal sequence alignment files like GATK. It uses a Bayesian probability approach to model the normal sample allele frequencies as diploid genotypes and tumor sample allele frequencies as a mixture of the normal sample with somatic variation. The approaches also apply some priors for strand bias, mapping qualities, somatic mutation rates, and estimated heterozygosity rates of the normal sample. SomaticSniper [27] uses a Bayesian probability model to compute the probability of all possible combined genotypes for the cancer–normal pair samples. The likelihood is given by the observed as well as prior information from the rates of population mutation, sequencing error, and somatic mutation. Each variant call in tumor samples is assigned a Phred-scaled score indicating the probability that the cancer and normal genotypes are different. JointSNVMix [47] utilizes a different Bayesian method with a mixed binomial model to call each variant in the tumor and normal samples. It analyzes the allelic count in paired cancer–normal samples using two probabilistic graphical models: JointSNVMix1 that assumes the base calls and read numbers and follows a perfect binomial distribution and JointSNVMix2 that weighs priors for base call and mapping quality.

1.2.4 Structural Variant Discovery

Structural variants (SVs) are widespread in human genomes and play important roles in the development of human diseases. As the growing number of SVs has been demonstrated to have clinical relevance, SV discovery is critical in precision medicine and cancer genomics. NGS technologies have revolutionized SV studies. Compared to traditional hybridization-based approaches such as array CGH and SNP microarrays, sequencing-based bioinformatics methods can detect multiple types of SVs at a wide size range [5]. Most of these approaches distinguish SVs based on two read mapping signatures including depth of coverage and paired-end mapping [39]. The first type of approaches searches the regions with abnormal read counts; the second type of tools investigates the configurations of the paired-end mappings [60]. In this section, we describe the computational approaches (Table 1.2) based on the two signatures below.

Depth of Coverage The approaches assume that read mapping follows a Poisson distribution and the divergence from this distribution indicates the SV signatures. The duplication has more reads mapping to region, and deletions show significantly reduced coverage. CNVnator [2] can detect the deletions and duplications using a statistical analysis of read mapping density for single-end and paired-end reads. It captures the read-depth signatures by dividing sequencing regions into equal-sized bins and computing the counts of reads in each bin. The partitioning of the signatures is based on a mean-shift approach with additional filters such as GC-bias correction. The statistical significance test is used to identify the regions with abnormal signals for detecting possible deletions or duplications. The read-depth approaches can predict the absolute copy numbers of genomic segments. However, they cannot detect the balanced SVs such as translocations and inversions.

Paired-End Mapping The approaches can be classified into two types of strategy: read pair and split read. Read-pair methods analyze the span and orientation of paired-end reads and identify the read pairs that are mapped with discordant separation distances or orientation. Read-pair approaches can detect all classes of SVs. BreakDancer [12] can detect read pairs with mapping span and orientation that are inconsistent with the control. It has two models: BreakDancerMax can identify five types of SVs including insertion, deletions, inversions, and intrachromosomal and interchromosomal translocations, while BreakDancerMini is used to detect INDELs. Split-read approaches are used to search split-read signatures to identify the breakpoints of SVs. The deletions and duplications can be identified from the continuous stretch of gaps in the sequence reads or references, respectively. Split-read methods are suitable for long reads, but some algorithms can use short reads to identify the breakpoints of large SVs. For example, Pindel [61] uses a pattern growth algorithm to find large deletions and medium insertions from short paired-end reads. The algorithm can align the gapped short sequences to reference sequences with local alignment, which can reduce memory and increase speed for searching potential split reads.

Structure variant discovery from targeted or whole-exome sequencing data is very challenging due to the noncontiguous reads in exons. The targeted sequencing results in some biases in sample collection, targeted genomic hybridization, and GC content. Multiple tools have been developed to overcome these biases. CONTRA [36] is a read-depth tool for CNV discovery. It uses BAM/SAM alignments as inputs and builds an average baseline across multiple samples as the control. CONTRA then computes the base-level log-ratios with corrections for imbalanced library size bias and GC content bias. It calculates two-tailed P-values to detect CNVs. XHMM [18] applies principal component analysis to normalize read depth in targets. It uses hidden Markov model (HMM) to detect CNVs across multiple samples (>50 samples). In addition to VCF files, Browser Extensible Data (BED) format files can be used to store and display large structural variants for further analysis.

1.3 Variant Analysis

Causal variant discovery is the key step in precision medicine informatics. Identifying the disease-related variants promises to dramatically expand current aspects of biomedical research in disease diagnostics and drug design. Multiple bioinformatics tools have been developed to distinguish the causal variants associated with human diseases from the massive number of nonfunctional variants detected by NGS variant callers. Annotation methods determine the possible functional impact of all identified variants. Association analyses connect the variants with complex diseases or clinical traits. Visualization tools provide the graphic views of identified causal variants. The disease-related casual variants can be identified by combining these approaches and stored in public variant databases such as ClinVar [25] and HGMD [54]. The Human Variome Project (http://www.humanvariomeproject.org/) has curated the gene-/disease-specific databases to collect the sequence variants and genes associated with diseases. In this section, we summarize the variant analysis approaches for identifying the most promising causal variants underlying human diseases.

1.3.1 Variant Annotation

Variant annotation can be used to determine the effects of sequence variants on genes and proteins and filter the functional important variants from a background of neutral polymorphisms. Coding mutations, such as nonsynonymous SNVs, could change amino acid sequences and affect protein structures and functions. They are more likely to be involved in the development of diseases. Regulatory variants located in noncoding regions could modulate the gene expressions and work as the causative modifiers of human diseases. Here, we describe the common computational tools for predicting the effects of coding mutations and regulatory variants. We also introduce the generally used annotation toolkits to access the prediction results generated from these tools.

Damaging Nonsynonymous Mutation Prediction With the advent of NGS technologies, particularly of exome sequencing, there is a significant need to interpret the coding variants. A number of tools have been developed to distinguish deleterious mutations from a large number of harmless nonsynonymous polymorphisms. Sorting Intolerant From Tolerant (SIFT) [41] is a commonly used method for predicting the effects of coding mutations on protein function. The algorithm assumes that important protein sites should be conserved throughout evolution and mutations located in these sites could alter protein functions. SIFT searches the target sequence in protein database and constructs the sequence alignments using closely related sequences. It computes the degree of conservation of protein residues to distinguish the deleterious and neutral coding mutations. Polymorphism Phenotyping v2 (PolyPhen2) [4] is another popular tool for predicting deleterious missense mutations. The PolyPhen2 prediction is based on sequence annotations, structural attributes, and comparative evolutionary considerations. PolyPhen2 uses an iterative greedy algorithm to extract sequence-based and structure-based features. Then, it constructs the supervised machine learning classifiers to predict missense variants as benign, possibly damaging, or probably damaging mutations. PolyPhen2 uses two data sets (HumDiv and HumVar) for training. HumDiv data set collects all damaging mutations associated with human Mendelian diseases from UniProtKB and non-damaging mutations between the proteins and their closely related mammalian homologs. HumDiv model can be used to analyze rare variants mildly deleterious at functionally important regions such as the regions involved in complex phenotypes or identified from genome-wide association studies (GWAS). HumVar data set uses all disease-causing mutations from UniProtKB as positive data and the common sequence variants not involved in disease as negative instances. HumVar model can be used to identify the damaging mutations with significant effects for Mendelian disease research. Other common in silico programs include likelihood ratio test (LRT) [13], which identifies the damaging mutations that disrupt significantly conserved amino acid positions within the human proteome, and MutationTaster [51] which evaluates the deleterious sequence variants using a naive Bayesian model constructed from features including splice-site alterations, mRNA changes, loss of protein, and evolutionary conservation.

Regulatory Variant Effect Prediction The majority of disease-related variant hits identified from GWAS fall in noncoding DNA region, which indicate the regulatory variants located in noncoding regions are critical in human disease. Regulatory variants play important roles in gene expression and protein modification. Several bioinformatics tools have been developed for predicting the functional effects of regulatory variants. Genome-wide annotation of variants (GWAVA) [45] uses a random forest algorithm to construct three classifiers to distinguish the functional sequence variants in regulatory regions from a background of neutral variants. The classifiers integrate genomic features such as evolutionary conservation and GC content and range of epigenomic annotations from the Encyclopedia of DNA

Elements (ENCODE) project [15]. Combined Annotation Dependent Depletion (CADD) [22] is a score that can be used to prioritize the functional variants including coding variants and regulatory variants. CADD tool constructs support vector machine classifiers to integrate various genomic and epigenomic annotations into a single measure (C score) for each sequence variant. Recently, deep learning algorithm has been applied for interpretation of regulatory variants. DeepSEA [62] is a deep learning-based tool for predicting the effects of noncoding variant and prioritizing regulatory variants. The software uses deep learning algorithms to learn regulatory sequence code from large-scale chromatin-profiling data and predict the effects of noncoding variants on chromatin accessibility such as DNase I sensitivities, transcription factor binding, and histone marks at regulatory elements.

General Variant Annotation Multiple annotation toolkits have been developed to determine the impacts of sequence variants on genes and proteins and access their functional effects from above predictors. ANNOVAR [58] is a command-line Perl software for annotating SNVs and INDELs based on genes, regions, or filters. In gene-based annotation, it can annotate whether sequence variants affect protein amino acid sequences (nonsense, missense, splice site, etc.). In region-based annotation, it can identify the variants located in ENCODE-annotated regions such as transcribed regions, enhancer regions, DNase I hypersensitivity sites, transcription factor binding site, and transcription factor ChIP-Seq data. In filter-based annotation, ANNOVAR can extract the information (allele frequency and identifier) of a sequence variant in public databases such as dbSNP [53], ClinVar [25], 1000 Genomes Project [1], and Exome Variant Server (http://evs.gs.washington.edu/EVS/). In addition, it can be used to access the annotations from damaging mutation predictors (SIFT, PolyPhen2, LRT, MutationTaster, etc.) for nonsynonymous mutations and CADD for regulatory variants. SnpEff [14] is another popular annotation package to estimate the functional effects of SNVs, INDELs, and multiple nucleotide polymorphisms. Based on the functional impacts of the sequence variants, SnpEff classifies the variants to four classes: high, moderate, low, and modifier. It also provides the annotations for regulatory variants. SnpEff provides a summary HTML page to display overall statistics for sequences and variants (Table 1.3).

1.3.2 Variant Association Testing

Understanding how genetic variants contribute to diseases is the key challenge in precision medicine. There are two hypotheses for interpreting the genetic contribution of sequence variants in complex diseases such as cancers and mental disorders [50]. The "common disease–common variant" hypothesis states that a few common variants, usually defined as the allele frequency greater than 1 % in the population, make the major contributions for the genetic variance in complex disease susceptibility. In contrast, the "common disease–rare variant" hypothesis argues that multiple risk variants, each of which has low frequency (e.g., allele frequency less than 1 %) in the population, are the major contributors to the genetic

Table 1.3 Variant annotation tools

Tool	Description	URL	Reference
Damaging nonsynonymous mutation prediction			
SIFT	Tool can predict deleterious and neutral mutations based on sequence homology	http://sift.jcvi.org/	[41]
PolyPhen2	Tool can predict probably damaging, possibly damaging, and benign mutations based on sequence and structure features	http://genetics.bwh.harvard.edu/pph2/	[4]
LRT	Tool can predict deleterious, neutral, or unknown mutations using likelihood ratio test	http://www.genetics.wustl.edu/jflab/lrt_query.html	[13]
MutationTaster	Tool can predict disease-causing and polymorphism mutations using naive Bayesian model	http://www.mutationtaster.org/	[51]
Regulatory variant effect prediction			
GWAVA	Tool can predict the regulatory variant effects using random forest algorithm	https://www.sanger.ac.uk/sanger/StatGen_Gwava	[45]
CADD	Tool can predict the effects of coding and noncoding variants using support vector machine algorithm	http://cadd.gs.washington.edu/	[22]
DeepSEA	Tool can predict the regulatory variant effects using deep learning algorithm	http://deepsea.princeton.edu/	[62]
General variant annotation			
ANNOVAR	Perl annotation toolkit based on genes, regions, and filters	http://annovar.openbioinformatics.org/	[58]
SnpEff	Java annotation package based on genes	http://snpeff.sourceforge.net/	[14]

susceptibility to complex diseases. NGS technologies can detect the full spectrum of sequence variants including the rare variants that are difficult to be captured by traditional genotyping arrays. Here, we describe the generally used case–control association approaches for common and rare variants.

Case–Control Data QC The first step in any case-control association analysis is the data quality control [6]. The samples and variants with poor quality should be removed to reduce the numbers of false-positive and false-negative associations. The samples with outlying heterozygosity rates, high missing data rates, and discordant sex information have poor quality and should be removed firstly. In addition, the related samples or samples from divergent ancestry should not be used for case-control analysis. If the variants showed a high rate of missing genotypes, departure from Hardy–Weinberg equilibrium, or a different missing genotype rate between cases and controls, these variants should be excluded from case-control analysis.

Common-Variant Association Analysis The genome-wide association study (GWAS) is a generally used approach to identify the common variants associated with complex diseases and traits. The common methods used in GWAS are carried out based on a single-variant level. The variants are tested individually, and multiple testing correction should be used to control the family-wise error rate (FWER). PLINK (http://pngu.mgh.harvard.edu/~purcell/plink/) is the most commonly used software package for GWAS analysis in large-scale studies [43]. It provides numerous useful tools for genetic data management, data quality control, and association tests. Multiple association tests implemented in PLINK can be used to identify the common variants associated with diseases based on their minor allele frequencies between cases and controls. Fisher's exact test can be used for case-control traits in small-sized samples; permutation methods should be applied to control for FWER. Linear regression test can be utilized for complex quantitative traits, and permutation approaches should be performed to generate empirical P-values to avoid issues with the test statistic distribution caused by the combination of variants and traits that deviate greatly from normality. Another popular association testing tool is PLINK/SEQ (https://atgu.mgh.harvard.edu/plinkseq/). The toolset performs Fisher's exact test for single-variant association, on the contrary, based on the alternate allele frequencies of variants in cases and controls.

Rare-Variant Association Analysis GWAS research has identified many common variants strongly implicated in complex diseases. However, most of the common variants have modest effects on the disease risk and much of the genetic contribution to complex diseases remains unexplained [37]. Recent sequencing studies revealed the rare genetic variants have large effects on the risk for complex diseases such as schizophrenia [44]. The rare-variant association tests are usually carried out on a gene, or gene set level due to single-variant analysis is underpowered for rare variants unless the sample sizes are very large. The general rare-variant burden test collapses the rare variants across all samples into a single variable and compares the cumulative effects in cases with controls within a gene to evaluate the significance of the difference. The sequence kernel association test (SKAT, https://cran.r-project.org/web/packages/SKAT/) is particularly designed for the rare-variant analysis from NGS data [59]. It uses a kernel machine regression approach to aggregate the associations between variants in a gene region and a continuous or dichotomous trait. SKAT-O [28] test applies a unified test to search the optimal linear combination of the general burden test and SKAT test to maintain the power in both scenarios. In addition, the SKAT package provides "SKAT_CommonRare" function to evaluate the combined effects of rare and common variants. The permutation method can be used in rare-variant association analysis to control FWER.

1.3.3 Variant Visualization

Visualizing the individual genomes and causal variants based on the existing knowledge provides critical supports for biomedical research. Various standalone visualization tools have been developed for interactive exploration of NGS data from public resources and researchers' own studies. Integrative Genomics Viewer

(IGV) is a high-performance tool that provides a rapid visualization for large genomic data sets [46]. IGV tool (https://www.broadinstitute.org/igv/) can load sequence alignment BAM files, annotation data, and reference genomes from local computers or remote sites. IGV includes tools for data tiling and file format sorting and indexing [55]. It provides both stand-alone GUI desktop version and command-line scripts for generating different image snapshots. IGV is capable of displaying various sequence variants including SNVs and INDELs. As shown in Fig. 1.2, IGV provides the views of chromosome ideogram, genomic coordinates, coverage plot, sequence reads, and gene annotation tracks. The base mismatches compared to reference are highlighted with color bars in coverage plot or color bases in read tracks. Figure 1.2a shows an intronic SNV (rs3812384) in Src-like-adaptor

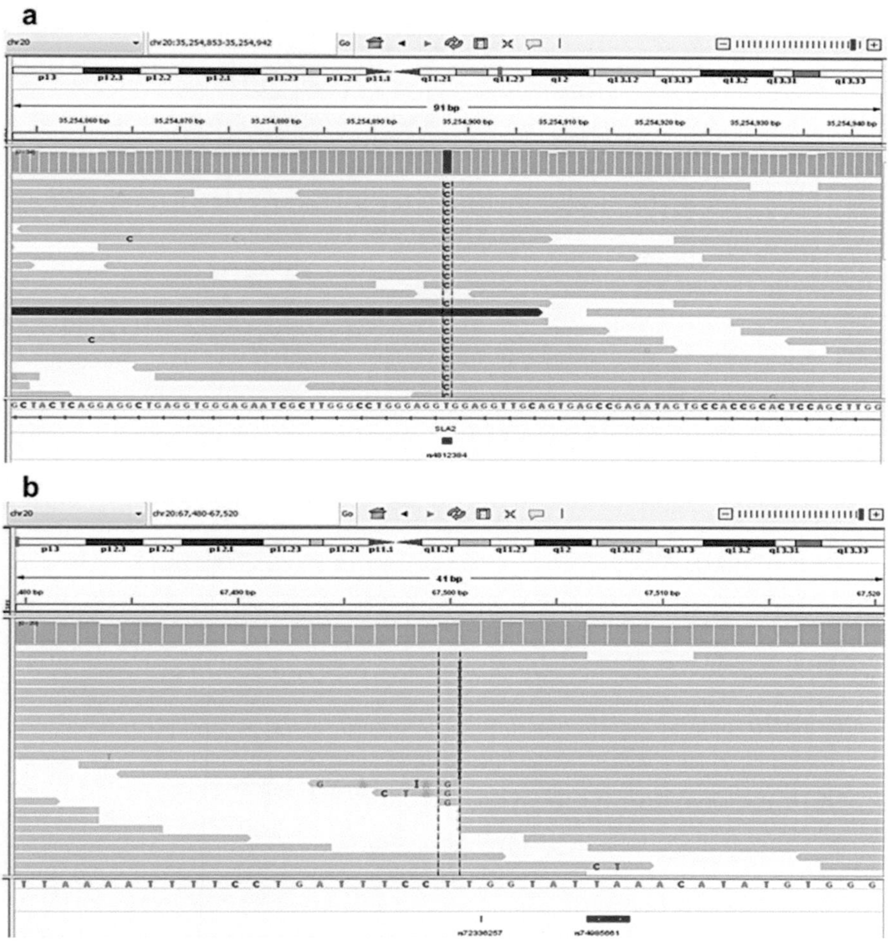

Fig. 1.2 IGV views of small sequence variants. Snapshots of IGV showing (**a**) an intronic SNV (rs3812384) in *SLA2* gene and (**b**) an insertion (rs72336257) in *DEFB125* gene

2 (*SLA2*) gene, and Fig. 1.2b displays an insertion (rs72336257) in defensin beta 125 (*DEFB125*) gene.

Several visualization tools have been developed for displaying the large SVs from NGS data. Sequence Annotation, Visualization, and ANalysis Tool (Savant) is a viewer for analyzing and visualizing sequence reads and variants [17]. Savant browser (http://genomesavant.com) uses a modular docking framework to show each module in a separate window. It provides the track module, bookmark module, and table view to analyze the NGS data. In particular, Savant provides multiple visualization modes to view and compare SVs in different samples (Fig. 1.3). For example, a 150 kb duplication presenting in case but not in control shows a higher coverage in zinc finger and AT-hook domain containing (*ZFAT*) gene in the

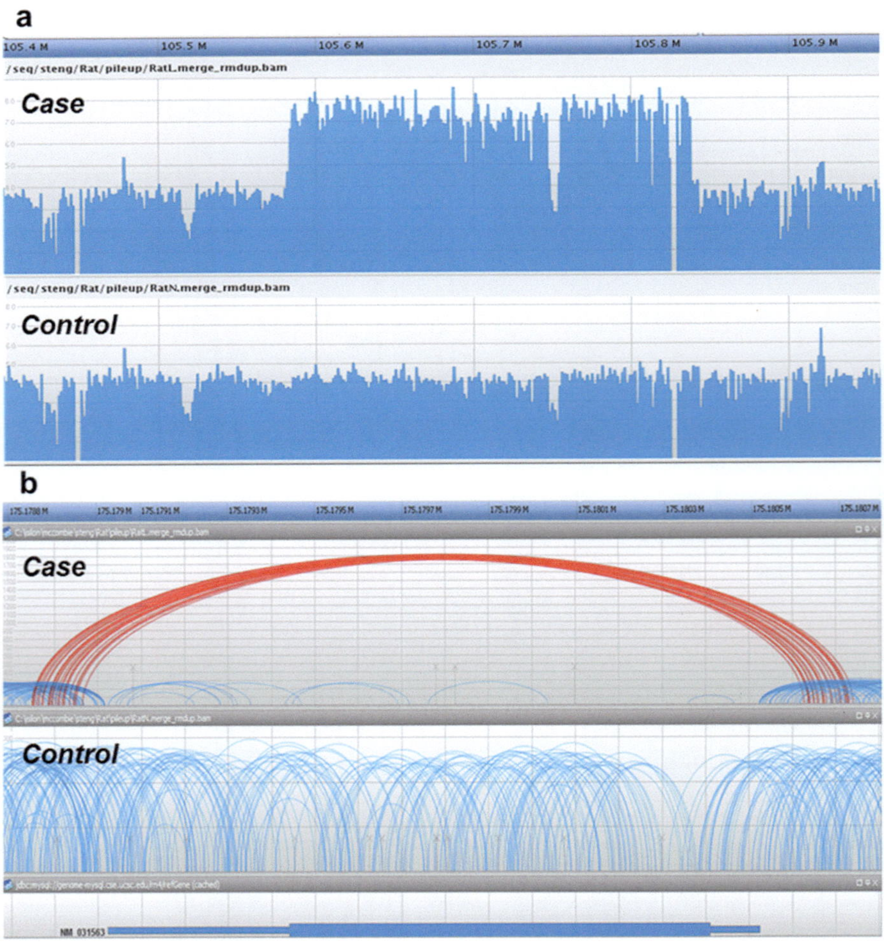

Fig. 1.3 Savant views of large SVs. Plots of Savant displaying the large SVs that present in case but not in control including (**a**) a 150 kb duplication in *ZFAT* gene and (**b**) a 1.5 kb deletion in *YBX1* gene

coverage mode (Fig. 1.3a). The Matepair (arc) mode displays the relative distance between paired-end reads. The red and taller arcs indicate a larger distance between the two reads of a pair, suggesting a 1.5 kb deletion in Y box binding protein 1 (*YBX1*) gene is only carried by the case sample (Fig. 1.3b).

1.4 Conclusions

NGS has significantly benefited the discovery of disease-related sequence variants, which greatly facilitated the improvement of diagnosis and treatment methods in precision medicine. Computational approaches have been developed to detect different types of sequence variants (SNVs, INDELs, somatic mutations, and structural variants) from NGS data. Bioinformatics methods have been applied to annotate, filter, and visualize the casual variants associated with complex diseases. There are limit standards regarding best practices in NGS variant detection and analysis. To meet the challenges in precision medicine, some international scientific organizations, such as Human Variome Project, are developing standardized workflows for analyzing sequence variants implicated in human diseases.

References

1. Abecasis GR, Altshuler D, Auton A, et al. A map of human genome variation from population-scale sequencing. Nature. 2010;467:1061–73. doi:10.1038/nature09534.
2. Abyzov A, Urban AE, Snyder M, Gerstein M. CNVnator: an approach to discover, genotype, and characterize typical and atypical CNVs from family and population genome sequencing. Genome Res. 2011;21:974–84. doi:10.1101/gr.114876.110.
3. Adekoya E, Ait-Zahra M, Allen N, et al. Initial sequencing and analysis of the human genome. Nature. 2001;409:860–921.
4. Adzhubei IA, Schmidt S, Peshkin L, et al. A method and server for predicting damaging missense mutations. Nat Methods. 2010;7:248–9. doi:10.1038/nmeth0410-248.
5. Alkan C, Coe BP, Eichler EE. Genome structural variation discovery and genotyping. Nat Rev Genet. 2011;12:363–76. doi:10.1038/nrg2958.
6. Anderson CA, Pettersson FH, Clarke GM, et al. Data quality control in genetic case–control association studies. Nat Protoc. 2010;5:1564–73. doi:10.1038/nprot.2010.116.
7. Andrews S. FastQC: a quality control tool for high throughput sequence data. babraham Bioinforma 1. 2010. doi: citeulike-article-id:11583827.
8. Barnett DW, Garrison EK, Quinlan AR, et al. BamTools: a C++ API and toolkit for analyzing and managing BAM files. Bioinformatics. 2011;27:1691–2. doi:10.1093/bioinformatics/btr174.
9. Blackwood DH, Fordyce A, Walker MT, et al. Schizophrenia and affective disorders-cosegregation with a translocation at chromosome 1q42 that directly disrupts brain-expressed genes: clinical and P300 findings in a family. Am J Hum Genet. 2001;69:428–33.
10. Bolger AM, Lohse M, Usadel B. Trimmomatic: a flexible trimmer for Illumina sequence data. Bioinformatics. 2014;30:2114–20. doi:10.1093/bioinformatics/btu170.
11. Burrows M, Wheeler D. A block-sorting lossless data compression algorithm. Algorithm, Data Compression 18. 1994. doi: 10.1.1.37.6774.

12. Chen K, Wallis JW, McLellan MD, et al. BreakDancer: an algorithm for high-resolution mapping of genomic structural variation. Nat Methods. 2009;6:677–81. doi:10.1038/nmeth.1363.
13. Chun S, Fay JC. Identification of deleterious mutations within three human genomes. Genome Res. 2009;19:1553–61. doi:10.1101/gr.092619.109.
14. Cingolani P, Platts A, le Wang L, et al. A program for annotating and predicting the effects of single nucleotide polymorphisms, SnpEff: SNPs in the genome of Drosophila melanogaster strain w1118; iso-2; iso-3. Fly (Austin). 2012;6:80–92. doi:10.4161/fly.19695.
15. Consortium TEP. The ENCODE (ENCyclopedia Of DNA Elements) project. Science. 2004;306:636–40. doi:10.1126/science.1105136.
16. DePristo MA, Banks E, Poplin R, et al. A framework for variation discovery and genotyping using next-generation DNA sequencing data. Nat Genet. 2011;43:491–8. doi:10.1038/ng.806.
17. Fiume M, Williams V, Brook A, Brudno M. Savant: genome browser for high-throughput sequencing data. Bioinformatics. 2010;26:1938–44. doi:10.1093/bioinformatics/btq332.
18. Fromer M, Moran JL, Chambert K, et al. Discovery and statistical genotyping of copy-number variation from whole-exome sequencing depth. Am J Hum Genet. 2012;91:597–607. doi:10.1016/j.ajhg.2012.08.005.
19. Hahn MK, Blakely RD. Monoamine transporter gene structure and polymorphisms in relation to psychiatric and other complex disorders. Pharmacogenomics J. 2002;2:217–35. doi:10.1038/sj.tpj.6500106.
20. Jiang H, Wong WH. SeqMap: mapping massive amount of oligonucleotides to the genome. Bioinformatics. 2008;24:2395–6. doi:10.1093/bioinformatics/btn429.
21. Joshi N, Fass J. Sickle: a sliding-window, adaptive, quality-based trimming tool for FastQ files (Version 1.33) [Software]. 2011. Available at https://github.com/najoshi/sickle2011.
22. Kircher M, Witten DM, Jain P, et al. A general framework for estimating the relative pathogenicity of human genetic variants. Nat Genet. 2014;46:310–5. doi:10.1038/ng.2892.
23. Koboldt DC, Zhang Q, Larson DE, et al. VarScan 2: somatic mutation and copy number alteration discovery in cancer by exome sequencing. Genome Res. 2012;22:568–76. doi:10.1101/gr.129684.111.
24. Koboldt DC, Steinberg KM, Larson DE, et al. The next-generation sequencing revolution and its impact on genomics. Cell. 2013;155:27–38.
25. Landrum MJ, Lee JM, Riley GR, et al. ClinVar: public archive of relationships among sequence variation and human phenotype. Nucleic Acids Res. 2014;42:D980–5. doi:10.1093/nar/gkt1113.
26. Langmead B, Trapnell C, Pop M, Salzberg S. Ultrafast and memory-efficient alignment of short DNA sequences to the human genome. Genome Biol. 2009;10:R25. doi:10.1186/gb-2009-10-3-r25.
27. Larson DE, Harris CC, Chen K, et al. SomaticSniper: identification of somatic point mutations in whole genome sequencing data. Bioinformatics. 2012;28:311–7. doi:10.1093/bioinformatics/btr665.
28. Lee S, Emond MJ, Bamshad MJ, et al. Optimal unified approach for rare-variant association testing with application to small-sample case–control whole-exome sequencing studies. Am J Hum Genet. 2012;91:224–37. doi:10.1016/j.ajhg.2012.06.007.
29. Levinson DF. The genetics of depression: a review. Biol Psychiatry. 2006;60:84–92. doi:10.1016/j.biopsych.2005.08.024.
30. Li H, Durbin R. Fast and accurate short read alignment with Burrows-Wheeler transform. Bioinformatics. 2009;25:1754–60. doi:10.1093/bioinformatics/btp324.
31. Li H, Homer N. A survey of sequence alignment algorithms for next-generation sequencing. Brief Bioinform. 2010;11:473–83. doi:10.1093/bib/bbq015.
32. Li H, Ruan J, Durbin R. Mapping short DNA sequencing reads and calling variants using mapping quality scores. Genome Res. 2008;18:1851–8. doi:10.1101/gr.078212.108.
33. Li R, Li Y, Kristiansen K, Wang J. SOAP: short oligonucleotide alignment program. Bioinformatics. 2008;24:713–4. doi:10.1093/bioinformatics/btn025.

34. Li H, Handsaker B, Wysoker A, et al. The sequence alignment/map format and SAMtools. Bioinformatics. 2009;25:2078–9. doi:10.1093/bioinformatics/btp352.
35. Li R, Yu C, Li Y, et al. SOAP2: an improved ultrafast tool for short read alignment. Bioinformatics. 2009;25:1966–7. doi:10.1093/bioinformatics/btp336.
36. Li J, Lupat R, Amarasinghe KC, et al. CONTRA: copy number analysis for targeted resequencing. Bioinformatics. 2012;28:1307–13. doi:10.1093/bioinformatics/bts146.
37. Manolio TA, Collins FS, Cox NJ, et al. Finding the missing heritability of complex diseases. Nature. 2009;461:747–53. doi:10.1038/nature08494.
38. McKenna A, Hanna M, Banks E, et al. The genome analysis toolkit: a MapReduce framework for analyzing next-generation DNA sequencing data. Genome Res. 2010;20:1297–303. doi:10.1101/gr.107524.110.
39. Medvedev P, Stanciu M, Brudno M. Computational methods for discovering structural variation with next-generation sequencing. Nat Methods. 2009;6:S13–20. doi:10.1038/nmeth.1374.
40. Mielczarek M, Szyda J. Review of alignment and SNP calling algorithms for next-generation sequencing data. J Appl Genet. 2015;57:1–9. doi:10.1007/s13353-015-0292-7.
41. Ng PC, Henikoff S. SIFT: predicting amino acid changes that affect protein function. Nucleic Acids Res. 2003;31:3812–4.
42. Pirooznia M, Kramer M, Parla J, et al. Validation and assessment of variant calling pipelines for next-generation sequencing. Hum Genomics. 2014;8:14. doi:10.1186/1479-7364-8-14.
43. Purcell S, Neale B, Todd-Brown K, et al. PLINK: a tool set for whole-genome association and population-based linkage analyses. Am J Hum Genet. 2007;81:559–75. doi:10.1086/519795.
44. Purcell SM, Moran JL, Fromer M, et al. A polygenic burden of rare disruptive mutations in schizophrenia. Nature. 2014;506:185–90. doi:10.1038/nature12975.
45. Ritchie GRS, Dunham I, Zeggini E, Flicek P. Functional annotation of noncoding sequence variants. Nat Methods. 2014;11:294–6. doi:10.1038/nmeth.2832.
46. Robinson JT, Thorvaldsdóttir H, Winckler W, et al. Integrative genomics viewer. Nat Biotechnol. 2011;29:24–6. doi:10.1038/nbt.1754.
47. Roth A, Ding J, Morin R, et al. JointSNVMix: a probabilistic model for accurate detection of somatic mutations in normal/tumour paired next-generation sequencing data. Bioinformatics. 2012;28:907–13. doi:10.1093/bioinformatics/bts053.
48. Sanger F, Nicklen S, Coulson AR. DNA sequencing with chain-terminating inhibitors. Proc Natl Acad Sci U S A. 1977;74:5463–7.
49. Saunders CT, Wong WSW, Swamy S, et al. Strelka: accurate somatic small-variant calling from sequenced tumor-normal sample pairs. Bioinformatics. 2012;28:1811–7. doi:10.1093/bioinformatics/bts271.
50. Schork NJ, Murray SS, Frazer KA, Topol EJ. Common vs. rare allele hypotheses for complex diseases. Curr Opin Genet Dev. 2009;19:212–9. doi:10.1016/j.gde.2009.04.010.
51. Schwarz JM, Rodelsperger C, Schuelke M, Seelow D. MutationTaster evaluates disease-causing potential of sequence alterations. Nat Methods. 2010;7:575–6. doi:10.1038/nmeth0810-575.
52. Shendure J, Ji H. Next-generation DNA sequencing. Nat Biotechnol. 2008;26:1135–45.
53. Sherry ST, Ward MH, Kholodov M, et al. dbSNP: the NCBI database of genetic variation. Nucleic Acids Res. 2001;29:308–11.
54. Stenson PD, Mort M, Ball EV, et al. The human gene mutation database: 2008 update. Genome Med. 2009;1:13. doi:10.1186/gm13.
55. Thorvaldsdóttir H, Robinson JT, Mesirov JP. Integrative Genomics Viewer (IGV): high-performance genomics data visualization and exploration. Brief Bioinform. 2013;14:178–92. doi:10.1093/bib/bbs017.
56. Van der Auwera G a., Carneiro MO, Hartl C, et al. From fastQ data to high-confidence variant calls: The genome analysis toolkit best practices pipeline. Curr Protoc Bioinfor. 2013;43:11.10.1–33. doi:10.1002/0471250953.bi1110s43.

57. van Dijk EL, Auger H, Jaszczyszyn Y, Thermes C. Ten years of next-generation sequencing technology. Trends Genet. 2014;30:418–26.
58. Wang K, Li M, Hakonarson H. ANNOVAR: functional annotation of genetic variants from high-throughput sequencing data. Nucleic Acids Res. 2010;38, e164. doi:10.1093/nar/gkq603.
59. Wu MC, Lee S, Cai T, et al. Rare-variant association testing for sequencing data with the sequence kernel association test. Am J Hum Genet. 2011;89:82–93. doi:10.1016/j.ajhg.2011.05.029.
60. Xi R, Kim T-M, Park PJ. Detecting structural variations in the human genome using next generation sequencing. Brief Funct Genomics. 2010;9:405–15. doi:10.1093/bfgp/elq025.
61. Ye K, Schulz MH, Long Q, et al. Pindel: a pattern growth approach to detect break points of large deletions and medium sized insertions from paired-end short reads. Bioinformatics. 2009;25:2865–71. doi:10.1093/bioinformatics/btp394.
62. Zhou J, Troyanskaya OG. Predicting effects of noncoding variants with deep learning-based sequence model. Nat Methods. 2015;12:931–4. doi:10.1038/nmeth.3547.

Chapter 2
RNA Bioinformatics for Precision Medicine

Jiajia Chen and Bairong Shen

Abstract The high-throughput transcriptomic data generated by deep sequencing technologies urgently require bioinformatics methods for proper data visualization, analysis, storage, and interpretation. The involvement of noncoding RNAs in human diseases highlights their potential as biomarkers and therapeutic targets to facilitate the precision medicine. In this chapter, we give a brief overview of the bioinformatics tools to analyze different aspects of RNAs, in particular ncRNAs. We first describe the emerging bioinformatics methods for RNA identification, structure modeling, functional annotation, and network inference. This is followed by an introduction of potential usefulness of ncRNAs as diagnostic, prognostic biomarkers and therapeutic strategies.

Keywords RNA • Precision medicine • Biomarkers • Bioinformatics • Cancer

2.1 Introduction

RNAs are polymeric molecules that carry genetic information and are implicated in protein synthesis. Recently, it is discovered that only a minor fraction of human genomes encode for proteins [10, 15], and the remaining large fraction of the transcripts are known as noncoding RNAs (ncRNAs).

ncRNAs could be broadly grouped into distinct classes based on the lengths. Some classes have been known for long, e.g., ribosomal RNAs, transport RNAs, and small nucleolar RNAs (snoRNAs). Several novel ncRNA classes have been discovered, e.g., microRNAs (miRNAs), small interfering RNAs (siRNAs), PIWI-interacting RNAs (piRNAs), hairpin RNAs (hpRNAs), and long noncoding RNAs (lncRNAs) [31, 36, 51, 56]. The repertoire of ncRNAs continues to expand.

J. Chen
School of Chemistry, Biology and Material Engineering, Suzhou University of Science and Technology, No1. Kerui road, Suzhou, Jiangsu 215011, China

B. Shen (✉)
Center for Systems Biology, Soochow University, No1. Shizi Street, 206, Suzhou, Jiangsu 215006, China
e-mail: bairong.shen@suda.edu.cn

B. Shen et al. (eds.), *Translational Biomedical Informatics*, Advances in Experimental Medicine and Biology 939, DOI 10.1007/978-981-10-1503-8_2

Each of the ncRNA class has its unique biogenesis routes, three-dimensional structure, and modes of action. Therefore, the functional spectrum of ncRNAs is richer than expected. RNAs function as critical players at the epigenetic, transcriptional, and posttranscriptional level [40]. They regulate gene expression through diverse mechanisms, e.g., by mediating imprinting [57], alternative splicing [61], and modification of other small noncoding RNAs.

The huge amount of biological data generated by high-throughput sequencing technologies have opened new possibilities for RNA research field. On the other hand, these data in turn give rise to an urgent demand for proper visualization, analysis, storage, and interpretation of the data. It's imperative to find efficient bioinformatics methods to utilize these rich data source for a better understanding of the RNA world.

Clinical transcriptomics will have great impact on the therapeutic strategy of human disorders. The involvement of the RNA in human diseases has widely been investigated for microRNAs. MicroRNAs, however, are just the tip of the iceberg. The heterogeneous family of long noncoding RNAs (lncRNAs) may also participate in the progression of human diseases.

In the year of 2015, US government invested $215 million to launch the Precision Medicine Initiative (PMI). Precision medicine (PM) is a customized healthcare model. This model takes into account the individual variability and tailors the medical treatment to the individual patient. Precision medicine is dependent on molecular diagnostics or other molecular and cellular analyses to select appropriate therapies based on the context of a patient's genetic content [45].

Bioinformatics analyses integrating high-throughput microarray, next-generation sequencing (NGS), and genotyping data in disease cases and matched healthy controls consistently reveal changes in gene expression of both protein-coding and regulatory noncoding RNAs. The strong correlations between deregulated lncRNAs and disease development and prognosis highlight the potential of ncRNA as biomarkers and therapeutic targets to facilitate the precision medicine.

In this chapter, we give a brief overview of the bioinformatics tool to analyze several different aspects of RNAs, in particular ncRNAs. We first describe the emerging bioinformatics methods for RNA identification, structure modeling, functional annotation, and network inference. This is followed by an introduction of potential usefulness of ncRNAs as diagnostic, prognostic biomarkers and therapeutic strategies.

2.2 RNA Detection

RNA-seq is now a rich source to discover new ncRNA transcripts. It is also an interesting topic to further validate any novel transcripts discovered.

2.2.1 Detection of Small ncRNAs

Most of the currently available methods or algorithms that investigate NGS-generated transcriptomic data aim at detecting, predicting, and quantifying small RNAs, in particular microRNAs.

miRDeep [19] is the first stand-alone tool that identifies both novel and known microRNAs from large-scale sRNA-seq data. miRDeep evaluates the possibility of a hairpin structure by using Bayesian probability controls during the processes of microRNA biogenesis. miRDeep successfully identifies new miRNA candidates by searching for the characteristic read profile covering the mature miRNA and its complement miRNA*. In miRDeep2, the well-conserved structure of ncRNA families is used as a supplementary step to confirm a novel or known ncRNA [49]. miRDeep* [1] employs a miRNA precursor prediction algorithm to minimize the putative range of the precursor loci. It outperforms both versions of miRDeep by reducing false negatives.

In addition to the microRNA-specific tools, other approaches or pipelines generally concentrate on the whole family of small RNAs.

The web service DARIO is a comprehensive approach designed to predict and analyze different types of small RNAs generated from any next-generation RNA sequencing experiments [17]. The web server CPSS [67] makes further improvement on DARIO. CPSS is able to analyze small RNA-seq data that originate from single or paired samples. CPSS also predicts target genes for interested miRNAs and performs functional annotation of the predicted target genes. Different from CPSS, ncPRO-seq [7] is an integrated pipeline that identifies regions significantly enriched with short reads which do not belong to any known ncRNA families, thus allowing the identification of novel ncRNA- or siRNA-producing regions.

2.2.2 Detection of lncRNAs

Compared with messenger RNAs or small ncRNAs, detection of long ncRNAs from assembled transcripts is more complicated, because more deep sequencing reads are needed for lncRNAs to achieve sufficient coverage.

The main problem in the lncRNA detection is to reconstruct the transcripts from the short sequencing reads. In order to meet the ever-increasing need for effective NGS data mining, a variety of bioinformatics tools have been developed for NGS-based RNA transcriptome investigation. Below we will give a brief introduction on the currently available tools.

Sun et al. [59] developed a computational pipeline lncRScan to predict novel lncRNA from transcriptome sequencing data. lncRScan uses expression level as a filter to preclude artifacts or mRNAs from the initially assembled candidate transcripts. iSeeRNA [60] trained a support vector machine-based classifier using conservation, open reading frame, and sequence information as features. iSeeRNA could identify putative lincRNAs from RNA-seq data with high accuracy and speed. More recently, Musacchia et al. [53] have provided a pipeline, Annocript, that distinguishes lncRNAs from coding RNAs by combining functional annotation databases and sequence analysis tools. Most recently, Jiang and colleagues [28] developed LncRNA2Function, an ontology-driven web service which annotates lncRNAs with the functional ontologies of the groups of protein-coding genes that are coexpressed with them.

2.3 RNA Structure Prediction

The ncRNAs have now been recognized as an abundant class of genes which often function through their structure. During the past decades, RNA structure prediction has always been a research focus. Understanding of RNA structure not only helps the biologists to investigate the biological of the RNA in vivo but also will facilitate the clinicians to design novel strategies against genetic disorders.

Currently, RNA secondary structure determination is a challenging task. Unfortunately, most RNAs are currently difficult to crystallize. Moreover, it is rather difficult and expensive to determine RNA structures experimentally at atomic level. Therefore, developing mathematical and computational algorithms to determine the secondary structure of RNA from a known RNA sequence is very necessary.

A number of computational tools for secondary structure prediction have been proposed. A large majority of the methods are based on the thermodynamic folding of linear nucleotide sequences. Among them the most well known are the ViennaRNA package [43], Mfold [69], Rfold [32], and RNAalifold [3]. Instead of using energy parameters, probabilistic approaches provide an alternative to the thermodynamic RNA folding. Probabilistic approaches provide a probability distribution of common secondary structures of the input alignment. Popular programs include Pfold [33], PETfold [58], McCaskill-MEA, and CentroidAlifold.

Methods based merely on single-sequence folding are far from sufficient to accurately predict RNA structure. Therefore, some implementations of thermodynamic folding, e.g., UNAfold and RNAfold, also incorporate structure information to improve prediction accuracy. These folding algorithms restrict the folding of RNA sequence to structures that are consistent with the experimental constraints. There are also some attempts that predict RNA structure from homologous sequences of conserved species to improve secondary structure prediction.

2.4 Functional Analysis

RNAs have proved to be a gold mine of novel functions necessary for the entire human genome. Given the ever-growing high-throughput sequencing data, understanding the function of ncRNAs relies on computational approaches to functionally characterize ncRNAs. There are some general methods for ncRNA clustering and function prediction from sRNA-seq data.

ncRNAs are highly structured and interact with DNAs, RNAs, and proteins. This versatility suggests that ncRNAs might serve as links between the proteins and nucleotide sequences encoded by the entire genome. Therefore one key step in the study of ncRNA function includes finding lncRNA associations with other molecules. For that purpose, one needs approaches to directly analyze physical interactions. Basically, we have three major classes of interactions: RNA-protein interaction, RNA-RNA interaction, and RNA-DNA interaction.

RNA-protein interactions can be determined by microarray (RIP-CHIP) or high-throughput sequencing (RIP-seq). The bioinformatics analysis of RIP-seq and CLIP-seq data usually uses peak callers to reduce the false-positive rate. For CHIP-seq data, standard CHIP-seq peak callers like MACS [66] are often used. For RIP-seq data, specialized peak callers like PARalyzer [11] have to be used.

There are a multitude of tools to discover the sequence motifs for transcription factor-binding sites, with popular examples being MEME, DREME, MatrixREDUCE [18], and DRIMust [39]. These approaches are often used to analyze RNA-protein interaction data from RIP-seq or CLIP-seq experiments. In addition to the RNA sequence, other methods also consider the structural features of the RNAs in binding motif detection, e.g., BioBayesNet [54], MEMERIS [24], and RNAcontext [29].

RNA-DNA interactions can be detected with the CHIRP-seq method. Bioinformatics analysis using sequence motif detection tools of triple helices in binding sites can be performed in this case.

For RNA-RNA interactions, there have been very successful approaches to predicting these interactions for small RNAs, especially for miRNA target prediction.

Computational tools for microRNA target gene prediction are usually based on the complementarity between microRNAs and their target mRNAs. Different methods have been developed for microRNA target prediction. The initial strategies for target prediction are mainly based on sequence complementarity. However, these methods generally suffer from low accuracy and high false-positive rate. As an improvement, conservation analysis reduces the false-positive rate by filtering out the microRNA-binding sites that are not conserved among species.

Recent methods have further improved the target prediction accuracy using CLIP-seq data, e.g., Piranha [62] and CLIPZ [30], and single nucleotide polymorphism (SNP) data, e.g., PolymiRTS [68], Patrocles [23] and miRdSNP, at the target regions.

In addition, several bioinformatics tool, e.g., InMiR [5], miRGen [42], miRGator [8], SigTerms [12], and TopKCEMC [41], have integrated target prediction and expression information to reveal microRNA-mRNA interactions.

2.5 RNA Interpretation at Pathway Level

The cross talk between RNAs and other biological molecules has generated a wide-ranging regulatory network at transcriptomic level. In order to investigate RNAs at systems level, numerous studies have emerged that converge ncRNAs and their interaction partners on specific biological pathways.

These systematic association analyses can be used for large numbers of lncRNAs simultaneously. They have integrated the predicted targets of ncRNAs of interest with other -omics data, e.g., the functional annotation and expression profiles, and thus provided global insights into the regulatory role of ncRNAs in a broader range of posttranscriptional pathway and network.

A number of pathway analysis pipelines have been developed for this purpose; popular examples include DAVID, WebGestalt [64], miR2Disease [27], and miReg [2].

2.6 ncRNA-Based Biomarkers

The discovery of disease-specific biomarkers may provide useful predictive parameters for early diagnosis, prognosis, and prediction of therapeutic response and also provide potential drug targets.

It has been shown that defects in lncRNA expression and epigenetics are not the passenger, but a hallmark of human diseases, particularly in cancer. Moreover, ncRNAs are stable and can be readily isolated and detected in biological fluids using qRT-PCR amplification. Therefore, ncRNAs are of great potential diagnostic, prognostic, and therapeutic value.

We searched PubMed for articles about cancer-related ncRNAs with key words "*ncRNA* [tiab] AND (cancer[tiab] or carcinoma[tiab])." The searching results were listed in Fig. 2.1.

Articles about ncRNAs implicated in cancer have been increasing steadily during the last 15 years, implying a growing interest during the next few years.

An ideal biomarker should be noninvasive, easy to detect, cost effective, and consistent across heterogeneous patient groups. In this sense, the use of noncoding RNAs as biomarkers has some intrinsic advantages over the protein-coding RNAs.

Currently the majority of biomarkers are proteins or glycoproteins. The major drawbacks of these protein-based biomarkers are relatively low sensitivity and specificity as well as high false-positive rates. A prominent example is given by prostate-specific antigen (PSA), which is a glycoprotein secreted from the prostate gland. PSA testing has been widely applied in prostate cancer screening. PSA testing lacks specificity for prostate cancer. Serum PSA is overexpressed not only in prostate cancer but also in other benign prostate diseases such as prostatitis and prostatic hyperplasia. It was reported that only 30 % of the patients with high PSA are diagnosed with prostate cancer after histopathological analysis. PSA-based

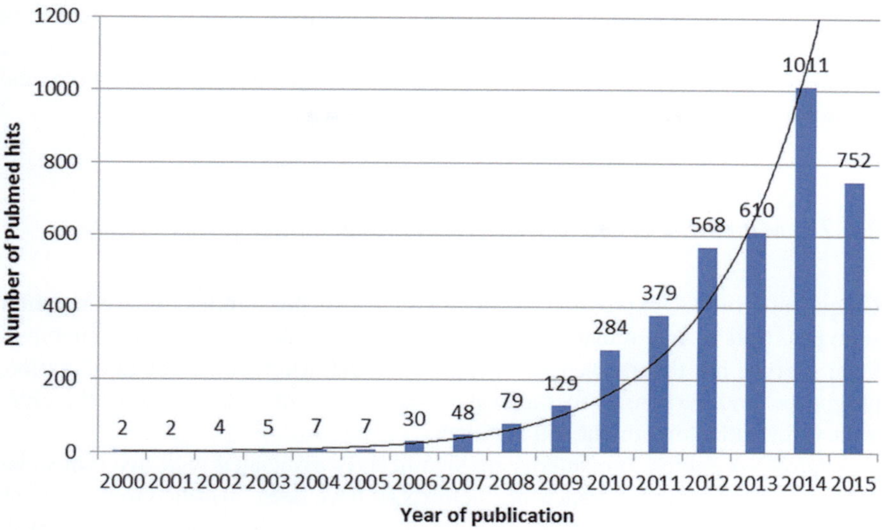

Fig. 2.1 The number of papers on cancer-related RNA during the past 15 years. The *bars* represent the number of PubMed hits for the keywords "*ncRNA* [tiab] AND (cancer[tiab] or carcinoma[tiab])"

screening may result in overdiagnosis and unnecessary treatment and does not help reduce PCa-associated mortality.

Different from protein-based biomarkers which are expressed from different tissues, most lncRNAs feature a tissue-specific expression pattern [14]. The tissue specificity of lncRNAs makes them ideal molecular signatures for clinical utilization. In the past decade, numerous studies have attempted to identify ncRNA-based molecular signatures for cancer diagnosis and prognosis. lncRNA and miRNA are two classes of the ncRNAs that have received special attention in oncologic research.

2.6.1 MicroRNAs in Cancer Diagnosis

MicroRNA is one of the most studied ncRNA classes in cancer research. MicroRNAs are short RNAs containing approximately 22 nucleotides. MicroRNAs silence target genes via partial base pairing with complementary regions of mRNAs and inhibit them from translation.

The target genes of microRNAs are found to play vital roles in many fundamental biological processes such as proliferation and apoptosis. Differential microRNA expression patterns have been extensively reported in many types of malignancies, in which they act either as oncogenes or as tumor suppressors.

Calin et al. [6] were the first to report microRNAs with differential expression in cancer samples. They found two microRNAs, hsa-miR15 and hsa-miR16, significantly downregulated in the majority of chronic lymphocytic leukemia (CLL) patients. Since then, many more reports have investigated aberrations in microRNA expression in a variety of cancer types including both hematological and solid neoplasms, summarized in a recent review [16].

2.6.2 MicroRNAs for Tumor Subtyping

Unique microRNA expression signatures can also differentiate between tumor subtypes. Cancer is a remarkably heterogeneous disease with multiple subtypes. Each subtype has diverse genetic backgrounds and different molecular alteration, prognosis, and responses to medical treatment. Tumor subtyping is therefore critical for patient treatment and survival.

MicroRNA expression patterns are significantly associated with five molecular subtypes of breast tumor. Panels of microRNAs have been identified to distinguish among breast cancer subtypes, e.g., luminal A [50], luminal B [44], HER2-enriched, basal-like, and normal-like breast cancer [4, 13, 63]. Recently, a microRNA classification system was developed to distinguish among four subtypes of renal cell carcinoma (RCC), clear cell RCC (ccRCC), papillary RCC (pRCC), chromophobe RCC (cRCC), and oncocytoma [65]. An eight-miRNA panel accurately distinguishes among the four subtypes of lung cancers, namely, small cell lung cancer (SCLC), NSCLC, squamous cell carcinoma, and carcinoid [20]. -hsa-miR21 and hsa-miR29b differentiate SCLC from NSCLC, while hsa-miR129 and hsa-miR205 are differentially expressed in squamous versus non-squamous NSCLC [37, 38]. A set of microRNAs was recently validated that differentiate among common subtypes of peripheral T-cell lymphoma (PTCL) [35].

2.6.3 MicroRNA-Based Therapeutic Strategies

Oncogenic microRNAs also provide new potential targets for anticancer therapies.

Inhibition of microRNA function can be achieved by using antisense oligonucleotides (ASOs). ASOs target microRNA molecules via base-pairing complementarity. In mammary cancer, administration of the antagonistic oligonucleotides targeting miR10b remarkably prevents lung metastases [47].

Alternatively, one could express "microRNA sponges" that competitively inhibit microRNA function. MicroRNA sponges contain artificial microRNA-binding sites and prevent microRNAs from binding with their natural targets [9]. This strategy has been adopted by Ma et al. to inhibit miR9 in breast cancer cells to reduce metastasis formation [48].

Another innovative strategy involves restoring the expression of tumor suppressor microRNAs that are downregulated in cancer. A microRNA replacement therapy was recently reported to restore miR26a expression in liver cancer. An adenoviral vector was used to deliver miR26a in murine model of hepatocellular carcinoma, leading to suppressed tumor growth and increased apoptosis [34].

2.6.4 lncRNAs in Diagnosis of Cancer

It's increasingly accepted that many cancer-associated risk loci are transcribed into lncRNAs and these noncoding transcripts are broadly involved in cancer transformation and progression.

These findings, along with tissue- and cancer-type-specific expression manner, make lncRNAs intriguingly potential diagnostic and prognostic markers. In the past decade, several large-scale studies have found lncRNA-based expression signatures in different types of cancer. A detailed list of the deregulated lncRNAs in cancer is provided in Table 2.1.

The PCA3 (prostate cancer antigen 3) is an established diagnostic marker in prostate cancer. First identified in 1999, it has already been translated into clinical practice. PCA3 is a prostate-specific gene markedly overexpressed in prostate cancer. It can be easily detected in urine or urine sediments. Compared with currently available biomarker prostate-specific antigen (PSA), PCA3 demonstrates higher prostate specificity and helps avoid unnecessary prostate biopsy. In addition, Progensa™, a commercialized PCA3 urine test, has been developed for general use in clinical testing.

Another interesting example for lncRNA-based diagnosis is the liver-specific RNA HULC (highly upregulated in liver cancer). HULC is detectable in primary liver tumors and hepatic metastasis of colorectal cancer, but absent in primary colon cancer or non-liver metastasis, indicating diagnostic potential of HULC.

2.6.5 lncRNAs in Prognosis of Cancer

Additionally, lncRNA expression was also found to be an independent prognostic parameter which predicts metastasis and patient outcome.

For example, HOTAIR can serve as a candidate prognostic marker of lymph node metastasis in HCC. Upregulation of HOTAIR also predicts recurrence in HCC patients who have received liver transplant. Moreover, HOTAIR expression positively correlates with primary breast cancer, stomach cancer, and colorectal cancers. The same is true for MALAT1 whose expression is an independent predictor of patient survival in early-stage lung cancer [26]. Furthermore, lncRNA expression also predicts patient response to chemotherapy. For instance, the expression of the XIST is strongly associated with the disease-free survival of cancer patients under Taxol therapy [25].

Table 2.1 Deregulated lncRNAs and their function in human cancers

Name	Cancer type	Location	Expression	Function	Clinical utility
3′aHIF-1α	Kidney, breast, paraganglioma	14q23.2	Up	Downregulates HIF-1α translation by binding to its 3′ UTR	Subtyping
5′aHIF-1α	Kidney	14q23.2	Up	Functions in membrane transport, regulates HIF-1α pathway	
ANRIL	Prostate, ALL, glioma, melanoma	9q21.3	Up	Epigenetically silences CDKN2B and adjacent tumor suppressor	
BC200	Breast, lung, ovary, tongue	2p21	Up	Protein binding	
CBR3-AS1	Prostate	21q22.1	Up	Reciprocal regulation with AR	
CTBP1-AS	Prostate	4p16.3	Up	Inhibits CTBP1 and other tumor suppressor genes in trans	
GAS5	Prostate	1q25.1	Down	Decoy of glucocorticoid receptor, induces apoptosis, suppresses proliferation	
GAS5	Kidney, breast	1q25.1	Down	Inhibits cell proliferation and invasion, induces cell apoptosis, arrests cell cycling	
HOTAIR	Prostate, breast, stomach, HCC, pancreatic, laryngeal, nasopharyngeal	12q13.13	Up	Guides PRC2 and LSD1/CoREST/REST complexes to HOXD locus for epigenetic silencing of tumor suppressor genes	
H19	Lung, cervix, esophagus, ovary, bladder, liver, breast, prostate, colorectal	11p15.5	Up	Paternally hypomethylated, silences IGF2 and NKD1, activates Wnt/β-catenin, promotes EMT	Prognosis
H19	Kidney	11p15.5	Down	Maternally methylated, causes biallelic expression of IGF2, precursor of miR675, promoted growth after hypoxia recovery and cell cycle progression and inhibit apoptosis	Risk prediction

HULC	Liver	6p24.3		MicroRNA decoy, mediated HBV-induced cell proliferation and anchorage-independent growth	
KCNQ1OT1	Kidney, breast, colon	11p15.5	Up	Binds PRC2 and G9a to promote H3K27me3 and H3K9me3, mediates the silencing of the flanking gene	
linc-UBC1	Bladder	1q32.1	Up	Recruits PRC2 complex and regulates histone methylation of target genes	Prognosis
Linc00963	Prostate	9q34.11	Up	Inhibits EGFR expression, attenuates motility and invasion, enhances apoptosis	Prognosis
MALAT1	Bladder, kidney, prostate, liver, lung, colorectal, colon, breast, uterus, pancreas, cervix, osteosarcoma, neuroblastoma	11q13.1	Up	Interacts with unmethylated Pc2, activates growth control gene, influences alternative splicing	Diagnosis and prognosis
MEG3	Bladder	14q32.3	Down	Hypomethylated, induces accumulation of p53 and stimulates p53-dependent transcription	
MEG3	Kidney, brain	14q32.3	Up	Hypermethylated, downregulates DLK1	
ncRNACCND1	Prostate	11q13	Up	Recruits TLS to inhibit CBP/p300 activities on the CCND1	
ncRAN	Bladder, colon, cervix, neuroblastoma	17q25.1	Up	Promotes tumor growth, invasion, and survival	
NEAT1	Prostate	11q13.1	Up	Induces H3K4Me3 and H3AcK9 at the promoter of PSMA and GJB1	
PCA3	Prostate	9q21-22	Up	Modulates AR signaling	Diagnosis
PCAN-R1	Prostate	1q32.1	Up	Inhibits cell growth and soft-agar colony formation	Prognosis

(continued)

Table 2.1 (continued)

Name	Cancer type	Location	Expression	Function	Clinical utility
PCAN-R2	Prostate	9q22.32	Up	Inhibits cell proliferation and soft-agar colony formation	Prognosis
PCAT1	Prostate	8q24.21	Up	Downregulates BRCA2, disrupts miR34a-mediated downregulation of cMyc	Prognosis
PCAT18	Prostate	18q11.2	Up	Inhibits apoptosis and migration	Prognosis
PCAT29	Prostate	15q23	Down	Suppresses growth and metastasis	Prognosis
PCGEM1	Prostate	2q32	Up	Promotes cell proliferation, delays p21 and p53 expression, inhibits apoptosis, miRNA decoy	Risk prediction; prognosis
PRNCR1	Prostate	8q24	Up	Increases viability in the androgen-dependent LNCaP cells	Risk prediction
PTENP1	Prostate	9p21	Down	Interferes with the regulation of PTEN by miRNAs	
RASSF1-AS1	Prostate	3p21.3	Up	Represses RASSF1A transcription via H3K27Me3	Prognosis
RMRP	Leukemia, lymphoma	9p21-p12		Mitochondrial RNA processing endoribonuclease, hTERT dependent	
SCHLAP1	Prostate	2q31.1	Up	Impairs SWI/SNF-mediated epigenetic regulation	Prognosis
SNHG16	Bladder	17q25.1	Up	Promotes cell growth, migration, and invasion	Prognosis
SPRY4-IT1	Kidney	5q31.3	Up	Enhances proliferation, migration, and invasion	Prognosis
SRA-1/SRA	Breast, uterus, ovary	5q31.3		Binding transcription	

TRPM2-AS	Prostate	21q22.3	Up	Silences TRPM2 expression	Prognosis Therapeutic target
TUG1	Bladder	22q12.2	Up	Represses genes involved in cell cycle regulation	
UCA1	Bladder, lung, thyroid, liver, breast, esophagus, stomach	19p13.12	Up	Regulates cell cycle distribution via CREB through PI3K-dependent pathway	Diagnosis
XIST	Prostate	Xq13.2		Hypomethylated in PCa, recruits PRC2 onto the Xic and initiates H3K27me3	Diagnosis

2.6.6 lncRNA-Based Therapeutic Strategies

lncRNAs also provide new potential targets for anticancer therapies. One could block the activity of oncogenic lncRNAs that are upregulated in cancer in several ways.

First, by decreasing oncogenic lncRNA expression, one of the most explored methods is to deliver small interfering RNAs (siRNAs) or longer antisense oligonucleotides (ASOs) that are complementary to the lncRNA targets. While siRNAs induce lncRNA degradation in RISC (RNA-induced silencing) complex, ASO function through RNase H1 catalyzed cleavage of the RNA molecules.

For example, inhibition of HOTAIR by siRNAs decreased invasiveness of breast cancer cells [21], attenuated proliferation of pancreatic cancer, and improved the chemotherapeutic sensitivity of liver cancer cell lines. siRNA-mediated depletion of HULC compromised HCC cell progression. Knockdown of MALAT1 by ASO attenuated malignant phenotypes in cervical cancer cells and blocked metastasis of lung cancer cells.

Taking into account their interaction with regulatory proteins, additional therapeutic strategies can be designed that disrupt lncRNA-protein interactions. Therapeutic strategies that block the protein-microRNA interactions could be applied to the lncRNA field. The strategy for inhibiting microRNAs is also applicable to lncRNAs. This could be achieved via the use of either antagonistic oligonucleotide that targets the ncRNA and blocks their binding. Alternatively, small-molecule inhibitors that block the binding site of lncRNA partners can be used to functionally silence lncRNAs.

The tissue-specific expression pattern of lncRNAs has translated them into preliminary clinical trials. For example, H19 is an extensively investigated lncRNA which has entered clinical trial making use of its high tissue-specific expression. Hochberg et al. have developed an expression vector BC-819 containing diphtheria toxin A, under control of the regulatory sequences of H19 genes [55]. This vector allows for a tissue-specific expression of diphtheria toxin A. When administered intratumorally, the vector significantly inhibited tumor proliferation without any toxicity to the surrounding cells. A phase 2 clinical trial of the BC-819 was conducted among unresectable pancreatic cancer patients [22]. Recent studies have yielded promising results of H19 in many other malignancies, e.g., bladder, colon, and ovarian cancers.

Alternatively, one could find a way to restore the expression level of tumor-suppressive lncRNAs such as GAS5 [52] and TERRA [46] or to administer synthetic lncRNA mimics.

2.7 Conclusions

The examples of ncRNAs provided in this chapter have showed the potential clinical utility of these novel transcripts. High-throughput human genomic information can be fully utilized only if these noncoding fractions of the human transcriptome are well understood. The bioinformatics tools discussed above will shed light on the regulatory role of ncRNAs and improve the diagnosis and prognosis of diseases toward the ultimate goal of precision medicine.

References

1. An J, Lai J, Lehman ML, Nelson CC. miRDeep*: an integrated application tool for miRNA identification from RNA sequencing data. Nucleic Acids Res. 2013;41(2):727–37.
2. Barh D, Bhat D, Viero C. miReg: a resource for microRNA regulation. J Int Bioinform. 2010;7 (1). doi:10.2390/biecoll-jib-2010-144.
3. Bernhart SH, Hofacker IL, Will S, Gruber AR, Stadler PF. RNAalifold: improved consensus structure prediction for RNA alignments. BMC Bioinf. 2008;9:474.
4. Blenkiron C, Goldstein LD, Thorne NP, Spiteri I, Chin SF, Dunning MJ, Barbosa-Morais NL, Teschendorff AE, Green AR, Ellis IO, et al. MicroRNA expression profiling of human breast cancer identifies new markers of tumor subtype. Genome Biol. 2007;8(10):R214.
5. Bruno AE, Li L, Kalabus JL, Pan Y, Yu A, Hu Z. miRdSNP: a database of disease-associated SNPs and microRNA target sites on 3'UTRs of human genes. BMC Genomics. 2012;13:44.
6. Calin GA, Dumitru CD, Shimizu M, Bichi R, Zupo S, Noch E, Aldler H, Rattan S, Keating M, Rai K, et al. Frequent deletions and down-regulation of micro- RNA genes miR15 and miR16 at 13q14 in chronic lymphocytic leukemia. Proc Natl Acad Sci U S A. 2002;99 (24):15524–9.
7. Chen CJ, Servant N, Toedling J, Sarazin A, Marchais A, Duvernois-Berthet E, Cognat V, Colot V, Voinnet O, Heard E, et al. ncPRO-seq: a tool for annotation and profiling of ncRNAs in sRNA-seq data. Bioinformatics. 2012;28(23):3147–9.
8. Cho S, Jun Y, Lee S, Choi HS, Jung S, Jang Y, Park C, Kim S, Lee S, Kim W. miRGator v2.0: an integrated system for functional investigation of microRNAs. Nucleic Acids Res. 2011;39 (Database issue):D158–62.
9. Cohen SM. Use of microRNA sponges to explore tissue-specific microRNA functions in vivo. Nat Methods. 2009;6(12):873–4.
10. Consortium EP, Birney E, Stamatoyannopoulos JA, Dutta A, Guigo R, Gingeras TR, Margulies EH, Weng Z, Snyder M, Dermitzakis ET, et al. Identification and analysis of functional elements in 1% of the human genome by the ENCODE pilot project. Nature. 2007;447(7146):799–816.
11. Corcoran DL, Georgiev S, Mukherjee N, Gottwein E, Skalsky RL, Keene JD, Ohler U. PARalyzer: definition of RNA binding sites from PAR-CLIP short-read sequence data. Genome Biol. 2011;12(8):R79.
12. Creighton CJ, Nagaraja AK, Hanash SM, Matzuk MM, Gunaratne PH. A bioinformatics tool for linking gene expression profiling results with public databases of microRNA target predictions. RNA. 2008;14(11):2290–6.
13. de Rinaldis E, Gazinska P, Mera A, Modrusan Z, Fedorowicz GM, Burford B, Gillett C, Marra P, Grigoriadis A, Dornan D, et al. Integrated genomic analysis of triple-negative breast cancers reveals novel microRNAs associated with clinical and molecular phenotypes and sheds light on the pathways they control. BMC Genomics. 2013;14:643.

14. Djebali S, Davis CA, Merkel A, Dobin A, Lassmann T, Mortazavi A, Tanzer A, Lagarde J, Lin W, Schlesinger F, et al. Landscape of transcription in human cells. Nature. 2012;489 (7414):101–8.
15. Elgar G, Vavouri T. Tuning in to the signals: noncoding sequence conservation in vertebrate genomes. Trends Genet. 2008;24(7):344–52.
16. Faruq O, Vecchione A. microRNA: diagnostic perspective. Front Med. 2015;2:51.
17. Fasold M, Langenberger D, Binder H, Stadler PF, Hoffmann S. DARIO: a ncRNA detection and analysis tool for next-generation sequencing experiments. Nucleic Acids Res. 2011;39: W112–7.
18. Foat BC, Tepper RG, Bussemaker HJ. TransfactomeDB: a resource for exploring the nucleotide sequence specificity and condition-specific regulatory activity of trans-acting factors. Nucleic Acids Res. 2008;36(Database issue):D125–31.
19. Friedlander MR, Chen W, Adamidi C, Maaskola J, Einspanier R, Knespel S, Rajewsky N. Discovering microRNAs from deep sequencing data using miRDeep. Nat Biotechnol. 2008;26(4):407–15.
20. Gilad S, Lithwick-Yanai G, Barshack I, Benjamin S, Krivitsky I, Edmonston TB, Bibbo M, Thurm C, Horowitz L, Huang Y, et al. Classification of the four main types of lung cancer using a microRNA-based diagnostic assay. J Mol Diagn. 2012;14(5):510–7.
21. Gupta RA, Shah N, Wang KC, Kim J, Horlings HM, Wong DJ, Tsai MC, Hung T, Argani P, Rinn JL, et al. Long non-coding RNA HOTAIR reprograms chromatin state to promote cancer metastasis. Nature. 2010;464(7291):1071–6.
22. Hanna N, Ohana P, Konikoff FM, Leichtmann G, Hubert A, Appelbaum L, Kopelman Y, Czerniak A, Hochberg A. Phase 1/2a, dose-escalation, safety, pharmacokinetic and preliminary efficacy study of intratumoral administration of BC-819 in patients with unresectable pancreatic cancer. Cancer Gene Ther. 2012;19(6):374–81.
23. Hiard S, Charlier C, Coppieters W, Georges M, Baurain D. Patrocles: a database of polymorphic miRNA-mediated gene regulation in vertebrates. Nucleic Acids Res. 2010;38(Database issue):D640–51.
24. Hiller M, Pudimat R, Busch A, Backofen R. Using RNA secondary structures to guide sequence motif finding towards single-stranded regions. Nucleic Acids Res. 2006;34(17):e117.
25. Huang KC, Rao PH, Lau CC, Heard E, Ng SK, Brown C, Mok SC, Berkowitz RS, Ng SW. Relationship of XIST expression and responses of ovarian cancer to chemotherapy. Mol Cancer Ther. 2002;1(10):769–76.
26. Ji P, Diederichs S, Wang W, Boing S, Metzger R, Schneider PM, Tidow N, Brandt B, Buerger H, Bulk E, et al. MALAT-1, a novel noncoding RNA, and thymosin beta4 predict metastasis and survival in early-stage non-small cell lung cancer. Oncogene. 2003;22 (39):8031–41.
27. Jiang Q, Wang Y, Hao Y, Juan L, Teng M, Zhang X, Li M, Wang G, Liu Y. miR2Disease: a manually curated database for microRNA deregulation in human disease. Nucleic Acids Res. 2009;37(Database issue):D98–104.
28. Jiang Q, Ma R, Wang J, Wu X, Jin S, Peng J, Tan R, Zhang T, Li Y, Wang Y. LncRNA2Function: a comprehensive resource for functional investigation of human lncRNAs based on RNA-seq data. BMC Genomics. 2015;16 Suppl 3:S2.
29. Kazan H, Ray D, Chan ET, Hughes TR, Morris Q. RNAcontext: a new method for learning the sequence and structure binding preferences of RNA-binding proteins. PLoS Comput Biol. 2010;6:e1000832.
30. Khorshid M, Rodak C, Zavolan M. CLIPZ: a database and analysis environment for experimentally determined binding sites of RNA-binding proteins. Nucleic Acids Res. 2011;39 (Database issue):D245–52.
31. Kim VN. Small RNAs just got bigger: piwi-interacting RNAs (piRNAs) in mammalian testes. Genes Dev. 2006;20(15):1993–7.
32. Kiryu H, Kin T, Asai K. Rfold: an exact algorithm for computing local base pairing probabilities. Bioinformatics. 2008;24(3):367–73.

33. Knudsen B, Hein J. Pfold: RNA secondary structure prediction using stochastic context-free grammars. Nucleic Acids Res. 2003;31(13):3423–8.
34. Kota J, Chivukula RR, O'Donnell KA, Wentzel EA, Montgomery CL, Hwang HW, Chang TC, Vivekanandan P, Torbenson M, Clark KR, et al. Therapeutic microRNA delivery suppresses tumorigenesis in a murine liver cancer model. Cell. 2009;137(6):1005–17.
35. Laginestra MA, Piccaluga PP, Fuligni F, Rossi M, Agostinelli C, Righi S, Sapienza MR, Motta G, Gazzola A, Mannu C, et al. Pathogenetic and diagnostic significance of microRNA deregulation in peripheral T-cell lymphoma not otherwise specified. Blood Cancer J. 2014;4:259.
36. Lau NC, Seto AG, Kim J, Kuramochi-Miyagawa S, Nakano T, Bartel DP, Kingston RE. Characterization of the piRNA complex from rat testes. Science. 2006;313(5785):363–7.
37. Lebanony D, Benjamin H, Gilad S, Ezagouri M, Dov A, Ashkenazi K, Gefen N, Izraeli S, Rechavi G, Pass H, et al. Diagnostic assay based on hsa-miR-205 expression distinguishes squamous from nonsquamous non-small-cell lung carcinoma. J Clin Oncol. 2009;27 (12):2030–7.
38. Lee JH, Voortman J, Dingemans AM, Voeller DM, Pham T, Wang Y, Giaccone G. MicroRNA expression and clinical outcome of small cell lung cancer. PLoS One. 2011;6(6):e21300.
39. Leibovich L, Paz I, Yakhini Z, Mandel-Gutfreund Y. DRIMust: a web server for discovering rank imbalanced motifs using suffix trees. Nucleic Acids Res. 2013;41(Web Server issue): W174–9.
40. Lewis BP, Burge CB, Bartel DP. Conserved seed pairing, often flanked by adenosines, indicates that thousands of human genes are microRNA targets. Cell. 2005;120(1):15–20.
41. Lin S, Ding J. Integration of ranked lists via cross entropy Monte Carlo with applications to mRNA and microRNA Studies. Biometrics. 2009;65(1):9–18.
42. Lindow M, Gorodkin J. Principles and limitations of computational microRNA gene and target finding. DNA Cell Biol. 2007;26(5):339–51.
43. Lorenz R, Bernhart SH, Honer Zu Siederdissen C, Tafer H, Flamm C, Stadler PF, Hofacker IL. ViennaRNA Package 2.0. Algorithms Mol Biol. 2011;6:26.
44. Lowery AJ, Miller N, Devaney A, McNeill RE, Davoren PA, Lemetre C, Benes V, Schmidt S, Blake J, Ball G, et al. MicroRNA signatures predict oestrogen receptor, progesterone receptor and HER2/neu receptor status in breast cancer. Breast Cancer Res. 2009;11(3):R27.
45. Lu YF, Goldstein DB, Angrist M, Cavalleri G. Personalized medicine and human genetic diversity. Cold Spring Harbor Perspect Med. 2014;4(9):a008581.
46. Luke B, Lingner J. TERRA: telomeric repeat-containing RNA. EMBO J. 2009;28 (17):2503–10.
47. Ma L, Reinhardt F, Pan E, Soutschek J, Bhat B, Marcusson EG, Teruya-Feldstein J, Bell GW, Weinberg RA. Therapeutic silencing of miR-10b inhibits metastasis in a mouse mammary tumor model. Nat Biotechnol. 2010;28(4):341–7.
48. Ma L, Young J, Prabhala H, Pan E, Mestdagh P, Muth D, Teruya-Feldstein J, Reinhardt F, Onder TT, Valastyan S, et al. miR-9, a MYC/MYCN-activated microRNA, regulates E-cadherin and cancer metastasis. Nat Cell Biol. 2010;12(3):247–56.
49. Mackowiak SD. Identification of novel and known miRNAs in deep-sequencing data with miRDeep2. In: Current protocols in bioinformatics/editorial board, Andreas D Baxevanis [et al]. 2011, Chapter 12:Unit 12 10.
50. McDermott AM, Miller N, Wall D, Martyn LM, Ball G, Sweeney KJ, Kerin MJ. Identification and validation of oncologic miRNA biomarkers for luminal A-like breast cancer. PLoS One. 2014;9(1):e87032.
51. Mitra SA, Mitra AP, Triche TJ. A central role for long non-coding RNA in cancer. Front Genet. 2012;3:17.
52. Mourtada-Maarabouni M, Pickard MR, Hedge VL, Farzaneh F, Williams GT. GAS5, a non-protein-coding RNA, controls apoptosis and is downregulated in breast cancer. Oncogene. 2009;28(2):195–208.

53. Musacchia F, Basu S, Petrosino G, Salvemini M, Sanges R. Annocript: a flexible pipeline for the annotation of transcriptomes able to identify putative long noncoding RNAs. Bioinformatics. 2015;31(13):2199–201.
54. Nikolajewa S, Pudimat R, Hiller M, Platzer M, Backofen R. BioBayesNet: a web server for feature extraction and Bayesian network modeling of biological sequence data. Nucleic Acids Res. 2007;35(Web Server issue):W688–93.
55. Ohana P, Bibi O, Matouk I, Levy C, Birman T, Ariel I, Schneider T, Ayesh S, Giladi H, Laster M, et al. Use of H19 regulatory sequences for targeted gene therapy in cancer. Int J Cancer. 2002;98(5):645–50.
56. Okamura K, Chung WJ, Ruby JG, Guo H, Bartel DP, Lai EC. The Drosophila hairpin RNA pathway generates endogenous short interfering RNAs. Nature. 2008;453(7196):803–6.
57. Redrup L, Branco MR, Perdeaux ER, Krueger C, Lewis A, Santos F, Nagano T, Cobb BS, Fraser P, Reik W. The long noncoding RNA Kcnq1ot1 organises a lineage-specific nuclear domain for epigenetic gene silencing. Development. 2009;136(4):525–30.
58. Seemann SE, Menzel P, Backofen R, Gorodkin J. The PETfold and PETcofold web servers for intra- and intermolecular structures of multiple RNA sequences. Nucleic Acids Res. 2011;39 (Web Server issue):W107–11.
59. Sun L, Zhang Z, Bailey TL, Perkins AC, Tallack MR, Xu Z, Liu H. Prediction of novel long non-coding RNAs based on RNA-Seq data of mouse Klf1 knockout study. BMC Bioinf. 2012;13:331.
60. Sun K, Chen X, Jiang P, Song X, Wang H, Sun H. iSeeRNA: identification of long intergenic non-coding RNA transcripts from transcriptome sequencing data. BMC Genomics. 2013;14 Suppl 2:S7.
61. Tripathi V, Ellis JD, Shen Z, Song DY, Pan Q, Watt AT, Freier SM, Bennett CF, Sharma A, Bubulya PA, et al. The nuclear-retained noncoding RNA MALAT1 regulates alternative splicing by modulating SR splicing factor phosphorylation. Mol Cell. 2010;39(6):925–38.
62. Uren PJ, Bahrami-Samani E, Burns SC, Qiao M, Karginov FV, Hodges E, Hannon GJ, Sanford JR, Penalva LO, Smith AD. Site identification in high-throughput RNA-protein interaction data. Bioinformatics. 2012;28(23):3013–20.
63. van Schooneveld E, Wildiers H, Vergote I, Vermeulen PB, Dirix LY, Van Laere SJ. Dysregulation of microRNAs in breast cancer and their potential role as prognostic and predictive biomarkers in patient management. Breast Cancer Res. 2015;17:21.
64. Wang J, Duncan D, Shi Z, Zhang B. WEB-based GEne SeT AnaLysis Toolkit (WebGestalt): update 2013. Nucleic Acids Res. 2013;41(Web Server issue):W77–83.
65. Youssef YM, White NM, Grigull J, Krizova A, Samy C, Mejia-Guerrero S, Evans A, Yousef GM. Accurate molecular classification of kidney cancer subtypes using microRNA signature. Eur Urol. 2011;59(5):721–30.
66. Zhang Y, Liu T, Meyer CA, Eeckhoute J, Johnson DS, Bernstein BE, Nusbaum C, Myers RM, Brown M, Li W, et al. Model-based analysis of ChIP-Seq (MACS). Genome Biol. 2008;9(9): R137.
67. Zhang Y, Xu B, Yang Y, Ban R, Zhang H, Jiang X, Cooke HJ, Xue Y, Shi Q. CPSS: a computational platform for the analysis of small RNA deep sequencing data. Bioinformatics. 2012;28(14):1925–7.
68. Ziebarth JD, Bhattacharya A, Chen A, Cui Y. PolymiRTS Database 2.0: linking polymorphisms in microRNA target sites with human diseases and complex traits. Nucleic Acids Res. 2012;40(Database issue):D216–21.
69. Zuker M. Mfold web server for nucleic acid folding and hybridization prediction. Nucleic Acids Res. 2003;31(13):3406–15.

Chapter 3
Exploring Human Diseases and Biological Mechanisms by Protein Structure Prediction and Modeling

Juexin Wang, Joseph Luttrell IV, Ning Zhang, Saad Khan, NianQing Shi, Michael X. Wang, Jing-Qiong Kang, Zheng Wang, and Dong Xu

Abstract Protein structure prediction and modeling provide a tool for understanding protein functions by computationally constructing protein structures from amino acid sequences and analyzing them. With help from protein prediction tools and web servers, users can obtain the three-dimensional protein structure models and gain knowledge of functions from the proteins. In this chapter, we will provide several examples of such studies. As an example, structure modeling methods were used to investigate the relation between mutation-caused misfolding of protein and human diseases including epilepsy and leukemia. Protein structure

J. Wang
Department of Computer Science, University of Missouri, Columbia, MO 65211, USA

Christopher S. Bond Life Sciences Center, University of Missouri, Columbia, MO 65211, USA

J. Luttrell IV • Z. Wang
School of Computing, University of Southern Mississippi, 118 College Drive, Hattiesburg, MS 39406, USA

N. Zhang • S. Khan
Christopher S. Bond Life Sciences Center, University of Missouri, Columbia, MO 65211, USA

Informatics Institute, University of Missouri, Columbia, MO 65211, USA

N. Shi
Department of Medicine, Division of Cardiovascular Medicine, University of Wisconsin, Room 8418, 1111 Highland Ave, Madison, WI 53706, USA

M.X. Wang
Department of Pathology and Anatomical Sciences, University of Missouri, Columbia, MO 65211, USA

J.-Q. Kang
Department of Neurology, Vanderbilt University Medical Center, Nashville, TN 37232, USA

D. Xu (✉)
Department of Computer Science, University of Missouri, Columbia, MO 65211, USA

Christopher S. Bond Life Sciences Center, University of Missouri, Columbia, MO 65211, USA

Informatics Institute, University of Missouri, Columbia, MO 65211, USA
e-mail: xudong@missouri.edu

B. Shen et al. (eds.), *Translational Biomedical Informatics*, Advances in Experimental Medicine and Biology 939, DOI 10.1007/978-981-10-1503-8_3

prediction and modeling were also applied in nucleotide-gated channels and their interaction interfaces to investigate their roles in brain and heart cells. In molecular mechanism studies of plants, rice salinity tolerance mechanism was studied via structure modeling on crucial proteins identified by systems biology analysis; trait-associated protein-protein interactions were modeled, which sheds some light on the roles of mutations in soybean oil/protein content. In the age of precision medicine, we believe protein structure prediction and modeling will play more and more important roles in investigating biomedical mechanism of diseases and drug design.

Keywords Protein structure modeling • Protein structure prediction • Biological mechanism • Protein misfolding • Sequence mutation • Human disease • GWAS • Plant breeding

3.1 Introduction

As the most versatile macromolecules in living organisms ranging from bacteria to human, proteins serve crucial functions in essentially all biological processes [55]. Folding from an amino acid sequence, three-dimensional structures of proteins often gave us informative knowledge of protein functions. However, only a limited number of three-dimensional structures of proteins were known experimentally (116, 258 in Protein Data Bank (PDB) [4] as of Feb. 25, 2016), in contrast to 60, 971, 489 known protein sequence entries in Release 2016_02 of Feb. 17, 2016, of the UniProtKB/TrEMBL database [2]. This huge gap between protein sequence and structure makes protein structure prediction and modeling more and more important, i.e., to computationally predict the protein three-dimensional structure from its amino acid sequence and analyze the structural model. With decades of efforts, numerous protein structure prediction methods, software tools, and web servers have been developed and deployed. Comparing with the experimental approaches, computational structural prediction and modeling are quicker, cheaper, and becoming more and more reliable. Researchers could use these tools to model the target protein that they are interested in, and the structure models could help them obtain further insight on the function of the target protein and its role and mechanism of the underlying biological process.

This book chapter starts from a brief overview of current protein structure prediction tools and mainstream protein function prediction methods (Sect. 3.1). Several case studies using structure modeling are presented. The first case study is how to use ligand binding prediction server to study the proposed protein function (Sect. 3.2). Then we demonstrate how structure modeling is used in human protein studies on exploring diseases (Sect. 3.3). Plant studies on abiotic stress and agricultural traits using protein structure models are also presented (Sect. 3.4).

3.1.1 Protein Structure Prediction Methods

Computational protein structure prediction methods can be generally classified into three categories: ab initio prediction [26, 37, 79], comparative modeling (CM) [8, 65], and threading [63, 78, 82]. Different from CM and threading using other known protein structures as templates, ab initio methods predict a protein structure by optimizing some scoring functions based on the physical/statistical properties of proteins. CM methods, based on the fact that evolutionarily related proteins typically share a similar structure, build models for the target protein by aligning the target sequence to evolutionarily related (i.e., homologous) template structures. Threading methods are designed to find and align the target sequence to templates of similar structural folds, where target and template sequences are not required to be evolutionary related. Theoretically, ab initio prediction could discover new structural folds with more computational resources, but it has not been consistently successful. Template-based methods often obtain high-resolution models with the available templates and accurate alignments.

To advance the development of protein structure prediction, Critical Assessment of Techniques for Protein Structure Prediction (CASP) [32] was set up in 1994 to provide an objective assessment of the state of the art in the field. Since then, 11 CASPs have been done on a biannual basis. Table 3.1 shows several protein structure prediction servers that achieved success in the CASPs.

Table 3.1 A sample of current tools in the field of protein structure prediction

Method/server	URL	Brief description
HHpred Soding et al. [65]	http://toolkit.tuebingen.mpg.de/hhpred	Structure prediction by sensitive HMM-HMM search of template
I-TASSER Zhang [82]	http://zhanglab.ccmb.med.umich.edu/I-TASSER/	A hierarchical approach using multiple threading results and conformation sampling
MUFOLD [83, 84]	http://mufold.org/	Graph-based model generation using MDS and comprehensive model quality assessment
MULTICOM Wang et al. [73]	http://sysbio.rnet.missouri.edu/multicom_cluster/	Model generation using multi-template comparative modeling and refinement
RaptorX Kallberg et al. [22]	http://raptorx.uchicago.edu/StructurePrediction/predict/	Highly sensitive method for remote homolog identification and alignment
ROBETTA/ROSETTA Kim et al. [26]	http://www.robetta.org/submit.jsp	Model generation using both ab initio and comparative models of protein domains

3.1.2 Protein Function Prediction via Protein-Ligand Binding Comparison

A related problem to protein structure prediction is protein function prediction, which occupies an interesting and challenging area of computational research. In order to understand the process behind predicting a protein's function, it is necessary to understand what is meant by the term "function." For many proteins, this is a computationally challenging problem and can lead to answers that reveal much more than a single biological effect [6, 14]. However, regardless of the difficulty of the question, the goal of studying protein function is typically to simply understand what a given protein "does" in nature [14]. Furthermore, even though the details surrounding the question of a protein's function typically share a stronger relationship with biological research, development of methods for computational prediction of protein function continues to be fueled by the abundance of possible functions and the relative difficulty of determining them experimentally [14].

Just as there are many different functional roles that a protein may have, there are many different ways to predict these functions. One interesting method currently being researched involves protein-ligand binding site comparison. In this kind of prediction, previously known or predicted binding sites are used to infer the functional similarity of two proteins. Essentially, the main argument behind the usefulness of these predictions is the fact that binding sites of proteins with similar biological functions are typically better conserved than many other structural elements in evolutionarily distant proteins [29]. In other words, being able to predict the regions of a protein where binding with a ligand takes place can be a helpful clue in determining the function of that protein by allowing comparison with proteins that have similar binding and classified functions.

As a result, a number of tools and databases have been developed using these methods. Even within this closely related group of tools, the methods used vary. For example, ProBiS predicts ligand binding sites that may exist on a given query protein by assessing the surface structure of that protein and then comparing it to a database of proteins to find structurally similar binding sites [28, 31]. Some other tools, such as CAST, focus on predicting binding sites through the automatic location and measurement of regions on the input protein known as pockets [38]. Table 3.2 lists a few more of the software packages and projects that are currently utilizing these techniques.

In addition to these servers, a number of databases also contain protein-ligand binding information. These services often distinguish themselves by including information from differing sources and by utilizing differing search algorithms. For example, the PoSSuM database can rapidly compare binding sites among structures with similar or differing global folds [21]. Databases like this may provide users with an idea of the function of similar structures without the need for computationally expensive predictions. In order to make use of the information available from these databases, projects like GIRAF continue to experiment with different data handling approaches [46].

Table 3.2 A sample collection of current tools in the field of protein function prediction that are based on protein-ligand binding comparison

Method/server	URL	Summary
CAST Liang et al. [38]	http://sts.bioe.uic.edu/castp/	Predicting binding sites through shape matching
CatSId Nilmeier et al. [48] and Kirshner et al. [27]	http://catsid.llnl.gov/	Predicting catalytic sites by using a structure matching algorithm
COFACTOR Roy et al. [56]	http://zhanglab.ccmb.med.umich.edu/COFACTOR	Structure-based function predictions (ligand binding sites, GO, and enzyme commission)
DoGSiteScorer Volkamer et al. [71]	http://dogsite.zbh.uni-hamburg.de/	Pocket detection on protein surface and druggability prediction
Nucleos Friedberg [14] and Gherardini et al. [16]	http://nucleos.bio.uniroma2.it/nucleos/	Binding site prediction for different types of nucleotides
SA-Mot Friedberg [14]	http://sa-mot.mti.univ-paris-diderot.fr/main/SA_Mot_Method	Using HMM-SA to extract and describe structural motifs from protein loop structures
SiteBinder	http://webchem.ncbr.muni.cz/Platform/App/SiteBinder	Building models of binding sites and superimposing an arbitrary number of small protein fragments
ProBiS Konc and Janezic [28] and Konc et al. [31]	http://probis.nih.gov/	Ligand binding site prediction by searching for protein surface with known binding of ligand

3.1.3 Protein Structure Modeling in the Era of Precision Medicine

Precision medicine often uses invaluable genetics information of many complex diseases. Researchers may explore the link between disease and nonsynonymous variation in a large scale, i.e., both population/family scale and individual whole genome-wide scale [15, 60]. Several tools such as SIFT and MutationTaster were developed to evaluate and predict the exon mutation effects on biological functions [33, 60]. However, the potential of using protein structure prediction and modeling for this purpose has been under-explored. There will be increasingly demands in accurately modeling comparative structures between wide-type and mutated proteins. These millions of sequencing data will also advance the method improvement of structure modeling, and the systems biology that incorporates different levels of biological information will expand the usefulness of protein structure information for more comprehensive understanding of biological mechanism. Furthermore, protein design and structure-based drug design will also benefit more and more from integrating the structural information and systems biology data for precision medicine.

3.2 Predicting the Protein Function: A Case Study

One way to start the process of predicting the function of a protein is to obtain protein-ligand binding site predictions from a web server such as ProBiS [31]. Using this software simply involves uploading a protein structure file. Also, advanced options include choosing which proteins from the database to compare with the query protein and choosing different limits for the scoring function that judges the results. In order to generate these predictions, ProBiS analyzes the solvent accessible surface of the query protein and compares "patches" of this surface with entries in a database of proteins known as the nr-PDB (nonredundant Protein Data Bank) [28]. Once the final predictions have been made, the user is sent an email with a link to a results page that contains visualization tools to aid in interpreting the results.

As an example of using ProBiS, Fig. 3.1 shows images generated with the tools available at the ProBiS website and depicts the process of obtaining ligand binding site predictions. In this case, the ProBiS web server was given the structure of CASP 9 target protein T0515 as a query for protein-ligand binding site prediction [45]. Using the aforementioned structural comparison process, ProBiS determined that the third top scoring similar model was PDB ID 2P3E (chain A). It is important to note that the coloring scheme is consistent across all of the images and was automatically applied by ProBiS.

After going through the process to obtain these results, biologists would be more informed in their search to identify the function of T0515. The ProBiS server even predicts what type of ligand may bind at each predicted binding site. However, this is just the first step toward predicting the function of a protein. At this point, any predictions made are heavily reliant on inference. Konc et al. performed predictions like these but also took their study a step further into more detailed function prediction using the information that they obtained from protein-ligand binding site comparison [30]. After they had used ProBiS to predict the binding sites on their query protein (Tm1631), they also noticed that the server detected binding site similarities between Tm1631 and another protein (PDB ID 2NQJ). Specifically, a similarity was detected between a predicted binding site on Tm1631 and the DNA-binding site of 2NQJ. Figure 3.2 depicts the structure of Tm1631 as represented in PDB ID 1VPQ and binding site predictions performed with the ProBiS server as an example illustration for this passage. By superimposing the structures of Tm1631 and 2NQJ, they were able to formulate a hypothetical model for the interaction of Tm1631 and DNA. In order to further validate this proposed interaction, a molecular dynamics simulation using Chemistry at Harvard Molecular Mechanics (CHARMM) was run on the hypothetical complex [9]. This process essentially tested the stability of the complex and determined that it could reasonably exist in a natural environment. With this as supporting evidence, they were able to draw even more detailed conclusions about Tm1631 and its functional role in binding to DNA. Starting with protein-ligand binding site predictions and

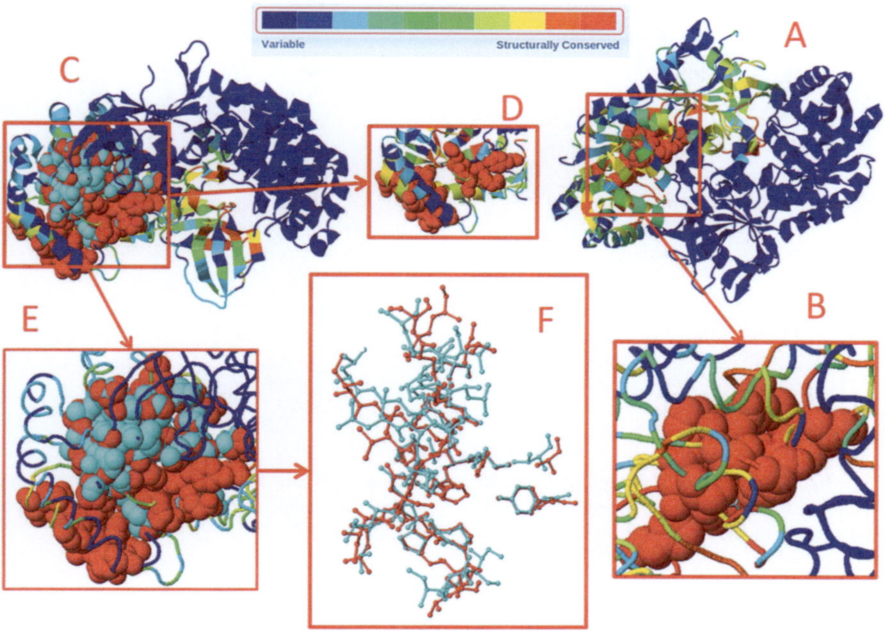

Fig. 3.1 An example of using the ProBiS web server to perform protein-ligand binding site prediction. Section (**a**) represents the template protein 2P3E and depicts its binding region (represented by *red spheres*) as identified by ProBiS. Section (**b**) shows an enlarged view of this binding region surrounded by a simplified view of the tertiary structure of 2P3E. Section (**c**) depicts the structure of T0515 with its predicted binding region and the aligned region of 2P3E. Section (**d**) provides a closer view of the predicted binding region of T0515 without the aligned region of 2P3E and surrounded by a simplified representation of the tertiary structure of T0515. Section (**e**) depicts the same view as section **c** but also includes the aligned region of 2P3E (represented as a mixture of *blue* and *red spheres*). Section (**f**) offers a more detailed view of the predicted binding region of T0515 and the aligned region of 2P3E

making comparisons with other proteins proved to be an effective strategy for this case of function prediction.

While these predictions are a useful starting point, it is important to remember that they do not perfectly reflect the natural conditions that proteins function in. Therefore, it may be beneficial to compare results derived from protein-ligand binding comparison with other computational methods of function prediction. One way to gain further verification of results in this situation is through the use of Gene Ontology term (GO term) prediction [1]. Essentially, GO terms are identifiers that establish a universal vocabulary for describing the function of a protein [1]. As an example of obtaining GO term predictions through methods other than protein-ligand binding comparison, the sequence for T0515 was submitted to the CombFunc server [76]. Using CombFunc from the web server page is a simple process that only involves entering a protein sequence and an email address for the results to be sent to. Once the sequence has been received, CombFunc utilizes an

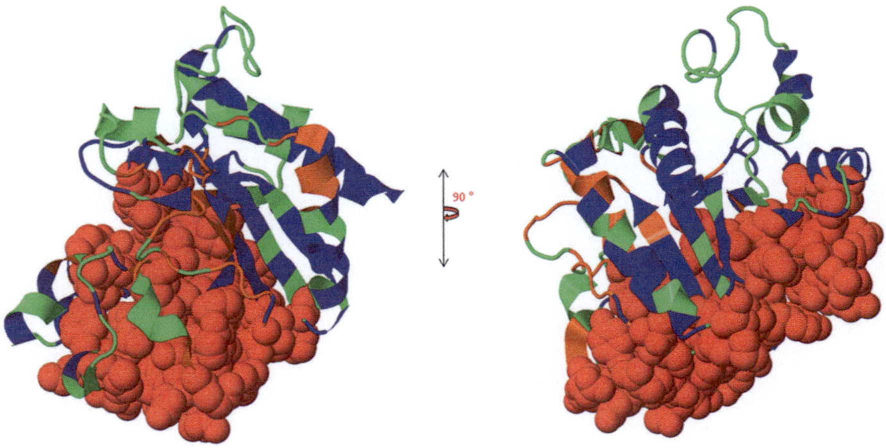

Fig. 3.2 ProBiS results depicting the tertiary structure of PDB ID 1VPQ (Tm1631) and its predicted binding region (*red spheres*). The *red spheres* represent the predicted binding region and the rest of the image follows the same coloring scheme as Fig. 3.1

algorithm that incorporates data from multiple protein function prediction sources. After the relevant data have been gathered, CombFunc produces and ranks the final predictions using a support vector machine (SVM) [69]. In this case, four biological process GO terms were predicted and listed in the order of increasing SVM probability. The first term (GO:0009089) represents the "lysine biosynthetic process via diaminopimelate." Second, the term GO:0008295 represents the "spermidine biosynthetic process." Third, the term GO:0006591 represents the "ornithine metabolic process." Finally, the fourth biological process prediction made by CombFunc was the term representing the "putrescine biosynthetic process."

Checking the CASP 9 target list revealed that the target T0515 is associated with PDB ID 3MT1. The PDB entry for 3MT1 contains records with two biological process annotations assigned to this protein. These two processes are identified with the GO terms GO:0008295 (described as the "spermidine biosynthetic process") and GO:0006596 (described as the "polyamine biosynthetic process"). Therefore, the second biological process prediction from CombFunc matched one of the function annotations in the PDB. In some cases, the top scoring predictions can be fairly close to the accepted annotations in terms of semantic similarity. Here, the results from CombFunc were compared with the annotations in the PDB using the "mgoSem" function of the GOSemSim library [81]. In this configuration, GOSemSim takes two lists of GO terms and returns a number (from 0 to 1) that indicates the percentage of similarity between the two lists. Specifically, GOSemSim looks at relationships between "ancestor" terms and the position of GO terms in the graph structure of Gene Ontology data [81]. Simply running the mgoSem function with the two lists (the two PDB GO terms and the four GO terms predicted by CombFunc) as input resulted in a semantic similarity of 0.723.

With all of these different pieces of information and methods of prediction, combining them is the first step to uncovering the story behind the function of a protein, which is highly challenging. One place where progress in protein function prediction can be seen is the Critical Assessment of Protein Function Annotation (CAFA) experiments [53]. In these experiments, researchers participate to develop the best function prediction methods using various methods including protein-ligand binding comparison and more. As techniques continue to be developed and evaluated, application of these concepts is becoming more feasible.

For example, the medical field can benefit from the ability of predicting ligand binding sites since many drugs operate by binding to these areas on proteins. This is referred to as druggability prediction and is a feature offered by fpocket and many other servers [58]. The general process behind druggability prediction often operates on the same principles of protein-ligand binding prediction. However, druggability prediction tends to focus more on applying knowledge from the field of drug development. Since the process of testing drugs can be very expensive, having the ability to make computational predictions about the interactions of a drug with a potential target protein is very promising for drug development and precision medicine [25]. Given the ability to predict the function of a protein and the types of ligands that it may bind with, it may become easier for health-care professionals to provide better care for patients on an individual basis. Essentially, this may lead to more efficient treatment because of an increase in the ability to predict the compatibility of a drug with specific cases of a disease [25].

3.3 Protein Structure Modeling and Human Disease

In this section, we will show three examples of using protein structure prediction and modeling in studying human diseases: (1) Three truncation mutations of $GABA_A$ receptor ($GABA_AR$) GABRG2 subunit were modeled, which reveals that these mutations caused protein misfolding that links to epilepsy (Sect. 3.3.1). (2) Structure of hyperpolarization-activated cyclic nucleotide-gated (HCN) channels and caveolin in cardiac tissues were modeled individually, their interactions were studied, and the possible binding interface was predicted associated with pacemakers in heart and brain cells (Sect. 3.3.2). (3) Whole-exome analysis identified various nonsynonymous mutations in juvenile myelomonocytic leukemia (JMML) patients, and structural modeling on these mutated proteins shed some light on the mechanism of this disease (Sect. 3.3.3).

3.3.1 Protein Modeling on Truncation Mutation of $GABA_A$ Receptor for Studying Epilepsy

Epilepsy is a central nervous system disorder (neurological disorder) in which the nerve cell activity in the brain is disrupted, causing seizures or periods of unusual behavior, sensations, and sometimes loss of consciousness [49]. Genetic epilepsies (GEs) are common neurological disorders that are associated frequently with mutations of ion channel genes. One of them is $GABA_A$ receptor ($GABA_AR$), which is an ionotropic receptor and ligand-gated ion channel. It has an endogenous ligand, γ-aminobutyric acid (GABA), which is the major inhibitory neurotransmitter in the central nervous system [18, 20, 24]. $GABA_AR$ causes an inhibitory effect on neurotransmission by diminishing the occurrence of a successful action potential. Mutations in $GABA_A$ receptor subunit genes (*GABRs*) are frequently associated with epilepsy, and nonsense mutations in *GABRG2* are associated with various types of epilepsy syndromes including the most severe form epileptic encephalopathy like Dravet syndrome [18]. The molecular basis for the phenotypic heterogeneity of *GABRG2* truncation mutations is still unclear, but evidences gathered suggest these mutations caused protein misfolding and abnormal receptor trafficking [23].

The first three-dimensional protein structure of $GABA_AR$ was resolved by X-ray diffraction (PDB ID, 4COF), and its $GABA_AR$-β3 homopentamer reveals its role as a pentamer in signal transduction [43]. However, mutants and truncation in different lengths at these subunits still lack structure-based explanation. We applied a structure modeling approach to investigate the structure details on three nonsense mutations in *GABRG2* (*GABRG2(R136X)*, *GABRG2(Q390X)*, and *GABRG2 (W429X)*) associated with epilepsies of different severities. We mainly used our in-house protein structure prediction tool MUFOLD [83] to construct protein models of mutant $GABA_A$ receptor subunits: (1) γ2 (R136X) subunit, in which all transmembrane regions are deleted and only part of the N-terminal domain remains; (2) γ2 (Q390X) subunit, in which the fourth hydrophobic transmembrane α-helix (YARIFFPTAFCLFNLVYWVSYLYL) is deleted and a new α-helix with many charged amino acids (KDKDKKKKNPAPTIDIRPRSATI) is found to assume its location; and (3) γ2 (W429X) subunit, in which the fourth hydrophobic transmembrane α-helix is truncated. Figure 3.3 presents structure models on wide-type γ2, γ2 (R136X), γ2 (Q390X), and γ2 (W429X) subunits.

Following the MUFOLD protocol, we identified a nicotinic acetylcholine receptor (PDB ID: 4COF and 2BG9) as the main template for $GABA_AR$-β3. Then multidimensional scaling (MDS) was used to construct multiple protein decoys based on the template and other minor templates. Then these decoys were clustered and evaluated. With several iterations of model generation and evaluation, one decoy was chosen as the predicted protein model and then refined by Rosetta [36]. For mutant $GABA_A$ receptor subunits, the original input subunits were split into different domains, and each domain was modeled individually and then assembled together.

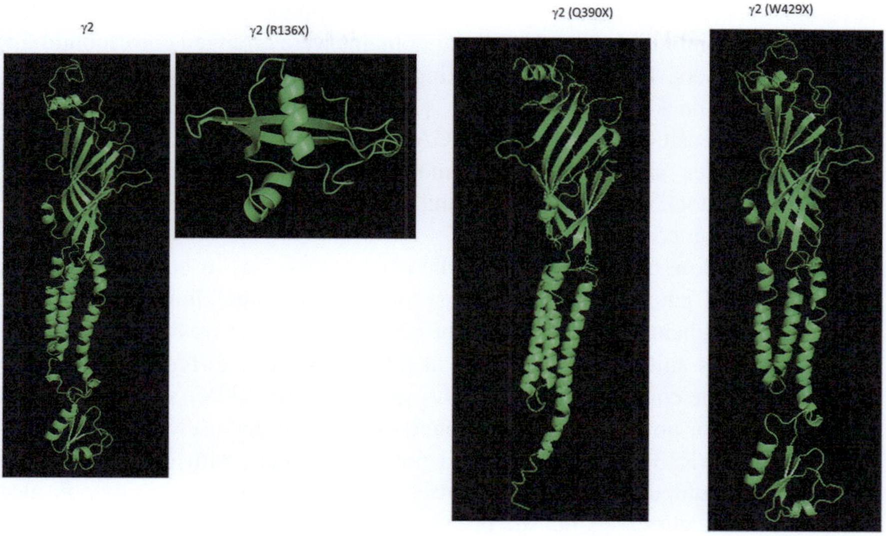

Fig. 3.3 Predicted protein structure modeling of the wild-type *γ2* and the mutant *γ2 (R136X)*, *γ2 (Q390X)*, and *γ2 (W429X)* subunits

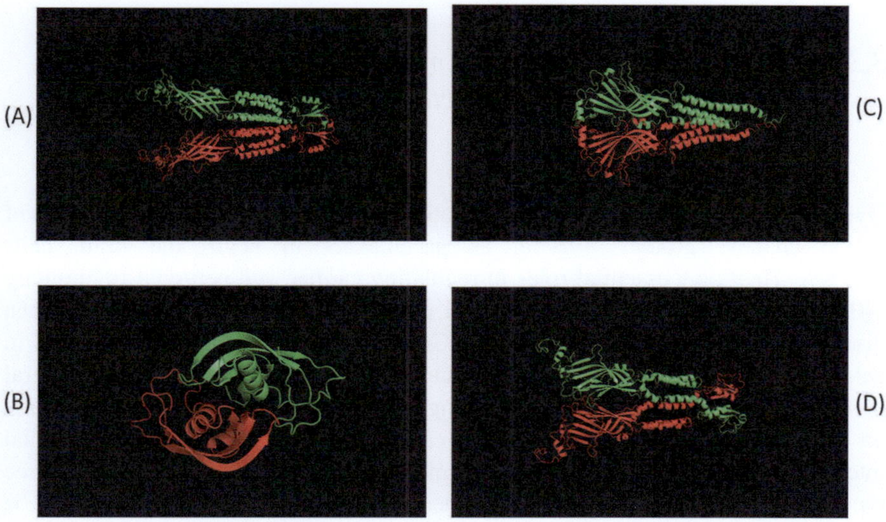

Fig. 3.4 Docking models for potential mutant γ2 subunit homodimers by SymmDock. In each panel, the two γ2 subunits are shown in *red* and *green*; (**a**) wide-type γ2 dimer, (**b**) γ2 (R136X) mutant dimer, (**c**) γ2 (Q390X) mutant dimer, (**d**) γ2 (W429X) mutant dimer

To further understand the stabilities of these wide-type and mutant subunits, a dimer structure was constructed between two subunits by symmetric docking of SymmDock [59], detailed in Fig. 3.4. As classified as transporter protein in the membrane, special filtering on dimer models was applied to make sure the intracellular, transmembrane, and extracellular domains were arranged correspondingly between monomers. Template-free dockings were performed in conjunction with template-based docking [66] between γ2 and α-subunits by mapping their corresponding positions to template $GABA_AR$-β3 homopentamer (PDB ID: 4COF). Pentamer and hypothetic homopentamer were also constructed by template-based docking. Chimera [51] and PyMOL [12] were used to display the protein structure models.

Along with flow cytometry and biochemical approaches in combination with lifted whole-cell patch clamp recordings, the structural modeling and structure-based analysis indicated that the wild-type γ2 subunit surface was naturally hydrophobic, which is suitable to be buried in membrane. The different γ2 subunits adopted different conformations, and the mutant γ2 (Q390X) subunits formed protein-specific or nonspecific stable protein dimers with themselves or other proteins, while γ2 (R136X) subunits could not form dimers with other partnering subunits but could dimerize with themselves. The γ2 (W429X) subunits also dimerized with themselves, but the protein conformation was similar to the wild-type γ2 subunit protein. Our modeling study provides good hypotheses to understand the mechanisms and effects of the *GABRG2* truncation mutations in epilepsy.

3.3.2 Exploring the Interaction of Caveolin-3 with Hyperpolarization-Activated Cyclic Nucleotide-Gated Channels 2 and 4

Hyperpolarization-activated cyclic nucleotide-gated (HCN) channels are a group of cation channel proteins serving as pacemakers in heart and brain cells [13, 50]. They play essential roles in regulating cardiac and neuronal rhythmicity [40, 57]. Currently, four types of HCN (HCN1 to HCN4) channels have been discovered. Among them, HCN4 and HCN2 are the main isoforms expressed in cardiac tissues. A type of related proteins to HCN is the caveolin family of integral membrane proteins, which are the building blocks of caveolae, a type of lipid rafts on cell membrane [52, 62]. In addition, caveolins act as scaffolding proteins and interact with a variety of proteins to form macromolecular complexes [7]. Three types of caveolin proteins have been identified so far (Cav1 to Cav3). Cav3 is reported to be specifically expressed in skeletal, smooth, and cardiac and muscle cells [67]. Interestingly, studies have shown that Cav3 is associated with HCN4 and affects its function [80]. However, the detailed interaction information is still largely unknown. In this work, we explored the interaction of Cav3 with HCN2/HCN4 and predicted the possible binding interface of Cav3 with HCN2/HCN4 using computational methods.

We started our analysis by searching for caveolin-binding motif on HCN2 and HCN4 protein sequences, respectively. The existence of caveolin-binding motif was not a definitive evidence of the binding interface, but it served as a reasonable

Table 3.3 Predicted caveolin-binding motif for HCN2 and HCN4

	Sequence	Pattern matched	Topology
HCN2-human-wt	WiihpYsdF (202–210)	[FWY]XXXX[FWY]XX [FWY]	Cytoplasmic (1–215)
HCN2-human-wt	WdFtmllF (214–221)	[FWY]X[FWY]XXXX [FWY]	Transmembrane (216–236)
HCN4-human-wt	WiihpYsdF (253–261)	[FWY]XXXX[FWY]XX [FWY]	Cytoplasmic (1–266)

starting point. We used ScanProsite [11] and targeted on two known motif patterns, i.e., [FWY]X[FWY]XXXX[FWY] and [FWY]XXXX[FWY]XX[FWY] [10]. We found two hits on HCN2 and one hit on HCN4 (Table 3.3). One hit (214–222) on HCN2 was located in the transmembrane region, making it less likely to be binding interface. The other hit on HCN2 (202–210) had the exactly the same motif sequence with HCN4 hit. Both hits were in the N-terminal cytoplasmic domain.

Meanwhile, to determine the possible binding interfaces of HCN2/HCN4 and Cav3, we performed correlated mutation analysis using i-COMS web server [19]. Correlated mutation aims to discover those coevolved pairs of amino acid residues between two proteins. Based on our motif search, we limited our search in the N-terminus of HCN2/HCN4. We predicted possible correlated amino acid residues between the ion transport protein N-terminal domain (PF08412) of HCN2/HCN4 and the caveolin domain (PF01146) of Cav3. We combined the prediction results from three algorithms of computing correlated mutations, including mutual information, pseudolikelihood maximization direct-coupling analysis, and mean-field direct-coupling analysis.

We found several amino acid pairs whose prediction scores ranked top 100 inter-protein links for all three methods: D209 in HCN2 vs. T66 in Cav3, D260 in HCN4 vs. T66 in Cav3, and S230 in HCN4 vs. F65 in Cav3. For the first two pairs (HCN2 D209 vs. Cav3 T66 and HCN4 D260 vs. Cav3 T66), the amino acid residues were in caveolin-binding motif regions previously determined. This further inferred that the possible binding sites on HCN2/HCN4 were located within the caveolin-binding motif.

Next we built a structural model showing the interaction between HCN2/HCN4 and Cav3. Results from protein disorder analysis indicated that most of the N-terminal sequences of HCN2/HCN4 were disordered regions. Therefore, the structure of the N-terminal domains could be highly variable, and it might be difficult to obtain a reliable and consistent structure. We selected the ordered region of HCN2 (186–225) and HCN4 (236–275) and used I-TASSER [82] to predict their 3D structures. On the other hand, since Cav3 is a transmembrane protein, we predicted the 3D structures using its cytoplasmic N-terminal sequences (1–85) based on I-TASSER. Next, the best predicted structures were docked using PatchDock [59] and FireDock [42] to determine and refine the possible binding conformations of the N-terminal structures of HCN2/HCN4 and Cav3 (Fig. 3.5). Among the top docking solutions, we could find interaction interface via caveolin-binding motif in HCN2/HCN4 with the N-terminus of Cav3.

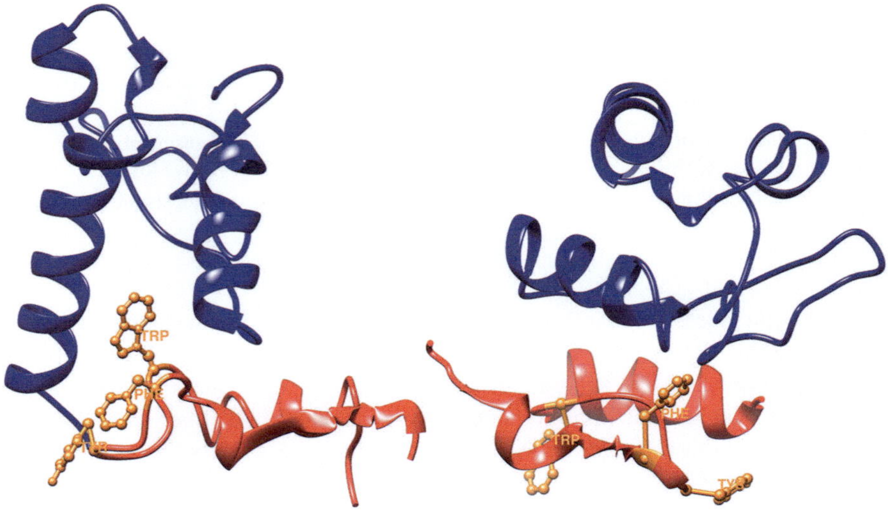

Fig. 3.5 A schematic representation of the interaction between HCN2/HCN4 and Cav3. HCN2 was shown in *left* and HCN4 in *right*. Cav3 N-terminal domain was shown in *blue*, and HCN2/HCN4 N-terminal ordered region was shown in *red*. The side chains of the three hydrophobic residuals (W, Y, F) in caveolin-binding motif were labeled in *orange*

3.3.3 Point Mutations Identified in Juvenile Myelomonocytic Leukemia (JMML) Patient's Exome and Their Effects on Protein Structure

Juvenile myelomonocytic leukemia (JMML) is a rare and chronic leukemia found to occur in 1.2 cases per million. JMML affects children in the age group of four and below. It is thought that JMML is a congenital disorder. The majority of mutations that have been found in JMML patients so far belong to RAS/MAPK signaling pathway; these include NRAS, KRAS, NF, and PTPN11 mutations. It has been established that JMML is fundamentally a disease of hyperactive RAS signaling, but targeted chemotherapy of this pathway has not been successful [3, 35].

We did whole-exome analysis of a 2-year-old JMML patient's bone marrow specimen using whole-exome sequencing and verified it using cancer panel sequencing. We were able to identify several novel mutations in NTRK1, HMGA2, MLH3, MYH9, and AKT1 genes. We were also able to confirm the already identified mutation of PTPN11 (exon 3 181G > T) [39] in JMML. Here, we discuss the methods we used to elucidate whether any of these novel/already identified mutations at DNA level affect the respective protein structure of their proteins.

Whole-exome analysis identified various nonsynonymous mutations, which were then confirmed by cancer panel sequencing. These include ITPR3 (chr6:33651070, G > A, exon 35, A1562 > T), PTPN11 (chr12:112888165, G > T, exon 3, D61 > Y), AKT1 (chr14:105239869, G > A, exon 9, R251 > C), MLH3 (chr14:75514537, A > T, exon 2, F608 > I), and MYH9 (chr22:36685257, C > A, exon 32, K1477 > N). In order to identify whether these mutations have any effect on their protein sequence, SIFT and PROVEAN predictions were used [33]. The already identified mutations in PTPN11 and the novel mutations in AKT1, MLH3, and MYH9 were found to be deleterious and damaging. In order to identify whether the particular mutation is likely to be associated with the disease, we used SuSPect web server (http://www.sbg.bio.ic.ac.uk/~suspect/). SuSPect web server indicated that out of all the mutations, two mutations, namely, AKT1 and already known PTPN11, were most likely disease-causing mutations. Protein sequences of the two mutations were further extracted from the .vcf file using customProDB [72]. Mutation taster [60] was used to detect that the AKT1 mutation occurs at an evolutionary conserved site by comparison against different species. A homology model of the AKT1 mutation was constructed using SWISS-MODEL [5] based on the template (PDB: 3O96) with 97 % identity as shown in Fig. 3.6a (superimposed with the native structure). The 3D structure shows that mutation R251 > C occurs at the

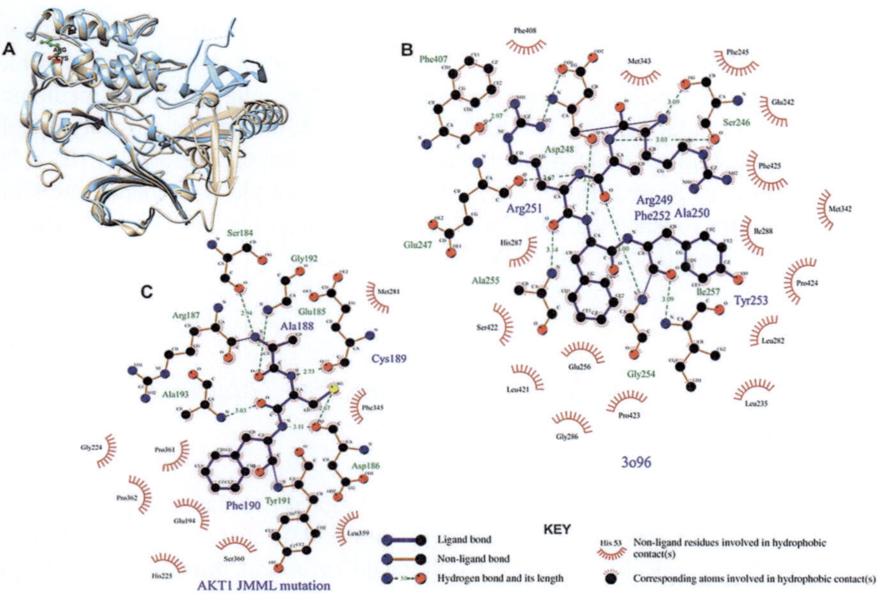

Fig. 3.6 Effect of AKT1 mutation on the protein structure. (**a**) Superimposed structures of AKT1 protein structure template (PDB ID = 3O96) and homology model of *AKT1* protein with point mutation (from *JMML* patient) (the *AKT1* protein sequence in template structure is shifted by 62 residues in the template sequence). (**b**) and (**c**) Polar contacts around the wild-type *ARG251* residue and its immediate neighbors were visualized using LigPlot + [34]; similarly, polar contacts around CYS189 and its immediate neighbors were also visualized using LigPlot+. There is a change in the electron density due to mutation as shown by change in positions of the hydrophobic contacts and loss of hydrogen bonding between *ASP248* and *PHE407*

surface of the 3D protein as indicated by ARG and CYS residues in the structure created visualized using UCSF Chimera [51].

Both Motif Scan and UniProt indicate that the R251 > C lies near or within multiple protein domains which can potentially act as active sites in the protein. In order to see if any active site lies near the mutation which can potentially have a damaging effect on overall structure of the protein HOPE [70], web server was used. Results from HOPE server point toward a structural disruption in an InterPro [44] domain, i.e., the protein kinase domain (IPR000719) [17].

In order to find if there is a change in the polar contacts of the neighboring residues of the mutated residue, we compared the polar contact maps of native ARG251 and the mutated CYS189 residues and their immediate neighbors using LigPlot+ [34]. As indicated by Fig. 3.6, there is a change in the electron density due to mutation as shown by change in positions of the hydrophobic contacts and loss of hydrogen bonding between mutated sites and ASP248 and PHE407.

3.4 Protein Modeling and Plant Analysis

In this section, we present two examples of integrating protein structure prediction/ modeling with other omics data for understanding plant protein mechanism. Although they represent agriculture research problems, applications in medicine can work in a similar fashion. The first example is to combine rice microarray and structure modeling to study proteins that may have important roles in the rice salinity resistance process (Sect. 3.4.1). The second example is to conduct protein modeling on trait-associated protein-protein interactions identified from soybean genome-wide association study (GWAS) (Sect. 3.4.2).

3.4.1 Structure Modeling in Exploring Rice Tolerance Mechanism

Under the pressure of global climate change and global population explosion, soil salinity causes rice production reduction in about 30 % of the rice-growing area worldwide. The study in exploring the mechanism of salt tolerance starts from selecting differential expressed genes in the whole gene set, then a putative mechanism network was built, and several network modules were identified upon information from protein-protein interaction. These modules were annotated and assessed by quantitative trait loci (QTL) [61], co-expression and regulatory binding motif analysis. The topological hub genes in these modules are considered as the most important genes dominating the inherent function [74].

Among all the genes in the module, one of the most interesting genes is LOC_Os01g52640.3, which is a hub gene in the largest module and overlaps with a QTL region. This gene corresponds to a hypothetical protein Os01g0725800, which interacts with 32 of the 51 proteins in the module. It contains four InterPro domains, namely, IPR000719, IPR001680, IPR011046, and IPR011009. IPR011009 domains can also be found in RIO kinase (IPR018935), a SPA1-related serine/threonine-specific and tyrosine-specific protein kinase. This protein also has an ortholog in *Arabidopsis thaliana* as SPA4 (SPA1-RELATED 4), which is a binding protein and a signal transducer. MUFOLD [83, 84] was applied to predict the structure for LOC_Os01g52640.3. Using the identified templates of 2GNQ, 3EMH, and 3DM0 in PDB, the model for the protein region of 196–627 for the protein with the length of 432 was constructed, as shown in Fig. 3.7. The protein structure model contains the WD40 structure motif repeats, each with a tryptophan-aspartic acid (WD) dipeptide termination. As WD40 proteins often play important roles in signal transduction and transcription regulation [47], the structure prediction suggests that this protein may be related to signal transduction in the salt resistance process.

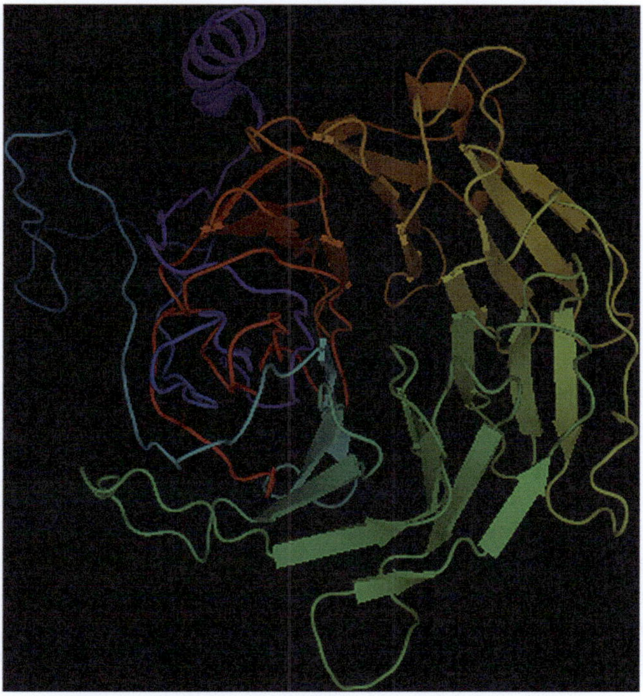

Fig. 3.7 Predicted structural model of protein Os01g0725800

3.4.2 Structure Modeling for Soybean Trait Improvement

Soybeans represent one of the most important agricultural crops providing nutrition and sustenance to humans and household animals and become an increasingly valuable feedstock for industrial applications [64]. Among hundreds of agricultural trials, seed oil content and seed protein content are both polygenic traits controlled by several gene loci in soybeans. Many of the QTL alleles with positive and negative effects on oil content are often dispersed among genotypes [85], which suggests that accumulation of the positive alleles from different genetic backgrounds could eventually lead to the development of genotypes with higher seed oil content or protein content [77]. To address the "missing heritability" problem in complex traits by the original GWAS analysis under the hypothesis of single SNP association with the phenotype [41], Bayesian High-Order Interaction Toolset (BHIT) [75] was applied to explore the SNP interactions associated with the phenotypes. The most interesting interactions identified were four loci across two chromosomes located in position 20,897,627; 20,954,490 of chromosome 8 and 8,642,446; 12,051,017 of chromosome 19 in soybean genome. Among them, protein Glyma08g26580.1 containing first SNP (SNP293) and protein Glyma19g07330.1 containing third SNP (SNP792) were computationally predicted to interact by ProPrInt [54].

The first SNP (named as SNP293) in the results is located in gene Glyma08g26580.1, which has an *Arabidopsis* homolog AT3G0140 (EC/6.3.2.19) and a ubiquitin-protein ligase. At the sequence level, this polymorphism makes the minor allele nucleotide adenine (A) replaced the major allele nucleotide guanine (G), which causes the 73th amino acid of the protein change from glycine (G) to arginine (R). The added positive-charged arginine may have significant impact on the protein conformation and function. The third SNP (named as SNP792) in the results is located in gene Glyma19g07330.1, which also causes amino acid change from glycine (G) to arginine (R). This gene has the *Arabidopsis* homolog AT3G48990.1, which encodes an oxalyl-CoA synthetase and is required for oxalate degradation and normal seed development processes.

MUFOLD [83, 84] was applied to predicted protein structures of gene Glyma08g26580.1 and gene Glyma19g07330.1. The two predicted structures were docked together using GRAMM-X [68]. Interestingly, the distance between the residue containing SNP293 and the residue containing SNP792 in the docking complex was 1.17 **Å**, shorter than 0.0052 % of all the paired distances between the two structures, as shown in Fig. 3.8. This result suggests that the epistatic interaction between the two SNPs may play a role in the interaction between the two proteins. And this interaction caused by amino acid changes may shed some light on the mechanism in controlling oil/protein contents in soybean.

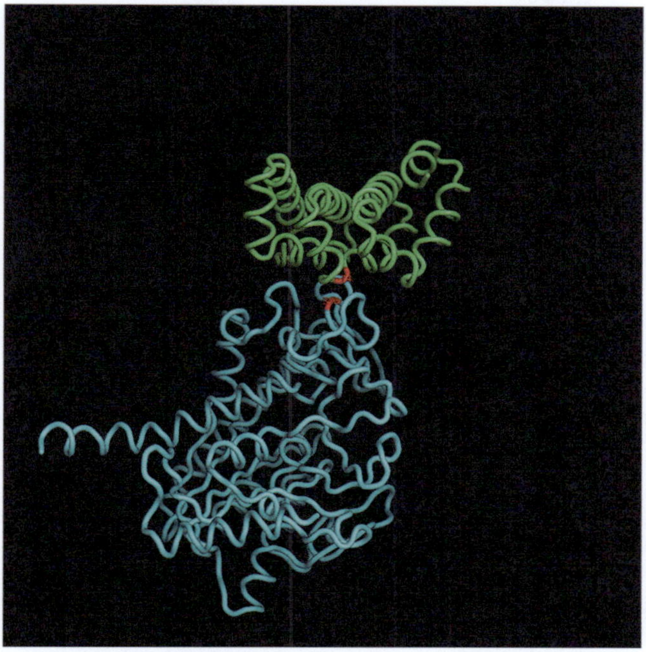

Fig. 3.8 Protein-protein interaction on predicted protein structures containing SNP locations. SNP293 is located in the protein Glyma08g26580.1 (*upper*, *green*), and SNP792 is located in the protein Glyma19g07330.1 (*lower*, *cyan*). The polymorphism sites (*red*) are located at the interface of the interaction

3.5 Conclusions

Protein structure modeling provides a tool to explore the mechanism of a biological process and the function of a protein. In the studies reviewed in this book chapter, we showed various use cases of combining with docking, protein-protein interaction prediction, GWAS analysis, systems biology, other analyses, and protein structure modeling in studying protein conformation, function, mutation, and disease/phenotype effects. The structure-based prediction and analysis expand our knowledge in biological mechanism and human disease and help design drug treatment in the age of precision medicine.

References

1. Ashburner M, et al. Gene Ontology: tool for the unification of biology. Nat Genet. 2000;25 (1):25–9.
2. Bairoch A, et al. The universal protein resource (UniProt) 2009. Nucleic Acids Res. 2009;37: D169–74.

3. Baumann I, Bennett JM, Niemeyer CM, Thiele J, Shannon K. Juvenile Myelomonocytic Leukemia (JMML). In: Swerdlow SH, I.A.f.R.o. Cancer, W.H. Organization, editors. WHO classification of tumours of haematopoietic and lymphoid tissues. Lyon: International Agency for Research on Cancer; 2008.
4. Berman HM, et al. The protein data bank. Nucleic Acids Res. 2000;28(1):235–42.
5. Biasini M, et al. SWISS-MODEL: modelling protein tertiary and quaternary structure using evolutionary information. Nucleic Acids Res. 2014;p. gku340.
6. Borgwardt KM, et al. Protein function prediction via graph kernels. Bioinformatics. 2005;21: I47–56.
7. Boscher C, Nabi IR. Caveolin-1: role in cell signaling. Adv Exp Med Biol. 2012;729:29–50.
8. Bowie JU, Luthy R, Eisenberg D. A method to identify protein sequences that fold into a known 3-dimensional structure. Science. 1991;253(5016):164–70.
9. Brooks BR, et al. CHARMM: the biomolecular simulation program. J Comput Chem. 2009;30 (10):1545–614.
10. Couet J, et al. Identification of peptide and protein ligands for the caveolin-scaffolding domain. Implications for the interaction of caveolin with caveolae-associated proteins. J Biol Chem. 1997;272(10):6525–33.
11. de Castro E, et al. ScanProsite: detection of PROSITE signature matches and ProRule-associated functional and structural residues in proteins. Nucleic Acids Res. 2006;34(Web Server issue):W362–5.
12. DeLano WL. The PyMOL molecular graphics system. Palo Alto: DeLano Scientific; 2002.
13. DiFrancesco D. Pacemaker mechanisms in cardiac tissue. Annu Rev Physiol. 1993;55:455–72.
14. Friedberg I. Automated protein function prediction – the genomic challenge. Brief Bioinform. 2006;7(3):225–42.
15. Gao M, Zhou HY, Skolnick J. Insights into disease-associated mutations in the human proteome through protein structural analysis. Structure. 2015;23(7):1362–9.
16. Gherardini PF, et al. Modular architecture of nucleotide-binding pockets. Nucleic Acids Res. 2010;38(11):3809–16.
17. Hanks SK, Quinn AM, Hunter T. The protein kinase family: conserved features and deduced phylogeny of the catalytic domains. Science. 1988;241(4861):42–52.
18. Harkin LA, et al. Truncation of the GABA(A)-receptor gamma2 subunit in a family with generalized epilepsy with febrile seizures plus. Am J Hum Genet. 2002;70(2):530–6.
19. Iserte J, et al. I-COMS: interprotein-COrrelated mutations server. Nucleic Acids Res. 2015;43 (W1):W320–5.
20. Ishii A, et al. Association of nonsense mutation in GABRG2 with abnormal trafficking of GABAA receptors in severe epilepsy. Epilepsy Res. 2014;108(3):420–32.
21. Ito JI, et al. PoSSuM: a database of similar protein-ligand binding and putative pockets. Nucleic Acids Res. 2012;40(D1):D541–8.
22. Kallberg M, et al. Template-based protein structure modeling using the RaptorX web server. Nat Protoc. 2012;7(8):1511–22.
23. Kang JQ, et al. Slow degradation and aggregation in vitro of mutant GABAA receptor gamma2 (Q351X) subunits associated with epilepsy. J Neurosci. 2010;30(41):13895–905.
24. Kang J-Q, et al. The human epilepsy mutation GABRG2 (Q390X) causes chronic subunit accumulation and neurodegeneration. Nat Neurosci. 2015;18(7):988–996.
25. Khoury MJ, et al. A population approach to precision medicine. Am J Prev Med. 2012;42 (6):639–45.
26. Kim DE, Chivian D, Baker D. Protein structure prediction and analysis using the Robetta server. Nucleic Acids Res. 2004;32:W526–31.
27. Kirshner DA, Nilmeier JP, Lightstone FC. Catalytic site identification-a web server to identify catalytic site structural matches throughout PDB. Nucleic Acids Res. 2013;41(W1):W256–65.
28. Konc J, Janezic D. ProBiS algorithm for detection of structurally similar protein binding sites by local structural alignment. Bioinformatics. 2010;26(9):1160–8.

29. Konc J, Janezic D. Binding site comparison for function prediction and pharmaceutical discovery. Curr Opin Struct Biol. 2014;25:34–9.
30. Konc J, et al. Structure-based function prediction of uncharacterized protein using binding sites comparison. Plos Comput Biol. 2013;9(11).
31. Konc J, et al. ProBiS-CHARMMing: web interface for prediction and optimization of ligands in protein binding sites. J Chem Inf Model. 2015;55(11):2308–14.
32. Kryshtafovych A, Fidelis K, Moult J. CASP10 results compared to those of previous CASP experiments. Proteins-Struct Funct Bioinf. 2014;82:164–74.
33. Kumar P, Henikoff S, Ng PC. Predicting the effects of coding non-synonymous variants on protein function using the SIFT algorithm. Nat Protoc. 2009;4(7):1073–82.
34. Laskowski RA, Swindells MB. LigPlot+: multiple ligand–protein interaction diagrams for drug discovery. J Chem Inf Model. 2011;51(10):2778–86.
35. Lauchle JH, Braun B. Targeting RAS signaling pathways in Juvenile Myelomonocytic Leukemia (JMML). In: Houghton PJ, Arceci RJ, editors. Molecularly targeted therapy for childhood cancer. New York: Springer; 2010. p. 123–38.
36. Leaver-Fay A, et al. ROSETTA3: an object-oriented software suite for the simulation and design of macromolecules. Methods Enzymol. 2011;487:545.
37. Li Z, Scheraga HA. Monte Carlo-minimization approach to the multiple-minima problem in protein folding. Proc Natl Acad Sci U S A. 1987;84(19):6611–5.
38. Liang J, Edelsbrunner H, Woodward C. Anatomy of protein pockets and cavities: measurement of binding site geometry and implications for ligand design. Protein Sci. 1998;7(9):1884–97.
39. Loh ML, Vattikuti S, Schubbert S, Reynolds MG, Carlson E, Lieuw KH, Ptpn T. Mutations in PTPN11 implicate the SHP-2 phosphatase in leukemogenesis. Blood. 2004;103(6):2325–32.
40. Ludwig A, et al. Two pacemaker channels from human heart with profoundly different activation kinetics. EMBO J. 1999;18(9):2323–9.
41. Mackay TFC. Epistasis and quantitative traits: using model organisms to study gene-gene interactions. Nat Rev Genet. 2014;15(1):22–33.
42. Mashiach E, et al. FireDock: a web server for fast interaction refinement in molecular docking. Nucleic Acids Res. 2008;36(Web Server issue):W229–32.
43. Miller PS, Aricescu AR. Crystal structure of a human GABAA receptor. Nature. 2014;18(7):988–996.
44. Mitchell A, et al. The InterPro protein families database: the classification resource after 15 years. Nucleic Acids Res. 2015;43(D1):D213–21.
45. Moult J, et al. Critical assessment of methods of protein structure prediction (CASP) – Round IX. Proteins-Struct Funct Bioinf. 2011;79:1–5.
46. Nagarajan N, Kingsford C. GiRaF: robust, computational identification of influenza reassortments via graph mining. Nucleic Acids Res. 2011;39(6):e34.
47. Neer EJ, et al. The ancient regulatory-protein family of WD-repeat proteins. Nature. 1994;371 (6495):297–300.
48. Nilmeier JP, et al. Rapid catalytic template searching as an enzyme function prediction procedure. Plos One. 2013;8(5):e62535.
49. Noebels JL. Exploring new gene discoveries in idiopathic generalized epilepsy. Epilepsia. 2003;44:16–21.
50. Pape HC. Queer current and pacemaker: the hyperpolarization-activated cation current in neurons. Annu Rev Physiol. 1996;58:299–327.
51. Pettersen EF, et al. UCSF Chimera—a visualization system for exploratory research and analysis. J Comput Chem. 2004;25(13):1605–12.
52. Philippova MP, et al. T-cadherin and signal-transducing molecules co-localize in caveolin-rich membrane domains of vascular smooth muscle cells. FEBS Lett. 1998;429(2):207–10.
53. Radivojac P, et al. A large-scale evaluation of computational protein function prediction. Nat Methods. 2013;10(3):221–7.
54. Rashid M, Ramasamy S, Raghava GPS. A simple approach for predicting protein-protein interactions. Curr Protein Pept Sci. 2010;11(7):589–600.

55. Rossmann MG, Moras D, Olsen KW. Chemical and biological evolution of nucleotide-binding protein. Nature. 1974;250(463):194–9.
56. Roy A, Yang JY, Zhang Y. COFACTOR: an accurate comparative algorithm for structure-based protein function annotation. Nucleic Acids Res. 2012;40(W1):W471–7.
57. Santoro B, et al. Identification of a gene encoding a hyperpolarization-activated pacemaker channel of brain. Cell. 1998;93(5):717–29.
58. Schmidtke P, Barril X. Understanding and predicting druggability. A high-throughput method for detection of drug binding sites. J Med Chem. 2010;53(15):5858–67.
59. Schneidman-Duhovny D, et al. PatchDock and SymmDock: servers for rigid and symmetric docking. Nucleic Acids Res. 2005;33(Web Server issue):W363–7.
60. Schwarz JM, et al. MutationTaster evaluates disease-causing potential of sequence alterations. Nat Meth. 2010;7(8):575–6.
61. Seaton G, et al. QTL Express: mapping quantitative trait loci in simple and complex pedigrees. Bioinformatics. 2002;18(2):339–40.
62. Simons K, Toomre D. Lipid rafts and signal transduction. Nat Rev Mol Cell Biol. 2000;1 (1):31–9.
63. Simons KT, et al. Assembly of protein tertiary structures from fragments with similar local sequences using simulated annealing and Bayesian scoring functions. J Mol Biol. 1997;268 (1):209–25.
64. Snyder CL, et al. Acyltransferase action in the modification of seed oil biosynthesis. N Biotechnol. 2009;26(1–2):11–6.
65. Soding J, Biegert A, Lupas AN. The HHpred interactive server for protein homology detection and structure prediction. Nucleic Acids Res. 2005;33:W244–8.
66. Szilagyi A, Zhang Y. Template-based structure modeling of protein–protein interactions. Curr Opin Struct Biol. 2014;24:10–23.
67. Tang Z, et al. Molecular cloning of caveolin-3, a novel member of the caveolin gene family expressed predominantly in muscle. J Biol Chem. 1996;271(4):2255–61.
68. Tovchigrechko A, Vakser IA. GRAMM-X public web server for protein-protein docking. Nucleic Acids Res. 2006;34:W310–4.
69. Vapnik VN. An overview of statistical learning theory. IEEE Trans Neural Netw. 1999;10 (5):988–99.
70. Venselaar H, et al. Protein structure analysis of mutations causing inheritable diseases. An e-Science approach with life scientist friendly interfaces. BMC Bioinf. 2010;11:548.
71. Volkamer A, et al. Combining global and local measures for structure-based druggability predictions. J Chem Inf Model. 2012;52(2):360–372.
72. Wang X, Zhang B. customProDB: an R package to generate customized protein databases from RNA-Seq data for proteomics search. Bioinformatics. 2013;29(24):3235–7.
73. Wang Z, Eickholt J, Cheng J. MULTICOM: a multi-level combination approach to protein structure prediction and its assessments in CASP8. Bioinformatics. 2010;26(7):882–8.
74. Wang J, et al. A computational systems biology study for understanding salt tolerance mechanism in rice. PLoS One. 2013;8(6):e64929.
75. Wang J, et al. A Bayesian model for detection of high-order interactions among genetic variants in genome-wide association studies. BMC Genomics. 2015;16(1):1011.
76. Wass MN, Barton G, Sternberg MJE. CombFunc: predicting protein function using heterogeneous data sources. Nucleic Acids Res. 2012;40(W1):W466–70.
77. Weselake RJ, et al. Increasing the flow of carbon into seed oil. Biotechnol Adv. 2009;27 (6):866–78.
78. Xu Y, Xu D. Protein threading using PROSPECT: design and evaluation. Proteins-Struct Funct Genet. 2000;40(3):343–54.
79. Xu D, Zhang Y. Ab initio protein structure assembly using continuous structure fragments and optimized knowledge-based force field. Proteins. 2012;80(7):1715–35.
80. Ye B, et al. Caveolin-3 associates with and affects the function of hyperpolarization-activated cyclic nucleotide-gated channel 4. Biochemistry. 2008;47(47):12312–8.

81. Yu GC, et al. GOSemSim: an R package for measuring semantic similarity among GO terms and gene products. Bioinformatics. 2010;26(7):976–8.
82. Zhang Y. I-TASSER server for protein 3D structure prediction. BMC Bioinf. 2008;9:40.
83. Zhang JF, et al. MUFOLD: a new solution for protein 3D structure prediction. Proteins-Struct Funct Bioinf. 2010;78(5):1137–52.
84. Zhang J, et al. Prediction of protein tertiary structures using MUFOLD. In: Functional genomics. New York: Springer; 2012. p. 3–13.
85. Zhao JY, et al. Oil content in a European x Chinese rapeseed population: QTL with additive and epistatic effects and their genotype-environment interactions. Crop Sci. 2005;45(1):51–9.

Chapter 4
Computational Methods in Mass Spectrometry-Based Proteomics

Sujun Li and Haixu Tang

Abstract This chapter introduces computational methods used in mass spectrometry-based proteomics, including those for addressing the critical problems such as peptide identification and protein inference, peptide and protein quantification, characterization of posttranslational modifications (PTMs), and data-independent acquisitions (DIA). The chapter concludes with emerging applications of proteomic techniques, such as metaproteomics, glycoproteomics, and proteogenomics.

Keywords Proteomics • Mass spectrometry • Protein identification • Protein quantification • Post-translational modifications • Algorithms

4.1 Introduction

4.1.1 Overview

Proteome [49] is defined as the entire set of proteins and their alternative forms in a specific species. Accordingly, the term "proteomics" [58] is defined as a large-scale, comprehensive study of a certain proteome. Objectives of such studies include the characterization of protein sequences, abundances, variants, and post-translational modifications (PTMs), as well as the interactions and pathways among proteins. Among those rapidly evolving proteomic techniques, for example, protein microarray [85] and cell flow cytometry [44], mass spectrometry (MS) is the dominant technology for accurate and high-throughput proteomics, specifically for identifying and quantifying the proteins in complex proteome samples with high sensitivity.

S. Li • H. Tang (✉)
School of Informatics and Computing, Indiana University, Bloomington, IN, USA
e-mail: hatang@indiana.edu

B. Shen et al. (eds.), *Translational Biomedical Informatics*, Advances in Experimental Medicine and Biology 939, DOI 10.1007/978-981-10-1503-8_4

4.1.2 Basic Mass Spectrometry

While there are many alternative procedures in mass spectrometry, shotgun proteomics, which combines liquid chromatography with tandem mass spectrometry (LC-MS/MS), is the most frequently used approach. Shotgun proteomic experiments are normally performed in the following steps: (1) proteins are extracted from lysed cells; (2) they are enzymatically digested; and (3) the resulting peptides are chromatographically separated prior to MS/MS analyses. The separated peptides are ionized, and the mass-to-charge (m/z) ratios of these ionized peptides are analyzed and recorded by mass spectrometry. The acquired mass spectra from the experimental samples contain multiple dimensions: different elution times of LC-MS scans, mass-to-charge (m/z) ratios, and their corresponding ion intensities. For identification purpose, certain selected ionized peptides are subjected to fragmentation in a secondary tandem mass spectrometry (MS/MS). The resulting MS/MS spectra are then used to identify the amino acid sequences associated with the ionized peptides (e.g., precursor ions). Relative and absolute quantification methods can be developed based on the quantitative information (e.g., ion intensities) embedded in MS or MS/MS spectra.

4.1.3 Emerging Applications

Over the past decade, mass spectrometric techniques have made great advancement on their throughput and sensitivity that enabled the high depth of protein identifications even in very complex proteome samples [3, 6, 82, 108, 135]. As a result, shotgun proteomics has been successfully applied to many fields in life sciences. With the further improvement of protein identifications by proteomic techniques (the coverage of the current draft of the human proteome has reached 92 % of human proteins [143] and is anticipated to have complete coverage of the human proteome eventually), shotgun proteomics has become to play a critical role in functional studies of proteins at the whole genome scale, such as the quantifying thousands of proteins in eukaryotic organisms [28], profiling dynamic change of protein phosphorylation in cancer cell lines [99], and identifying cross-linked peptides in complex samples [145].

4.1.4 Computational Challenges

Due to the complexity and large volume of mass spectrometric data, computational methods played an essential role in the data analysis. In the last decade, since the very first few peptide identification algorithms were developed, there was a significant burst of computational methods for proteomic data analysis. The efforts include but are not limited to the fields of peptide identification, protein inference, protein quantification, identification of PTMs, and cross-linked peptides. To ensure to deliver reasonable and reliable results, it is necessary to understand the basic underlying assumptions and theory of these computational proteomic methods, as well as the pipeline to combine these different methods for various applications of shotgun proteomics. In this chapter, we will introduce the basic concepts of the computational methods in proteomics and provide an overview of existing methods to address specific computational challenges in the field.

4.1.5 Integrative Pipelines

It is desirable by experimental researchers to have integrative computational framework or software packages such that they can conveniently analyze their data. We summarized the commonly used computational frameworks, currently available and actively maintained in proteomics in Table 4.1. The details for each

Table 4.1 Available integrative pipelines for the general purpose proteomic data analyses

Name	Type	Description	Availability
OpenMS [120]	Open source	An open-source C++ library for LC-MS data management and analyses. It offers an infrastructure for the rapid development of mass spectrometry-related software	open-ms.sourceforge.net/about/
Trans-Proteomic Pipeline (TPP) [30]	Free software	TPP includes all steps in the Institute of Systems Biology (ISB) MS/MS analysis pipeline after the peptide identification	tools.proteomecenter.org/software.php
PeptideShaker [131]	Open source	A search-engine-independent platform for interpretation of proteomic identification results from multiple search engines, currently supporting X!Tandem, MSGF+, MS Amanda, OMSSA, MyriMatch, Comet, Tide, Mascot, Andromeda, and mzIdentML	compomics.github.io/projects/peptide-shaker.html
Compomics-utilities [9]	Open source	As a user-friendly, well-documented, and open-source library, compomics-utilities greatly simplifies the implementation of the basic features needed in most proteomic tools. Implemented in Java	compomics.github.io/projects/compomics-utilities.html
ProteoWizard [63]	Open source	The ProteoWizard software project provides a modular and extensible set of open-source, cross-platform tools and libraries. The tools perform proteomic data analyses	proteowizard.sourceforge.net/
PRIDE-Toolsuite [133]	Open source	PRIDE-Toolsuite comprises a selection of mass spectrometry-related tools	github.com/PRIDE-Toolsuite

step are laid out in each of the following sections. These frameworks consist of modules for the purpose of file format conversion, spectra preprocessing, and interface to different peptide algorithms for protein identification, quantification, or visualization.

4.2 Peptide Identification and Protein Inference

4.2.1 Peptide Identification

After acquiring the MS/MS spectra, the very first step is to identify the amino acid sequences corresponding to the spectra. To address this problem, numerous algorithms have been developed. Depending upon if the process is involved with theoretical database or pre-existing spectral library, those algorithms are divided into three categories: protein database searching, de novo sequencing, and spectral library searching. The predominant method used here is the protein database searching based on the matching between experimental and theoretical spectra of peptides in the database. Essentially, the database searching process compares the experimental spectra with the theoretical spectra generated from a protein database, which could be predicted from all putative genes in a genome, and then reported the scores of the top peptide-spectrum matches (PSMs). Based on the scores and quality assessment of these PSMs, final identification results are reported. In short, these database searching engines take spectra and a protein database as the input and then output likely true peptide-spectrum matches (PSMs). Recently, spectral library searches draw much attention because the advancement of mass spectrometry instruments improved the experimental throughput and precision; thus, a large amount of high-quality spectra have been deposited to the spectra library. In contrast, de novo sequencing methods only use spectra itself to predict the peptide sequences without any prior knowledge of in silico digested peptides. This section will briefly introduce the protein database searching algorithms and provide an overview of the publicly available software tools for peptide/protein identifications.

4.2.2 Protein Database Searching

The protein database searching algorithms were pioneered by Mann and Wilm's "peptide sequence tag" method [81] and the SEQUEST algorithm developed by Yates and colleagues [36]. Currently, many well-established peptide searching engines have been successfully applied to proteomics, including commonly used tools such as Mascot [20], X!Tandem [24], OMSSA [46], InsPecT [123],

MyriMatch [121], Protein Prospector [14, 19], COMPASS [141], Andromeda [22], Morpheus [140], Comet [37], Peppy [109], MS Amanda [33], and MSGF+ [64]. Protein database searching has become a routine practice in computational proteomics [119], similar as the protein homologous search by using BLAST in bioinformatics. Despite the different algorithms and features employed by the database searching engines, all of them follow the same procedure comparing the experimental spectra with spectrum generated from theoretical peptides. Because some software tools are commercially available and others may not open source yet, we will use SEQUEST [36] and MSGF+ [64] as examples to illustrate the basic concept of database searching algorithms.

SEQUEST first preprocesses the experimental spectrum x into a vector $\widehat{x}$. Given a candidate peptide y, a theoretical spectrum $\widehat{y}$ is constructed from y. The length of $\widehat{y}$ is equal to the length of $\widehat{x}$. Cross correlation is then calculated between these two vectors based on the XCorr score function. In SEQUEST, XCorr score represents the correlation between the theoretical spectrum and the experimental spectrum with the consideration of the background noise. The other popular score to use in SEQUEST is DeltaCN, which represents the difference between the XCorr of the top-ranked PSM and the other PSMs.

MSGF+ [64] is a recently developed tool toward a universal database search engine for proteomics. Due to the applicability and speed, it has attracted growing attention. MSGF+ uses a simple but robust dot-product scoring Score(P, S) = P*·S* after converting peptide P and spectrum S into peptide vector P* and spectral vector S*. Conversion of a spectrum S into a spectral vector S* uses a probabilistic model that ensures that the resulting dot-product scoring is adequate and makes the scoring and the computation of accurate E-values fast. After the dot-product scoring, MSGF+ then uses E-values to evaluate statistical significance of individual PSMs and the target-decoy approach to estimate false discovery rates (see the following sections).

4.2.3 *Available Software Tools for Protein Database Searching*

We summarize available and actively maintained software tools in Table 4.2.

4.2.4 *De Novo Peptide Sequencing*

Besides the protein database searching method, another approach for peptide identification is de novo sequencing, which requires no prior knowledge of

Table 4.2 Available software tools for peptide identification by database searching

Name	Type	Algorithm	Availability
SEQUEST [36]	Commercial	Comparing the experimental spectra with theoretical spectrum by cross correlation	fields.scripps.edu/
Comet [37]	Open source	An open-source tandem mass spectrometry (MS/MS) sequence database search tool. The open-source version of SEQUEST	comet-ms.sourceforge.net/
Tide [31]	Free software	An independent reimplementation of the SEQUEST algorithm, which identifies peptides by comparing the observed spectra to a catalog of theoretical spectra derived in silico from a database of known proteins	noble.gs.washington.edu/proj/tide/
OMSSA [46]	Free software	Probability-based scoring based on Poisson distribution	No longer maintained
Mascot [103]	Commercial	The most popular searching engine. Performs searching through a statistical evaluation of matches between observed and theoretical peptide fragments	www.matrixscience.com/
X!Tandem [24]	Open source	Calculate statistical confidence (expectation values) for all of the individual spectrum-to-sequence assignments	www.thegpm.org/tandem/
MSGF+ [64]	Open source	Based on inner product and compute rigorous E-values (using the generating function approach)	omics.pnl.gov/software/ms-gf
pFind [137]	Free upon request	Machine learning based	pfind.ict.ac.cn/
MassWiz [144]	Open source	A novel empirical scoring function that gives appropriate weights to major ions, continuity of b-y ions, intensities, and the supporting neutral losses based on the instrument type	sourceforge.net/projects/masswiz/
PEAKS DB [149]	Commercial	Integrate algorithm to validate the searching results	www.bioinfor.com
ProteinPilot software [118]	Commercial	Easy-to-use ProteinPilot software streamline protein identification and quantitation, enabling you to identify hundreds of peptide modifications and non-tryptic cleavages simultaneously	sciex.com/products/software/
Protein Prospector [14]	Free software	Proteomic tools for mining sequence databases in conjunction with mass spectrometry experiments	prospector.ucsf.edu/prospector/mshome.htm
SimTandem [95]	Free software	Employs the parameterized Hausdorff distance as a mass spectra similarity function	siret.ms.mff.cuni.cz/novak/simtandem/

(continued)

Table 4.2 (continued)

Name	Type	Algorithm	Availability
SQID [72]	Open source	Intensity-incorporated protein identification algorithm for tandem mass spectrometry	research.cbc.osu.edu/wysocki.11/group-home/bioinformatics/
Andromeda [22]	Free software	A novel peptide search engine using a probabilistic scoring model. It is included in MaxQuant software suites	www.maxquant.org
MyriMatch [121]	Open source	A statistical model to score peptide matches that is based upon the multivariate hypergeometric distribution. This scorer, part of the "MyriMatch" database search engine, places greater emphasis on matching intense peaks	medschool.vanderbilt.edu/msrc/
Morpheus [140]	Open source	A database search algorithm designed from the ground up for high-resolution tandem mass spectra	cwenger.github.io/Morpheus/
MS Amanda [33]	Free software	This algorithm is especially designed for high-resolution and high-accuracy tandem mass spectra	ms.imp.ac.at/?goto=msamanda
Byonic [11]	Commercial	Advanced method searches for tens or even hundreds of modification types simultaneously without a prohibitively large combinatorial explosion. Support database search for glycopeptides	www.proteinmetrics.com/products/byonic/

potential peptide sequences. The de novo sequencing can be considered as a peptide searching in a searching space containing all possible peptides. In general, the methods of de novo sequencing are divided into four categories based on the algorithms they adopt: [2] the naive approach, the graph theoretical approach, the probabilistic approach, and the combinatorial approach. The basic principle of the naive approach is to use the mass difference between two fragment ions to predict the potential amino acid residues of the peptide backbone. Many de novo sequencing methods use a graph theoretical model to compute the longest path in the spectrum graph by employing a dynamic programming algorithm [15, 27]. When the low-resolution MS instruments were widely used, the de novo sequencing algorithms were deemed to give less accurate results. With the advancement of high-resolution MS instruments, the performance of de novo sequencing methods has been significantly improved. Multiple methods have also been developed to utilize a combination of different types (e.g., CID (collision-induced dissociation) and ETD (electron transfer dissociation)) of high-resolution MS/MS data to achieve accurate de novo peptide sequencing [16, 60]. The existing methods for de novo peptide sequencing are summarized in Table 4.3.

Table 4.3 Available software for de novo peptide identification

Name	Type	Algorithm	Availability
PepNovo [41]	Open source	PepNovo uses a Bayesian network to model the peptide fragmentation events in a mass spectrometer. In addition, it uses a likelihood ratio hypothesis test to determine if the peaks observed in the mass spectrum resulted from the fragmentation of a peptide	proteomics.ucsd.edu/software-tools/
PEAKS [149]	Commercial	First released in 2002, PEAKS Studio software has become the industrial standard software for automated de novo sequencing and is well known for its accuracy, speed, and ease of use	www.bioinfor.com/peaks/
CycloBranch [96]	Open source	CycloBranch is a stand-alone, cross-platform, and open-source de novo peptide search engine for identification of non-ribosomal peptides (NRPs) from MS/MS spectra. Currently, the identification of linear, cyclic, branched, and branch-cyclic NRPs is supported	ms.biomed.cas.cz/cyclobranch/docs/html/
Lutefisk [62]	Open source	Lutefisk is a software tool for the de novo interpretation of CID spectra of peptides	www.hairyfatguy.com/Lutefisk/
pNovo + [16]	Free upon request	A de novo peptide sequencing algorithm using complementary HCD and ETD tandem mass spectra	pfind.ict.ac.cn/software/pNovo/
UniNovo [60]	Open source	A universal de novo peptide sequencing tool that works well for various types of spectra and pairs of spectra (e.g., from CID, ETD, HCD, CID/ETD, etc.)	proteomics.ucsd.edu/Software/UniNovo/
Novor [78]	Free for academia	Novor's scoring functions are based on two large decision trees built from a peptide spectral library	www.rapidnovor.com/

4.2.5 *Spectral Library Search*

Spectral library search approach has become widely used because the mass spectrometry data quality has significantly improved. The earliest study for spectral library search appeared in 1998 [146]. Based on the observation of reproducibility of mass spectrum generation, this approach constructed a spectral library first and then matched the incoming peptides to the library. Comparing to the database searching and de novo sequencing methods, the spectral library search approach is limited to the searching space of peptides with available high-quality experimental spectra, and the searching procedure can be slower. The current software for this purpose include BiblioSpec [42], X!Hunter [25], and SpectraST [67].

4.2.6 Validation of Peptide Identification

Because of the complexity of MS/MS spectra, the bias from the scoring functions of the computational methods, the potential bias from the theoretical background protein database and other reasons, the best scored PSMs are not always true. Therefore, in order to get the validated identification, controlling the false discovery rate (FDR) of the identification results is necessary and is even now mandated by most proteomic journals for reporting proteomic results. This FDR estimation is usually addressed by using target-decoy approach (TDA) [34], which becomes the standard in high-throughput MS studies because of its robustness, simplicity, and applicability.

Provided as input to the protein database search engines are a set of spectra and a protein database (i.e., the target database); TDA requires that spectra are searched not only against the target database but also against a decoy database. To construct a decoy database, reversing, shuffling, or other ways of randomizing the target database can be used, as long as the amino acid composition and length distribution of peptides in the target and decoy databases are similar. We then define the PSMs identified from the decoy database (e.g., decoy PSMs) are false identifications, while the PSMs from target database (e.g., target PSMs) are positive (potentially true but maybe false) identifications.

TDA is based on the assumption that the chance of false PSMs identified from the decoy database is equal to the false-positive PSMs identified from the target database because the constructed decoy database exhibits similar statistical properties as the target database. Therefore, the false discovery rate can be estimated based on the numbers of decoy PSMs (Nd) and target PSMs (Nt) that are identified at a certain score cutoff. By adjusting the score cutoff, one can obtain the peptide identification with a desirable FDR.

Although TDA is simple and intuitive, there is no uniform formula to calculate the FDRs by TDA. Sometimes the formula 2Nd/(Nd + Nt) is preferred over the formula Nd/Nt, sometimes vice versa. The discussion of the best way to compute FDR is beyond the scope of this chapter. However, researchers have provided a series of recommendations for better FDR evaluations [59].

4.2.7 Protein Inference

After a reliable set of peptides is identified, the next step is to assemble a reliable list of proteins from these identified peptides. This process is often referred to as the protein inference problem. Protein inference is a crucial and nontrivial procedure because some identified peptides are shared by two or more proteins in the target database, known as the degenerate peptides. It has been estimated that two million

Table 4.4 Available software tools for protein inference

Name	Type	Algorithm	Availability
ProteinProphet [93]	Open source	Automatically validates protein identifications made on the basis of peptides assigned to MS/MS spectra by database search engines	proteinprophet.sourceforge.net/
ProteinLP [56]	Open source	A linear programming model for protein inference	sourceforge.net/projects/prolp/
MSBayesPro [71]	Open source	A Bayesian protein inference algorithm for LC-MS/MS proteomic experiment	darwin.informatics.indiana.edu/yonli/DQmodel/
Fido [115]	Open source	A graph theoretical model for protein inference. Now it is integrated into the percolator	percolator.ms/

out of 3.8 million fully tryptic peptides are degenerated peptides [86]. The problem of determining which of the proteins are indeed present in the sample is an ongoing research topic, but has multiple solutions. Nesvizhskii et al. [93] first addressed this challenge by using a probabilistic model, but different problem formulations and new solutions have recently been proposed as well [56, 71]. The input of the protein inference can be modeled as a bipartite graph, in which one part is identified peptides and the other part is the proteins containing these peptides. Multiple features need to be considered in this type of graph, for example, score of the PSMs, peptide detectability, spectral counts, etc. Some existing algorithms exploit the quality of identified PSMs and the parsimonious rule to rank all potential proteins, while the other methods also exploit the quantification information of proteins.

While there are some articles summarizing the existing algorithms [53, 69], we provided a list of executable software tools for protein inference here as practical solutions in Table 4.4.

4.3 Protein Quantification

The next step in mass spectrometry proteomics is to quantify the identified peptides or inferred proteins in the sample. Quantitative proteomics is built upon a routine shotgun proteomic experiment, in which complex proteome samples are subject to proteolytic digestion followed by an LC-MS/MS analysis [1]. Multiple quantitative experimental strategies have been developed in recent years as reviewed in several articles [35, 116, 138]. Protein quantification provides information about the protein abundances in the sample and thus can be used as a tool to monitor the changes of protein expression under different conditions [83, 114], e.g., before and after viral infection [32] or across samples from healthy and diseased patients [29].

4.3.1 Protein Quantification Methods

Depending upon the experimental setting, the quantification methods differ in the forms of label-free and labeled quantification. The label-free approaches do not use any chemical modification to label the samples; instead this approach infers the quantification directly from the shotgun proteomic experimental data. It is used for the direct comparison of protein abundances across multiple LC-MS analyses of different samples using peptide peak areas (precursor intensity) [87] or spectral counts [18, 148] attributable to the proteins of interest. In addition, there was a recently proposed protein quantification method [47] that measures peptide abundance based on the ion signals of multiple fragments of a given peptide in the tandem mass spectrum. The labeled quantification methods utilized the isotope labeling techniques, such as isotope-coded affinity tags (ICAT) [52] and stable isotope labeling by amino acids in cell culture (SILAC) [101] and isobaric tags for relative and absolute quantification (iTRAQ) [142, 152]. It can be used to estimate the relative abundances of proteins in multiple samples through a single LC-MS analysis in which protein quantities from different samples can be distinguished based on specific isotopically labeled amino acids. Generally speaking, labeling techniques yield more accurate estimation of relative peptide/protein abundance in multiple samples, but require extra steps in sample preparation. Label-free quantification can achieve much higher throughput, while the quantification accuracy may be lower.

The tools for the quantitative proteomics are listed in Table 4.5. The list is somewhat incomplete because the tools are actively being developed in a very fast pace. Depending upon the experimental procedure for quantitative proteomics, the appropriate software tools can be different. Some of them may be specialized in labeled proteomic methods, while the others are with the focus of label-free methods. Even within the same category of labeling-based or label-free proteomics, the methods are very different in terms of the quantitative information they exploit. Users need to understand the experimental details to carefully choose the best-fit software.

4.3.2 Relative and Absolute Quantification

Quantitative proteomics [100] is invaluable when the quantitative information is used to study specific biological research problems. But in reality, the quantities provided by most of the quantification methods, labeling-based or label-free approaches, are primarily representing the relative protein quantification, i.e., the comparison of the protein abundances across multiple samples (e.g., under different conditions). The determination of the absolute abundances of different proteins in the same sample, i.e., the absolute protein quantification, is useful for many other important biological applications such as the mapping of protein expression patterns in a whole proteome scale [5]. In contrast to the same-protein-different-sample

Table 4.5 Available software tools for quantitative proteomics

Name	Type	Algorithm	Availability
MaxQuant [21]	Free software	A quantitative proteomic software package designed for analyzing large mass spectrometric datasets. It specifically aimed at high-resolution MS data. Several labeling techniques as well as label-free quantification are supported	coxdocs.org/doku.php?id=maxquant:start.
MZmine [104]	Open source	An open-source software for mass spectrometry data processing, with the main focus on LC-MS data	mzmine.github.io/
PEAKS Q	Commercial	The module of PEAKS for protein quantification analysis based on mass spectrometry data. PEAKS Q supports both labeling-based and label-free methods	www.bioinfor.com/
VEMS [110]	Free software	A program for analysis of MS-based proteomic data. VEMS can furthermore analyze iTRAQ, Mass Tag, SILAC, and labeled samples and also support label-free quantification	portugene.com/vems.html
PeakView	Commercial	A stand-alone software application that is compatible with all SCIEX mass spectrometer systems for the quantitative review of LC-MS and MS/MS data	sciex.com/products/software/peakview-software
Protein Quantifier [139]	Free software	An automated pipeline for high-throughput label-free quantitative proteomics and is integrated in OpenMS [120] suite	ftp.mi.fu-berlin.de/pub/OpenMS
ProteoSuite [48]	Open source	An open-source framework for the analysis of quantitative proteomic data	www.proteosuite.org/
MFPaQ [12]	Open source	Web application dedicated to parse, validate, and quantify proteomic data. It allows fast and user-friendly verification of Mascot result files, as well as data quantification using isotopic labeling methods or label-free approaches	mfpaq.sourceforge.net/

scenario, the quantitative measures such as spectral counts or peptide peak areas of different proteins in the same sample are not directly comparable. For example, two proteins of the same abundance may have distinct spectral counts because the peptides from one protein are more easily identified by LC-MS instruments than those from another protein. To address this issue, some computational methods are introduced to correct the bias. For example, emPAI [57] exponentially modified the protein abundance index (PAI) [106] value to determine the absolute quantification of proteins. A concept of peptide detectability was proposed to model the identification bias of peptides in a standard proteomic experiment [122]. As a result, the absolute protein abundances can be estimated from each protein's spectral counts corrected by the detectabilities of peptides in the protein [76, 134].

4.3.3 Label-Free Protein Quantification

Label-free quantification [97] is widely used because they are easy to adopt and can be integrated in most workflows without modification of the experimental protocols. But the intensity values derived directly from mass spectrometry are not recommended to be directly used as the indicator of peptide quantity because the intensity value is confounding by the ionization efficiency of peptides, which depends on the amino acid compositions. The algorithm for label-free quantification has to correct the bias by considering the chemical properties and the length of peptides.

The label-free quantification methods [23, 50] have two major categories: those based on the peptide peak intensity and those based on the spectral counts. The former method extracted information from the ion chromatography, i.e., XIC (extracted ion chromatograms), for given peptides or proteins; the latter simply counts the number of identified PSMs in a protein. Similar to the spectral counts, the detected peak areas belonging to different peptides are not directly comparable due to the detection bias caused by different chemical properties of peptides and thus need to be properly calibrated when used for absolute protein quantification [57]. The very first two label-free quantification methods based on spectral counts are emPAI [57] and absolute protein expression (APEX) [13]. The emPAI method quantifies a protein by the ratio between the number of observed peptides and the number of theoretical peptides, which is also implemented in Mascot [103]. The APEX uses the total observed spectral count normalized by the in silico predicted (i.e., expected) count of PSMs from each protein. The expected PSM count is computed by summing the predicted detectabilities of peptides from the proteins.

4.3.4 Labeling-Based Protein Quantification

Labeling-based techniques for protein quantification include isotope-coded affinity tagging (ICAT) [88], cleavable isotope-coded affinity tagging (cICAT) [70], isobaric tags for relative and absolute quantification (iTRAQ) [142], and stable isotope labeling by amino acids in cell culture (SILAC) [101]. All these techniques except iTRAQ rely on the search of a pair of MS (precursor) peaks corresponding to the same peptide with predefined mass difference, which result from the unlabeled peptide and isotopically labeled peptide (each from one sample to be compared, respectively). These methods work on two samples in most cases, but can also work on more samples with multiple isotopic labeling. iTRAQ, isobaric tags for relative and absolute quantification, uses isobaric tags to label N-terminal and lysine of peptides. There are currently two main reagents: 4-plex and 8-plex. Thus, the iTRAQ method can simultaneously label four or eight samples. The fragmentation of the isobaric tags produces the reporter ions in the low-mass region that give relative quantification of the peptides and the proteins from which the peptides are derived. Although the labeling-based quantification followed the same principle, specific algorithm is still needed for different kinds of labeling techniques. For example, for chemical labeling techniques, the mass difference between unlabeled and labeled peptides depends on the number of labeled amino acids. For iTRAQ data, it is needed to extract the intensities of reporter ions in MS/MS spectra with experimentally designated masses. Thus, when using the algorithms or software tools, one needs to set the parameters appropriately according to the labeling technique and other experimental settings.

4.4 Post-translational Modification

4.4.1 Principles for PTM Identification via Mass Spectrometry

PTM, or post-translational modification, is defined as modification of proteins occurring during or after protein biosynthesis, covalently or enzymatically, by introducing new chemical groups, such as the phosphate or methyl group, on amino acid residues (often on their side chains) in a protein. Recent proteomic studies [43, 51, 99] have revealed many novel PTMs that were previously unknown and thus significantly expanded the scope of PTMs. It is now estimated that there are on average 8–12 modified forms for each unmodified peptide [94].

Most of these modifications imposed characteristic mass shift to the unmodified peptide/proteins, and thus mass spectrometry can be used to detect the mass shifts corresponding to these modifications. Moreover, the modification sites can be localized [10] through tandem mass spectrometry. However, there is still considerable difficulty for implementing this simple principle [129], due to various aspects of PTMs: (1) the large number of potential modifications, (2) the low abundance of modified proteins, (3) the dynamic change of modifications, (4) the stability of modifications, and (5) the effect of modification on peptide ionization efficiency. Regardless of these considerations, the identification of PTMs has been improved to a great extent in the past few years [65, 90, 91, 124]. Thousands of PTM sites on proteins now can be confidently identified and localized thanks to the technical advancement of experimental enrichment of modified peptides [98]. From the quantification perspective, the label-free methods are particularly convenient for PTMs. It is also possible to determine the stoichiometry of PTMs at a large scale [99]. In this section, we will provide a brief yet comprehensive review of the computational methods for PTM identification based on the abovementioned proteomic techniques.

4.4.2 Database Searching for Peptides Containing PTM

For unmodified peptides, the database searching is done by comparing each experimental MS/MS spectrum with the theoretical spectrum. The database searching process is very similar when attempting to identify the peptide containing PTMs except that the mass shift due to the PTM on certain amino acid residues needs to be specified. For example, phosphorylation can be specified as an 80 Da or 98 Da mass shift on the S, T, and Y residues. Nonrestricted (blind) PTM searching [17, 66, 91, 129, 147], aiming at any modification on any amino acid, is not yet generally practical because such searching takes extremely long time, although bioinformatic algorithms are available. For the restricted PTM searching, conventional database searching engines, as listed in the above section, can be utilized. Because of the fragmentation characteristics of certain modifications (e.g., the neutral loss from phosphorylated peptides), some algorithms specifically designed for the post-processing [73, 77] of the database searching results have also been developed to improve the identification for peptides containing these PTMs.

4.4.3 Localization of Modification Sites from MS/MS

When the modification is sufficiently stable to withstand on the amino acid from the fragmentation energy of MS/MS analysis, the resulting fragment ions will retain the same mass shift from the precursor. In other words, these modifications can be identified not only based on the mass shift of the precursor mass but also the mass shift of the fragment ions containing the modified residue. As a result, the characteristics of these modifications can be incorporated into the post-processing of the database searching results to pinpoint the location of the modification group [10]. Many algorithms have been developed to localize modification sites, specifically for phosphorylation. Those algorithms are divided into two categories: one is based on the probability of fragments that matches with modification [7, 10, 40, 79, 99, 125, 127, 136], e.g., Ascore [10] and phosphoRS [125]; the other is based on the identification score difference between different modification sites, including MD-Score [112] and SLIP [8]. In general, PTM localization is not yet a solved problem, especially for PTMs other than phosphorylation.

4.4.4 Computational Methods for Detecting PTMs

From the abovementioned points of view, we summarized the currently available computational methods and resources specifically designed for PTMs in Table 4.6. Note that the general purpose database searching engines are not included in this table. The predominant types of PTMs of interest are phosphorylation, acetylation, ubiquitination, and oxidation. Other types of PTMs such as glycosylation have also become research subjects in the field (see below).

Table 4.6 Available software tools for posttranslational modification analysis

Name	Type	Algorithm	Availability
Ascore [10]	Open source	Ascore measures the probability of correct phosphorylation site localization based on the presence and intensity of site-determining ions in MS/MS spectra	ascore.med.harvard.edu/
PhosphoSitePlus [54]	Database	An online systems biology resource providing comprehensive information and tools for the study of protein PTMs including phosphorylation, ubiquitination, acetylation, and methylation	www.phosphosite.org/
phosphoRS [125]	Free software	This tool enables automated and confident localization of phosphorylation sites within validated peptide sequences and can be applied to all commonly used fragmentation techniques (CID, ETD, and HCD)	ms.imp.ac.at/?goto=phosphors
SysPTM [74]	Database	The database provides a systematic and sophisticated platform for proteomic PTM research, equipped not only with a knowledge base of manually curated multi-type modifications but also with four fully developed, in-depth data mining tools	www.biosino.org/SysPTM/
SLoMo [7]	Open source	A software tool enabling researchers to localize PTM sites on peptides identified in MS data	massspec.bham.ac.uk/slomo/

4.5 Data-Independent Acquisition (DIA)

4.5.1 Overview of DIA

Most shotgun proteomic studies adopt the data-dependent acquisition (DDA), and targeted proteomics used selected reaction monitoring (SRM) [4]. Some mass spectrometers offer an alternative operation mode, data-independent acquisition (DIA) [47, 68]. In the DIA mode, the mass spectrometer fragments all precursor ions within a window of retention time and mass-to-charge ratio in the sample, instead of each isolated precursor ion. Comparing to the DDA mode in shotgun proteomics, DIA has potential advantages [126]. Specifically, the data acquired in DIA mode is continuous in the time frame and records all fragment ions in the fragmentation chamber, which in turn retained much more information than the DDA data that selects precursor ions depending on their intensities (dynamic exclusion [45]). However, because of the same reasons, DIA data are much more complex and thus pose new bioinformatic challenges in data analysis.

4.5.2 Data Analysis Strategies for DIA

Unlike the data analysis pipelines that are well established in DDA shotgun proteomics, DIA data is more difficult to be analyzed due to their higher complexity. One typical issue is that the MS/MS spectra are always from the mixture of peptides because the DIA systematically fragmented all precursors by within a relatively wide mass window. Interpretation of such mixture spectra is nontrivial because they are not one to one corresponding to the peptide precursors [128]. Currently, DIA data analysis primarily adopts the strategy of synthesizing tandem mass spectra from the mixture spectra in silico by extracting precursor and fragment ions resulting from the same peptide based on their common chromatographic features. The conventional shotgun proteomic data analysis pipelines can then be applied to these synthetic MS/MS spectra for both identification and quantification. An alternative approach is proposed by the methods derived from SRM-based targeted data analysis procedure [47, 111], where the extracted ion chromatograms (XIC) of the most intense transitions of a targeted peptide from the assay library are generated from all corresponding MS/MS spectra, followed by the retention time alignment, peak grouping, and statistical analysis.

4.5.3 Methods in DIA

We summarized widely used software packages for DIA data analysis in Table 4.7.

Table 4.7 Available software tools for DIA data analysis

Name	Type	Algorithm	Availability
Skyline [80]	Open source	Windows client application for building and analyzing selected reaction monitoring (SRM)/multiple reaction monitoring (MRM), parallel reaction monitoring (PRM – targeted MS/MS and DIA/SWATH), and targeted DDA data	skyline.gs.washington.edu
OpenSWATH [111]	Free software	Proteomic software allowing the analysis of LC-MS/MS DIA data	www.openswath.org/
DIA-Umpire [128]	Open source	A Java program enabling untargeted peptide/protein identification and quantification from DIA data	diaumpire.sourceforge.net/
Group-DIA [75]	Open source	Group-DIA combines the elution profiles of precursor ions and the fragment ions from multiple data files to determine precursor-fragment pairs. Those pairs can be used to synthesize MS/MS spectra in silico that can be analyzed using conventional sequence database searching engines	yuanyueli.github.io/group-dia/

4.6 Emerging Applications

4.6.1 Proteogenomics

Proteogenomics [92] is an emerging research field at the intersection of proteomics and genomics, owning to the rapid advance of the sequencing technologies, such as RNA-seq, and the mass spectrometry-based proteomics. The data analysis pipelines in proteogenomics exploit customized protein sequence databases built upon genomic and transcriptomic sequence data to identify novel or mutated peptides from MS data. The corresponding proteomic data can also be used to improve gene annotations of the corresponding genome and to investigate protein expression levels.

When generating customized protein database, six-frame translation of genome can be directly used, as well as *ab initio* gene predicted from a genome by using gene prediction software. Moreover, transcriptomic data may provide novel splicing variants that should be incorporated into the target database. Some other alternative and integrative methods are also proposed recently, such as MSProGene [151]. Another focus of proteogenomics is to search for protein variants in a given variant database constructed from either short insertion and deletions (implicated in RNA-seq data) or single-amino-acid variants (e.g., from NCBI dbSNP database [117]). The variants discovered from matched genomic/transcriptomic data can be appended to the reference database and can be identified by following the routine proteomic protocols.

4.6.2 Glycoproteomics

Glycoproteomics [102] is a branch of proteomics aiming to identify and characterize site-specific glycosylations in proteins containing glycans as a PTM (i.e., the glycoproteins). Computational methods for glycoproteomics are challenging due to the inherent structural complexity of glycans as well as the microheterogeneity associated with each glycosylation site. There are two different common types of protein glycosylations: N linked (where the glycans are attached to the Asn residue) and O linked (where the glycans are attached to the Ser/Thr residues). According to the linked type, the glycoproteomics experimental design differs. For example, the identification of N-linked glycopeptides may require the combination of collision-induced dissociation (CID), higher-energy collision dissociation (HCD), and electron transfer dissociation (ETD) fragmentation methods. While it benefits the determination of glycan structure [130], the combination of different types of fragmentation data imposes additional challenges for the algorithm development. On the other hand, because the computational methods highly relied on the experimental setting, most software tools are designed for specific experimental protocols, and thus, no common computational pipeline has been widely adopted in the field. Nevertheless, several computational methods have been developed [55]. For example, GlycoFragwork [84] is an integrative computational framework to analyze multiple pre-aligned LC-MS/MS datasets and reports a glycomap of identified intact glycopeptides with their mass, elution time, and abundances.

4.6.3 Peptidomics

Peptidomics is another proteomic branch aiming at the study of endogenously produced protein fragments [26] by employing the conventional proteomic techniques. Endogenous protein fragments can be generated in multiple ways: catabolized dietary proteins, peptides released from food proteins, and peptides created from protease or protein substrates. With subtle modification of the computational methods in proteomics, most of these tools can be directly used in peptidomics. Because of the large variations in endogenous protein fragments, de novo sequencing methods may be favored over the database searching methods. Nonetheless, there are several comprehensive peptidomics databases, including SwePep [39] and Peptidome [61], available for database searching.

4.6.4 Metaproteomics

Similar as the experimental protocols, metaproteomic projects also follow the bioinformatics approaches used in bottom-up proteomics. Specifically, metaproteomic data analysis starts from the peptide identification, achieved by searching MS/MS spectra from an LC-MS/MS experiment against the tryptic peptides *in silico* digested from a target database of proteins that are potentially present in the metaproteomic sample. The conventional peptide search engines as summarized above can be used for this purpose. Their applications to metaproteomics rely on the preparation of a target protein database. Early metaproteomic studies used the collection of proteins encoded by fully sequenced bacterial genomes that are likely present in a specific environment (e.g., human gut) as the target database [38, 132]. This collection may be largely incomplete [105, 113]. Therefore, more recent metaproteomic studies employed a metagenome-guided approach, in which complete or fragmental coding genes were first predicted from metagenomic sequences (i.e., contigs or scaffolds), acquired from the matched community samples, and corresponding protein sequences were used in peptide identification [89]. Several software tools were developed for the purpose of gene prediction in metagenomic sequences including MetaGeneMark [150] and FragGeneScan [107].

Acknowledgment This work was supported by the grants R01 AI108888 and R01 GM103725 from National Institutes of Health (NIH).

References

1. Aebersold R, Mann M. Mass spectrometry-based proteomics. Nature. 2003;422:198–207.
2. Allmer J. Algorithms for the de novo sequencing of peptides from tandem mass spectra. Expert Rev Proteomics. 2011;8:645–57.
3. Altelaar AM, Munoz J, Heck AJ. Next-generation proteomics: towards an integrative view of proteome dynamics. Nat Rev Genet. 2013;14:35–48.
4. Anderson L, Hunter CL. Quantitative mass spectrometric multiple reaction monitoring assays for major plasma proteins. Mol Cell Proteomics. 2006;5:573–88.
5. Anderson NL, Anderson NG, Pearson TW, Borchers CH, Paulovich AG, Patterson SD, Gillette M, Aebersold R, Carr SA. A human proteome detection and quantitation project. Mol Cell Proteomics. 2009;8:883–6.
6. Angel TE, Aryal UK, Hengel SM, Baker ES, Kelly RT, Robinson EW, Smith RD. Mass spectrometry-based proteomics: existing capabilities and future directions. Chem Soc Rev. 2012;41:3912–28.
7. Bailey CM, Sweet SM, Cunningham DL, Zeller M, Heath JK, Cooper HJ. SLoMo: automated site localization of modifications from ETD/ECD mass spectra. J Proteome Res. 2009;8:1965–71.
8. Baker PR, Trinidad JC, Chalkley RJ. Modification site localization scoring integrated into a search engine. Mol Cell Proteomics. 2011;10:M111008078.
9. Barsnes H, Vaudel M, Colaert N, Helsens K, Sickmann A, Berven FS, Martens L. - Compomics-utilities: an open-source Java library for computational proteomics. BMC Bioinf. 2011;12:1.
10. Beausoleil SA, Vill´en J, Gerber SA, Rush J, Gygi SP. A probability-based approach for high-throughput protein phosphorylation analysis and site localization. Nat Biotechnol. 2006;24:1285–92.
11. Bern M, Kil YJ, Becker C. Byonic: advanced peptide and protein identification software. Current Protoc Bioinf. 2012:13–20. doi:10.1002/0471250953.bi1320s40.
12. Bouyssi´e D, de Peredo AG, Mouton E, Albigot R, Roussel L, Ortega N, Cayrol C, Burlet-Schiltz O, Girard J-P, Monsarrat B. MFPaQ, a new software to parse, validate, and quantify proteomic data generated by ICAT and SILAC mass spectrometric analyses: application to the proteomic study of membrane proteins from primary human endothelial cells. Mol Cell Proteomics. 2007;6(9):1621–37.
13. Braisted JC, et al. The APEX quantitative proteomics tool: generating protein quantitation estimates from LC-MS/MS proteomics results. BMC Bioinf. 2008;9:529.
14. Chalkley RJ, Baker PR, Medzihradszky KF, Lynn AJ, Burlingame A. In-depth analysis of tandem mass spectrometry data from disparate instrument types. Mol Cell Proteomics. 2008;7:2386–98.
15. Chen T, Kao M-Y, Tepel M, Rush J, Church GM. A dynamic programming approach to de novo peptide sequencing via tandem mass spectrometry. J Comput Biol. 2001;8:325–37.
16. Chi H, et al. pNovo+: de novo peptide sequencing using complementary HCD and ETD tandem mass spectra. J Proteome Res. 2012;12:615–25.
17. Chick JM, Kolippakkam D, Nusinow DP, Zhai B, Rad R, Huttlin EL, Gygi SP. A mass-tolerant database search identifies a large proportion of unassigned spectra in shotgun proteomics as modified peptides. Nat Biotechnol. 2015;33:743–9.
18. Choi H, Fermin D, Nesvizhskii AI. Significance analysis of spectral count data in label-free shotgun proteomics. Mol Cell Proteomics. 2008;7:2373–85.
19. Clauser KR, Baker P, Burlingame AL. Role of accurate mass measurement ($\pm$10 ppm) in protein identification strategies employing MS or MS/MS and database searching. Anal Chem. 1999;71:2871–82.
20. Cottrell JS, London U. Probability-based protein identification by searching sequence databases using mass spectrometry data. Electrophoresis. 1999;20:3551–67.

21. Cox J, Mann M. MaxQuant enables high peptide identification rates, individualized p.p.b.-range mass accuracies and proteome-wide protein quantification. Nat Biotechnol. 2008;26:1367–72.
22. Cox J, Neuhauser N, Michalski A, Scheltema RA, Olsen JV, Mann M. Andromeda: a peptide search engine integrated into the MaxQuant environment. J Proteome Res. 2011;10:1794–805.
23. Cox J, Hein MY, Luber CA, Paron I, Nagaraj N, Mann M. Accurate proteome-wide label-free quantification by delayed normalization and maximal peptide ratio extraction, termed MaxLFQ. Mol Cell Proteomics. 2014;13:2513–26.
24. Craig R, Beavis RC. TANDEM: matching proteins with tandem mass spectra. Bioinformatics. 2004;20:1466–7.
25. Craig R, Cortens J, Fenyo D, Beavis RC. Using annotated peptide mass spectrum libraries for protein identification. J Proteome Res. 2006;5:1843–9.
26. Dallas DC, Guerrero A, Parker EA, Robinson RC, Gan J, German JB, Barile D, Lebrilla CB. Current peptidomics: applications, purification, identification, quantification, and functional analysis. Proteomics. 2015;15:1026–38.
27. Dancik V, Addona TA, Clauser KR, Vath JE, Pevzner PA. De novo peptide sequencing via tandem mass spectrometry. J Comput Biol. 1999;6:327–42.
28. De Godoy LM, Olsen JV, Cox J, Nielsen ML, Hubner NC, Fr¨ohlich F, Walther TC, Mann M. Comprehensive mass-spectrometry-based proteome quantification of haploid versus diploid yeast. Nature. 2008;455:1251–4.
29. DeSouza L, Diehl G, Rodrigues MJ, Guo J, Romaschin AD, Colgan TJ, Siu KW. Search for cancer markers from endometrial tissues using differentially labeled tags iTRAQ and cICAT with multidimensional liquid chromatography and tandem mass spectrometry. J Proteome Res. 2005;4:377–86.
30. Deutsch EW, et al. A guided tour of the Trans-Proteomic Pipeline. Proteomics. 2010;10:1150–9.
31. Diament BJ, Noble WS. Faster SEQUEST searching for peptide identification from tandem mass spectra. J Proteome Res. 2011;10:3871–9.
32. Diamond DL, Jacobs JM, Paeper B, Proll SC, Gritsenko MA, Carithers RLJ, Larson AM, Yeh MM, Camp DG, Smith RD, Katze MG. Proteomic profiling of human liver biopsies: hepatitis C virus–induced fibrosis and mitochondrial dysfunction. Hepatology. 2007;46:649–57.
33. Dorfer V, Pichler P, Stranzl T, Stadlmann J, Taus T, Winkler S, Mechtler K. MS Amanda, a universal identification algorithm optimized for high accuracy tandem mass spectra. J Proteome Res. 2014;13:3679–84.
34. Elias JE, Gygi SP. Target-decoy search strategy for increased confidence in largescale protein identifications by mass spectrometry. Nat Methods. 2007;4:207–14.
35. Elliott MH, Smith DS, Parker CE, Borchers C. Current trends in quantitative proteomics. J Mass Spectrom. 2009;44:1637–60.
36. Eng JK, McCormack AL, Yates JR. An approach to correlate tandem mass spectral data of peptides with amino acid sequences in a protein database. J Am Soc Mass Spectrom. 1994;5:976–89.
37. Eng JK, Jahan TA, Hoopmann MR. Comet: an open-source MS/MS sequence database search tool. Proteomics. 2013;13:22–4.
38. Erickson AR, et al. Integrated metagenomics/metaproteomics reveals human hostmicrobiota signatures of Crohn's disease. PLoS One. 2012;7:e49138.
39. F¨alth M, Sk¨old K, Norrman M, Svensson M, Feny¨o D, Andren PE. SwePep, a database designed for endogenous peptides and mass spectrometry. Mol Cell Proteomics. 2006;5:998–1005.
40. Fermin D, Walmsley SJ, Gingras A-C, Choi H, Nesvizhskii AI. LuciPHOr: algorithm for phosphorylation site localization with false localization rate estimation using modified target-decoy approach. Mol Cell Proteomics. 2013;12:3409–19.

41. Frank A, Pevzner P. PepNovo: de novo peptide sequencing via probabilistic network modeling. Anal Chem. 2005;77:964–73.
42. Frewen BE, Merrihew GE, Wu CC, Noble WS, MacCoss MJ. Analysis of peptide MS/MS spectra from large-scale proteomics experiments using spectrum libraries. Anal Chem. 2006;78:5678–84.
43. Fu Y. Data analysis strategies for protein modification identification. Stat Anal Proteomics. 2016;1362:265–75.
44. Fulwyler MJ. Electronic separation of biological cells by volume. Science. 1965;150:910–1.
45. Gatlin CL, Eng JK, Cross ST, Detter JC, Yates JR. Automated identification of amino acid sequence variations in proteins by HPLC/microspray tandem mass spectrometry. Anal Chem. 2000;72:757–63.
46. Geer LY, Markey SP, Kowalak JA, Wagner L, Xu M, Maynard DM, Yang X, Shi W, Bryant SH. Open mass spectrometry search algorithm. J Proteome Res. 2004;3:958–64.
47. Gillet LC, Navarro P, Tate S, Rost H, Selevsek N, Reiter L, Bonner R, Aebersold R. Targeted data extraction of the MS/MS spectra generated by data-independent acquisition: a new concept for consistent and accurate proteome analysis. Mol Cell Proteomics. 2012;11: O111.016717.
48. Gonzalez-Galarza FF, Lawless C, Hubbard SJ, Fan J, Bessant C, Hermjakob H, Jones AR. A critical appraisal of techniques, software packages, and standards for quantitative proteomic analysis. Omics. 2012;16:431–42.
49. Gooley AA, Hughes G, Humphery-Smith I, Williams KL, Hochstrasser DF. From proteins to proteomes: large scale protein identification by two—dimensional electrophoresis and amino acid analysis. Biotechnology. 1996;14:1.
50. Griffin NM, Yu J, Long F, Oh P, Shore S, Li Y, Koziol JA, Schnitzer JE. Label-free, normalized quantification of complex mass spectrometry data for proteomic analysis. Nat Biotechnol. 2010;28:83–9.
51. Gupta N. Whole proteome analysis of post-translational modifications: applications of mass-spectrometry for proteogenomic annotation. Genome Res. 2007;17:1362–77.
52. Gygi SP, Rist B, Gerber SA, Turecek F, Gelb MH, Aebersold R. Quantitative analysis of complex protein mixtures using isotope-coded affinity tags. Nat Biotechnol. 1999;17:994–9.
53. He Z, Huang T, Liu X, Zhu P, Teng B, Deng S. Protein inference: a protein quantification perspective. Comput Biol Chem. 2016 (in press).
54. Hornbeck PV, Kornhauser JM, Tkachev S, Zhang B, Skrzypek E, Murray B, Latham V, Sullivan M. PhosphoSitePlus: a comprehensive resource for investigating the structure and function of experimentally determined post-translational modifications in man and mouse. Nucleic Acids Res. 2011;40:D261–70.
55. Hu H, Khatri K, Zaia J. Algorithms and design strategies towards automated glycoproteomics analysis. Mass Spectrom Rev. 2015. http://dx.doi.org/10.1002/mas.21487.
56. Huang T, He Z. A linear programming model for protein inference problem in shotgun proteomics. Bioinformatics. 2012;28:2956–62.
57. Ishihama Y, Oda Y, Tabata T, Sato T, Nagasu T, Rappsilber J, Mann M. Exponentially modified protein abundance index (emPAI) for estimation of absolute protein amount in proteomics by the number of sequenced peptides per protein. Mol Cell Proteomics. 2005;4:1265–72.
58. James P. Protein identification in the post-genome era: the rapid rise of proteomics. Q Rev Biophys. 1997;30:279–331.
59. Jeong K, Kim S, Bandeira N. False discovery rates in spectral identification. BMC Bioinf. 2012;13:1.
60. Jeong K, Kim S, Pevzner PA. UniNovo: a universal tool for de novo peptide sequencing. Bioinformatics. 2013;29(16):1953–62.
61. Ji L, Barrett T, Ayanbule O, Troup DB, Rudnev D, Muertter RN, Tomashevsky M, Soboleva A, Slotta DJ. NCBI Peptidome: a new repository for mass spectrometry proteomics data. Nucleic Acids Res. 2010;38:D731–5.

62. Johnson RS, Taylor JA. Searching sequence databases via de novo peptide sequencing by tandem mass spectrometry. Mol Biotechnol. 2002;22:301–15.
63. Kessner D, Chambers M, Burke R, Agus D, Mallick P. ProteoWizard: open source software for rapid proteomics tools development. Bioinformatics. 2008;24:2534–6.
64. Kim S, Pevzner PA. MS-GF+ makes progress towards a universal database search tool for proteomics. Nat Commun. 2014;5:5277.
65. Kim S, Na S, Sim JW, Park H, Jeong J, Kim H, Seo Y, Seo J, Lee K-J, Paek E. MODi: a powerful and convenient web server for identifying multiple posttranslational peptide modifications from tandem mass spectra. Nucleic Acids Res. 2006;34:W258–63.
66. Kim M-S, Zhong J, Pandey A. Common errors in mass spectrometry-based analysis of post-translational modifications. Proteomics. 2015;16(5):700–14.
67. Lam H, Deutsch EW, Eddes JS, Eng JK, King N, Stein SE, Aebersold R. Development and validation of a spectral library searching method for peptide identification from MS/MS. Proteomics. 2007;7:655–67.
68. Law KP, Lim YP. Recent advances in mass spectrometry: data independent analysis and hyper reaction monitoring. Expert Rev Proteomics. 2013;10:551–66.
69. Li YF, Radivojac P. Computational approaches to protein inference in shotgun proteomics. BMC Bioinf. 2012;13:S4.
70. Li J, Steen H, Gygi SP. Protein profiling with cleavable isotope-coded affinity tag (cICAT) reagents the yeast salinity stress response. Mol Cell Proteomics. 2003;2:1198–204.
71. Li YF, Arnold RJ, Li Y, Radivojac P, Sheng Q, Tang H. A Bayesian approach to protein inference problem in shotgun proteomics. J Comput Biol. 2009;16:1183–93.
72. Li W, Ji L, Goya J, Tan G, Wysocki VH. SQID: an intensity-incorporated protein identification algorithm for tandem mass spectrometry. J Proteome Res. 2011;10:1593–602.
73. Li S, Arnold RJ, Tang H, Radivojac P. Improving phosphopeptide identification in shotgun proteomics by supervised filtering of peptide-spectrum matches. In: Proceedings of the international conference on bioinformatics, computational biology and biomedical informatics. 2013;316.
74. Li J, et al. SysPTM 2.0: an updated systematic resource for post-translational modification. Database. 2014;2014. bau025.
75. Li Y, Zhong C-Q, Xu X, Cai S, Wu X, Zhang Y, Chen J, Shi J, Lin S, Han J. Group-DIA: analyzing multiple data-independent acquisition mass spectrometry data files. Nat Methods. 2015;12:1105–6.
76. Lu P, Vogel C, Wang R, Yao X, Marcotte EM. Absolute protein expression profiling estimates the relative contributions of transcriptional and translational regulation. Nat Biotechnol. 2007;25:117–24.
77. Lu B, Ruse C, Xu T, Park SK, Yates J. Automatic validation of phosphopeptide identifications from tandem mass spectra. Anal Chem. 2007;79:1301–10.
78. Ma B. Novor: real-time peptide de Novo sequencing software. J Am Soc Mass Spectrom. 2015;26:1885–94.
79. MacLean D, Burrell MA, Studholme DJ, Jones AM. PhosCalc: a tool for evaluating the sites of peptide phosphorylation from mass spectrometer data. BMC Res Notes. 2008;1:30.
80. MacLean B, Tomazela DM, Shulman N, Chambers M, Finney GL, Frewen B, Kern R, Tabb DL, Liebler DC, MacCoss MJ. Skyline: an open source document editor for creating and analyzing targeted proteomics experiments. Bioinformatics. 2010;26:966–8.
81. Mann M, Wilm M. Error-tolerant identification of peptides in sequence databases by peptide sequence tags. Anal Chem. 1994;66:4390–9.
82. Mann M, Kulak NA, Nagaraj N, Cox J. The coming age of complete, accurate, and ubiquitous proteomes. Mol Cell. 2013;49:583–90.
83. Marguerat S, Schmidt A, Codlin S, Chen W, Aebersold R, Bahler J. Quantitative analysis of fission yeast transcriptomes and proteomes in proliferating and quiescent cells. Cell. 2012;3:671–83.

84. Mayampurath A, Yu C-Y, Song E, Balan J, Mechref Y, Tang H. Computational framework for identification of intact glycopeptides in complex samples. Anal Chem. 2013;86:453–63.
85. Melton L. Protein arrays: proteomics in multiplex. Nature. 2004;429:101–7.
86. Meyer-Arendt K, Old WM, Houel S, Renganathan K, Eichelberger B, Resing KA, Ahn NG. IsoformResolver: a peptide-centric algorithm for protein inference. J Proteome Res. 2011;10:3060–75.
87. Monroe ME, Shaw JL, Daly DS, Adkins JN, Smith RD. MASIC: a software program for fast quantitation and flexible visualization of chromatographic profiles from detected LC-MS(/MS) features. Comput Biol Chem. 2008;32:215–7.
88. Moseley MA. Current trends in differential expression proteomics: isotopically coded tags. TRENDS Biotechnol. 2001;19:10–6.
89. Muller O, Emilie EL. Community-integrated omics links dominance of a microbial generalist to fine-tuned resource usage. Nat Commun. 2014;5:5603.
90. Na S, Jeong J, Park H, Lee K-J, Paek E. Unrestrictive identification of multiple post-translational modifications from tandem mass spectrometry using an error-tolerant algorithm based on an extended sequence tag approach. Mol Cell Proteomics. 2008;7:2452–63.
91. Na S, Bandeira N, Paek E. Fast multi-blind modification search through tandem mass spectrometry. Mol Cell Proteomics. 2012;11:M111–010199.
92. Nesvizhskii AI. Proteogenomics: concepts, applications and computational strategies. Nat Methods. 2014;11:1114–25.
93. Nesvizhskii AI, Keller A, Kolker E, Aebersold R. A statistical model for identifying proteins by tandem mass spectrometry. Anal Chem. 2003;75:4646–58.
94. Nielsen ML, Savitski MM, Zubarev RA. Extent of modifications in human proteome samples and their effect on dynamic range of analysis in shotgun proteomics. Mol Cell Proteomics. 2006;5:2384–91.
95. Nov´ak J, Galgonek J, Hoksza D, Skopal T. Similarity search and applications. Berlin/Heidelberg: Springer; 2012. p. 242–3.
96. Nov´ak J, Lemr K, Schug KA, Havl´ıˇcek V. CycloBranch: de novo sequencing of nonribosomal peptides from accurate product ion mass spectra. J Am Soc Mass Spectrom. 2015;26:1780–6.
97. Old WM, Meyer-Arendt K, Aveline-Wolf L, Pierce KG, Mendoza A, Sevinsky JR, Resing KA, Ahn NG. Comparison of label-free methods for quantifying human proteins by shotgun proteomics. Mol Cell Proteomics. 2005;4:1487–502.
98. Olsen JV, Mann M. Status of large-scale analysis of post-translational modifications by mass spectrometry. Mol Cell Proteomics. 2013;12:3444–52.
99. Olsen JV, Blagoev B, Gnad F, Macek B, Kumar C, Mortensen P, Mann M. Global, in vivo, and site-specific phosphorylation dynamics in signaling networks. Cell. 2006;127:635–48.
100. Ong S-E, Mann M. Mass spectrometry–based proteomics turns quantitative. Nat Chem Biol. 2005;1:252–62.
101. Ong SE, Blagoev B, Kratchmarova I, Kristensen DB, Steen H, Pandey A, Mann M. Stable isotope labeling by amino acids in cell culture, SILAC, as a simple and accurate approach to expression proteomics. Mol Cell Proteomics. 2002;1:376–86.
102. Pan S, Chen R, Aebersold R, Brentnall TA. Mass spectrometry based glycoproteomics from a proteomics perspective. Mol Cell Proteomics. 2011;10:R110–003251.
103. Perkins DN, Pappin DJ, Creasy DM, Cottrell JS. Probability-based protein identification by searching sequence databases using mass spectrometry data. Electrophoresis. 1999;20:3551–67.
104. Pluskal T, Castillo S, Villar-Briones A, Oreˇsiˇc M. MZmine 2: modular framework for processing, visualizing, and analyzing mass spectrometry-based molecular profile data. BMC Bioinf. 2010;11:1.
105. Qin J, et al. A human gut microbial gene catalogue established by metagenomic sequencing. Nature. 2010;464:59–65.

106. Rappsilber J, Ryder U, Lamond AI, Mann M. Large-scale proteomic analysis of the human spliceosome. Genome Res. 2002;12:1231–45.
107. Rho M, Tang H, Ye Y. FragGeneScan: predicting genes in short and error-prone reads. Nucleic Acids Res. 2010;38(20):e191.
108. Richards AL, Merrill AE, Coon JJ. Proteome sequencing goes deep. Curr Opin Chem Biol. 2015;24:11–7.
109. Risk BA, Edwards NJ, Giddings MC. A peptide-spectrum scoring system based on ion alignment, intensity, and pair probabilities. J Proteome Res. 2013;12:4240–7.
110. Rodr´ıguez-Su´arez E, Gubb E, Alzueta IF, Falc´on-P´erez JM, Amorim A, Elortza F, Matthiesen R. Virtual expert mass spectrometrist: iTRAQ tool for database-dependent search, quantitation and result storage. Proteomics. 2010;10:1545–56.
111. Rost HL. OpenSWATH enables automated, targeted analysis of data-independent acquisition MS data. Nat Biotechnol. 2014;32:219–23.
112. Savitski MM, Lemeer S, Boesche M, Lang M, Mathieson T, Bantscheff M, Kuster B. Confident phosphorylation site localization using the Mascot Delta Score. Mol Cell Proteomics. 2011;10:M110–003830.
113. Segata N, Waldron L, Ballarini A, Narasimhan V, Jousson O, Huttenhower C. Metagenomic microbial community profiling using unique clade-specific marker genes. Nat Methods. 2012;9:811–4.
114. Selbach M. Widespread changes in protein synthesis induced by microRNAs. Nature. 2008;455:58–63.
115. Serang O, MacCoss MJ, Noble WS. Efficient marginalization to compute protein posterior probabilities from shotgun mass spectrometry data. J Proteome Res. 2010;9:5346–57.
116. Shenoy A, Geiger T. Super-SILAC: current trends and future perspectives. Expert Rev Proteomics. 2015;12:13–9.
117. Sherry ST, Ward M-H, Kholodov M, Baker J, Phan L, Smigielski EM, Sirotkin K. dbSNP: the NCBI database of genetic variation. Nucleic Acids Res. 2001;29:308–11.
118. Shilov IV, Seymour SL, Patel AA, Loboda A, Tang WH, Keating SP, Hunter CL, Nuwaysir LM, Schaeffer DA. The Paragon Algorithm, a next generation search engine that uses sequence temperature values and feature probabilities to identify peptides from tandem mass spectra. Mol Cell Proteomics. 2007;6:1638–55.
119. Steen H, Mann M. The ABC's (and XYZ's) of peptide sequencing. Nat Rev Mol Cell Biol. 2004;5:699–711.
120. Sturm M, et al. OpenMS–an open-source software framework for mass spectrometry. BMC Bioinf. 2008;9:163.
121. Tabb DL, Fernando CG, Chambers MC. MyriMatch: highly accurate tandem mass spectral peptide identification by multivariate hypergeometric analysis. J Proteome Res. 2007;6:654–61.
122. Tang H, Arnold RJ, Alves P, Xun Z, Clemmer DE, Novotny MV, Reilly JP, Radivojac P. A computational approach toward label-free protein quantification using predicted peptide detectability. Bioinformatics. 2006;22:481–8.
123. Tanner S, Shu H, Frank A, Wang L-C, Zandi E, Mumby M, Pevzner PA, Bafna V. InsPecT: identification of posttranslationally modified peptides from tandem mass spectra. Anal Chem. 2005;77:4626–39.
124. Tanner S, Pevzner PA, Bafna V. Unrestrictive identification of post-translational modifications through peptide mass spectrometry. Nat Protoc. 2006;1:67.
125. Taus T, Kocher T, Pichler P, Paschke C, Schmidt A, Henrich C, Mechtler K. Universal and confident phosphorylation site localization using phosphoRS. J Proteome Res. 2011;10:5354–62.
126. Ting YS, Egertson JD, Payne SH, Kim S, MacLean B, K¨all L, Aebersold R, Smith RD, Noble WS, MacCoss MJ. Peptide-centric proteome analysis: an alternative strategy for the analysis of tandem mass spectrometry data. Mol Cell Proteomics. 2015;14:2301–7.

127. Trudgian DC, Singleton R, Cockman ME, Ratcliffe PJ, Kessler BM. ModLS: post-translational modification localization scoring with automatic specificity expansion. J Proteomics Bioinf. 2012;5:283–9.
128. Tsou C-C, Avtonomov D, Larsen B, Tucholska M, Choi H, Gingras AC, Nesvizhskii AI. - DIA-Umpire: comprehensive computational framework for data independent acquisition proteomics. Nat Methods. 2015;12:258–64.
129. Tsur D, Tanner S, Zandi E, Bafna V, Pevzner PA. Identification of posttranslational modifications via blind search of mass-spectra. In: Proceedings IEEE computational systems bioinformatics conference. 2005;157–166.
130. V´ekey K, Ozohanics O, T´oth E, Jek˝o A, R´ev´esz A, Kreny´acz J, Drahos L. Fragmentation characteristics of glycopeptides. Int J Mass Spectrom. 2013;345:71–9.
131. Vaudel M, Burkhart JM, Zahedi RP, Oveland E, Berven FS, Sickmann A, Martens L, Barsnes H. PeptideShaker enables reanalysis of MS-derived proteomics data sets. Nat Biotechnol. 2015;33:22–4.
132. Verberkmoes NC, et al. Shotgun metaproteomics of the human distal gut microbiota. ISME J. 2009;3:179–89.
133. Vizcaíno JA, et al. The PRoteomics IDEntifications (PRIDE) database and associated tools: status in 2013. Nucleic Acids Res. 2013;41:D1063–9.
134. Vogel C, Marcotte EM. Calculating absolute and relative protein abundance from mass spectrometry-based protein expression data. Nat Protoc. 2008;3:1444–51.
135. Walther TC, Mann M. Mass spectrometry–based proteomics in cell biology. J Cell Biol. 2010;190:491–500.
136. Wan Y, Cripps D, Thomas S, Campbell P, Ambulos N, Chen T, Yang A. PhosphoScan: a probability-based method for phosphorylation site prediction using MS2/MS3 pair information. J Proteome Res. 2008;7:2803–11.
137. Wang L-h, Li D-Q, Fu Y, Wang H-P, Zhang J-F, Yuan Z-F, Sun RX, Zeng R, He S-M, Gao W. pFind 2.0: a software package for peptide and protein identification via tandem mass spectrometry. Rapid Commun Mass Spectrom. 2007;21:2985–91.
138. Wasinger VC, Zeng M, Yau Y. Current status and advances in quantitative proteomic mass spectrometry. Int J Proteomics. 2013;2013:180605.
139. Weisser H, et al. An automated pipeline for high-throughput label-free quantitative proteomics. J Proteome Res. 2013;12:1628–44.
140. Wenger CD, Coon JJ. A proteomics search algorithm specifically designed for high-resolution tandem mass spectra. J Proteome Res. 2013;12:1377–86.
141. Wenger CD, Phanstiel DH, Lee M, Bailey DJ, Coon JJ. COMPASS: a suite of pre- and post-search proteomics software tools for OMSSA. Proteomics. 2011;11:1064–74.
142. Wiese S, Reidegeld KA, Meyer HE, Warscheid B. Protein labeling by iTRAQ: a new tool for quantitative mass spectrometry in proteome research. Proteomics. 2007;7:340–50.
143. Wilhelm M, et al. Mass-spectrometry-based draft of the human proteome. Nature. 2014;509:582–7.
144. Yadav AK, Kumar D, Dash D. MassWiz: a novel scoring algorithm with target decoy based analysis pipeline for tandem mass spectrometry. J Proteome Res. 2011;10:2154–60.
145. Yang B, et al. Identification of cross-linked peptides from complex samples. Nat Methods. 2012;9:904–6.
146. Yates JR, Morgan SF, Gatlin CL, Griffin PR, Eng JK. Method to compare collision-induced dissociation spectra of peptides: potential for library searching and subtractive analysis. Anal Chem. 1998;70:3557–65.
147. Ye D, Fu Y, Sun R-X, Wang H-P, Yuan Z-F, Chi H, He S-M. Open MS/MS spectral library search to identify unanticipated post-translational modifications and increase spectral identification rate. Bioinformatics. 2010;26:i399–406.
148. Zhang B, VerBerkmoes NC, Langston MA, Uberbacher E, Hettich RL, Samatova NF. Detecting differential and correlated protein expression in label-free shotgun proteomics. J Proteome Res. 2006;5:2909–18.

149. Zhang J, Xin L, Shan B, Chen W, Xie M, Yuen D, Zhang W, Zhang Z, Lajoie GA, Ma B. PEAKS DB: de novo sequencing assisted database search for sensitive and accurate peptide identification. Mol Cell Proteomics. 2012;11:M111–010587.
150. Zhu W, Lomsadze A, Borodovsky M. Ab initio gene identification in metagenomic sequences. Nucleic Acids Res. 2010;38:e132.
151. Zickmann F, Renard BY. MSProGene: integrative proteogenomics beyond six-frames and single nucleotide polymorphisms. Bioinformatics. 2015;31:i106–15.
152. Zieske LR. A perspective on the use of iTRAQTM reagent technology for protein complex and profiling studies. J Exp Bot. 2006;57:1501–8.

Chapter 5
Informatics for Metabolomics

Kanthida Kusonmano, Wanwipa Vongsangnak, and Pramote Chumnanpuen

Abstract Metabolome profiling of biological systems has the powerful ability to provide the biological understanding of their metabolic functional states responding to the environmental factors or other perturbations. Tons of accumulative metabolomics data have thus been established since pre-metabolomics era. This is directly influenced by the high-throughput analytical techniques, especially mass spectrometry (MS)- and nuclear magnetic resonance (NMR)-based techniques. Continuously, the significant numbers of informatics techniques for data processing, statistical analysis, and data mining have been developed. The following tools and databases are advanced for the metabolomics society which provide the useful metabolomics information, e.g., the chemical structures, mass spectrum patterns for peak identification, metabolite profiles, biological functions, dynamic metabolite changes, and biochemical transformations of thousands of small molecules. In this chapter, we aim to introduce overall metabolomics studies from pre- to post-metabolomics era and their impact on society. Directing on post-metabolomics era, we provide a conceptual framework of informatics techniques for metabolomics and show useful examples of techniques, tools, and databases for metabolomics data analysis starting from preprocessing toward functional interpretation. Throughout the framework of informatics techniques for metabolomics provided, it can be further used as a scaffold for translational biomedical research which can thus lead to reveal new metabolite biomarkers, potential metabolic targets, or key metabolic pathways for future disease therapy.

Keywords Data acquisition and analysis • Informatics techniques • Metabolomics • Metabolite biomarkers • Data mining

K. Kusonmano
Bioinformatics and Systems Biology Program, School of Bioresources and Technology, King Mongkut's University of Technology Thonburi, Bangkhuntien, Bangkok 10150, Thailand

W. Vongsangnak • P. Chumnanpuen (✉)
Department of Zoology, Faculty of Science, Kasetsart University, Bangkok 10900, Thailand

Computational Biomodelling Laboratory for Agricultural Science and Technology, Faculty of Science, Kasetsart University, Bangkok 10900, Thailand
e-mail: pramote.c@ku.ac.th

B. Shen et al. (eds.), *Translational Biomedical Informatics*, Advances in Experimental Medicine and Biology 939, DOI 10.1007/978-981-10-1503-8_5

5.1 Introduction

The word "metabolism" comes from the Greek word "metabolé" which means transformation or change, while the word "metabolome" is commonly defined as the measuring metabolites in a biological system. The study of metabolome usually relies on two different approaches, which are targeted and nontargeted metabolomics (or metabolite profiles). The targeted metabolomics focuses on the quantification of known compound, while the nontargeted approach aims for screening the patterns of the whole set of unknown metabolites. The behaviors of different metabolites are very dynamic in cellular regulatory and metabolic processes [21]. Practically, the levels of most metabolites change with half time of minutes and seconds or even faster either arising from natural fluctuations or response to the environmental change or external perturbations. To capture overall physiological status, metabolomics is considered for a "downstream" process in a molecular central dogma.

Metabolomics captures global biochemical events by assaying thousands of small molecules in cells, tissues, organs, or biological fluids followed by the applications of informatics techniques to define metabolite biomarkers. Currently, metabolomics can lead to reveal disease mechanisms, identify new diagnostic or prognostic markers, and also enhance understanding in drug response phenotypes. In this chapter, we aim to introduce overall metabolomics studies from pre- to post-metabolomics era. We initially describe the history of metabolomics and its impact on society. Emphasizing on post-metabolomics era, we provide a basic conceptual framework of informatics techniques for metabolomics that are used to study the metabolomics data as illustrated in Fig. 5.1, for instance, acquisition and preprocessing of the metabolomics data, analysis of the metabolomics data, and functional interpretation of the metabolomics data. Focusing on the analysis of metabolomics data in details, we describe three different approaches and show example researches that apply these techniques based on biomedical data, (1) unsupervised learning for viewing patterns and grouping the metabolomics data, (2) supervised learning for building a model for classifying the metabolomics data, and (3) feature selection for identifying candidate metabolite biomarkers. Additionally, we also list useful examples of tools and databases that are used for metabolomics data analysis toward functional interpretation. At the end, we highlight perspectives on potential metabolomics applications toward translational research and biomedical informatics.

5.2 History of Metabolomics and Impact on Society

Metabolic biochemists have arguably been "doing metabolomics" for decades. The earliest use of body fluids to determine a biological condition can be considered as the first uses of metabolomics, which can be traced back to the ancient Chinese

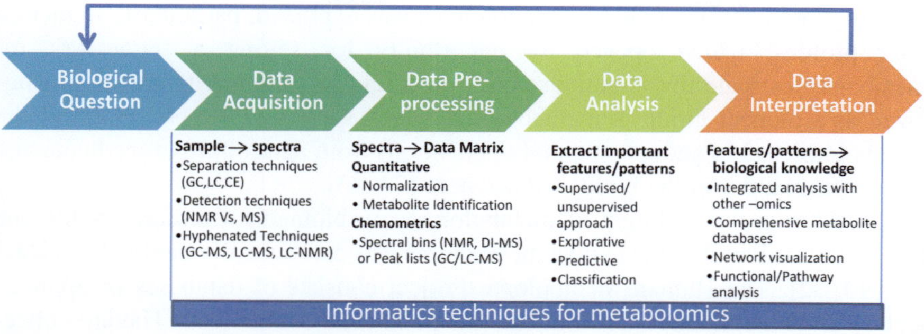

Fig. 5.1 A conceptual basic framework of informatics techniques for metabolomics studies

cultures (2000–1500 BC). At that time, there were some traditional Chinese doctors who began to evaluate the glucose level in urine of diabetic patients using ants. There is not only ancient China but also ancient Egypt and Greece that traditionally determine the urine taste to diagnose human diseases (300 BC). During this pre-metabolomics era, the chromatographic separation technique that made the initial detection of metabolites possible was developed in the late 1960s, which marked the technical origin of the metabolomics field. One of the medical luminaries of this early metabolomics revolution was Santorio Sanctorius who was the founding father of metabolic balance studies. His invaluable contributions were on insensible perspiration and useful invention published under the title of *Ars De Statica Medicina* in 1614 [28].

Joseph John Thomson and Francis William Aston were the first pioneers of mass spectrometry (MS) which was used to determine the nature of positively charged particles and evidence for isotopes of a stable (nonradioactive) element in 1913. With this powerful instrument, later on, MS has been greatly developed and improved regarding separation and sensitivity which led to the development of the mass spectrograph for metabolite quantification, metabolite profiles, and also structural elucidation.

Nuclear magnetic resonance (NMR) spectroscopy was afterward discovered in 1946 by Felix Bloch and Edward Purcell. NMR could be used to detect metabolites in unmodified biological samples [37]. Once instrumental sensitivity improved with the evolution of higher magnetic field strengths and magic angle spinning, NMR continues to be a leading analytical tool to investigate cellular metabolism [60, 84].

A development of metabolomics began in 1971 by Pauling's research team [65]. Although it was not called as metabolomics at that time, the first paper was "Quantitative Analysis of Urine Vapor and Breath by Gas–Liquid Partition Chromatography." Their studies investigated biological variability that could be explained by wider ranges of nutritional requirements than what was recognized. In analyzing the complicated chromatographic patterns of urine from vitamin B_6-loaded subjects, they also realized that the patterns of hundreds or thousands of chemical constituents in urine contained much useful information.

Few years later, the field of metabolomics was exploded, particularly in metabolite profiles, which opened an opportunity for setting a framework for metabolomics-scale investigations. Willmitzer and his research team were recognized as a pioneer group in metabolomics which suggested the promotion of the metabolomics field and its potential applications from agriculture to medicine and other related areas in the biological sciences [74, 83].

Considering the highlight of metabolomics on biomedical research fields, the first completed draft of the human metabolome was consequently published in 2007 [89, 90, 92]. The Human Metabolome Project consists of databases of approximately 2500 metabolites, 1200 drugs, and 3500 food components. Having collective and freely available database like the Human Metabolome Database (HMDB), the research fields on human metabolites related to disease diagnosis and advanced drug design can be rapidly modulated.

Nowadays, not only research on metabolomics related to human but also other model organisms have been increased during the post-metabolomics era, e.g., yeasts, fungi, insects, and plants [10, 19–21, 38, 56, 76, 87, 99, 102]. Such as in plants, the number of published paper related to metabolomics presented in several species, most notably *Medicago truncatula* and *Arabidopsis thaliana*, has been greatly increased for many years [1, 2, 17, 30, 44, 46, 47, 57, 58, 70, 86].

In the post-metabolomics era, the metabolomics data has been dramatically increased from the rapid development in high-throughput analytical techniques during pre-metabolomics era. Definitely, the informatics techniques and metabolomics tools and databases are required. A more developing area is the tool-aided functional interpretation of metabolite profiles. For instance, metabolomics tools aided for visual overlay of metabolite profiles onto biochemical network diagram and detection of statistical evidence for perturbation of particular pathways (e.g., enrichment analysis), identifying metabolite profile pattern that reliably indicates specific biological states and further using these patterns to diagnose the states of biological systems with an overall aim for metabolite biomarker identification. To illustrate the metabolomics timeline, Fig. 5.2 shows the highlight on instrumental and methodological development during pre-metabolomics era and informatics techniques during post-metabolomics era.

5.3 Informatics Techniques for Metabolomics

In order to extract information and knowledge from metabolomics data, informatics techniques are needed to analyze the derived data. Here we first introduce a data-centric overview of informatics techniques for metabolomics studies in three different approaches which include (1) acquisition and preprocessing of the metabolomics data, (2) analysis of the metabolomics data, and (3) functional interpretation of the metabolomics data.

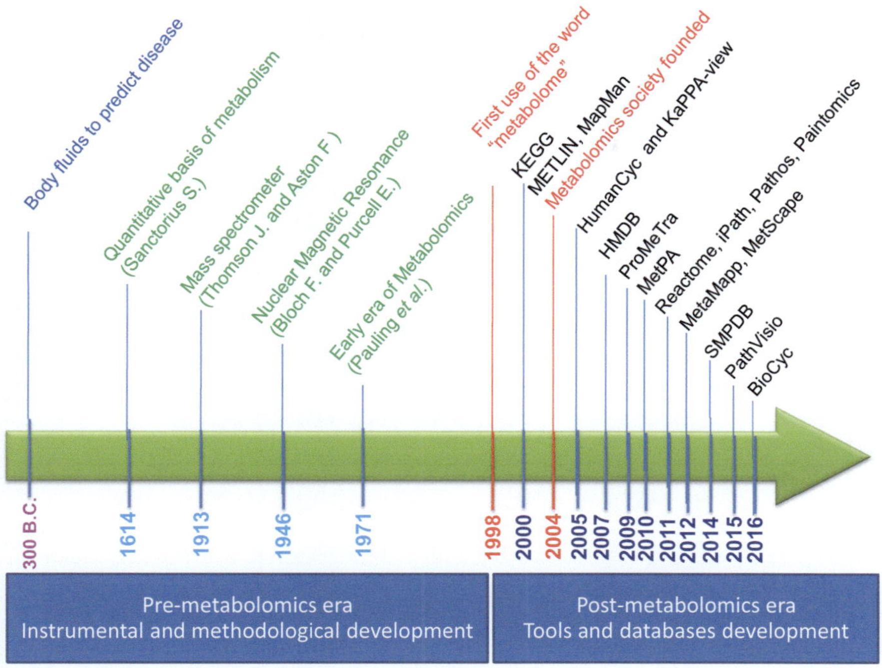

Fig. 5.2 Metabolomics timeline during pre- and post-metabolomics era

5.3.1 Acquisition and Preprocessing of the Metabolomics Data

Two of the most crucial and challenging steps for metabolome analysis are (1) how to acquire the raw (instrumental) data from the chosen high-throughput analytical techniques based on either MS or NMR and (2) how to convert them into the form of “extracted” data (e.g., peak tables) that can be easily further identified for specific metabolites and processed by statistical and data mining tools. So far, there are several statistical and data mining tools for acquisition and preprocessing of the metabolomics data as listed in Table 5.1.

5.3.1.1 Metabolomics Data Acquisition

For data acquisition in metabolome analysis, MS and NMR are the most frequently employed methods of detection. The chromatographic principles aim for the separation of the components by chromatographic instrumental, i.e., liquid chromatography (LC), gas chromatography (GC), or capillary electrophoresis (CE). The aim of this process is to collect and generate the spectral datasets from the designed experimental samples. NMR is particularly very useful for structure

Table 5.1 List of different tools used for acquisition and preprocessing of the metabolomics data

Tools		URL
AMDIS/NIST	AMDIS	http://www.amdis.net/
MASCOT	MATRIX SCIENCE	http://www.matrixscience.com/search_form_select.html
XCMS		https://metlin.scripps.edu/xcms/
MathDAMP	MathDAMP	http://mathdamp.iab.keio.ac.jp/
MetAlign	WAGENINGEN UR	http://www.wageningenur.nl/en/Expertise-Services/Research-Institutes/rikilt/show/MetAlign-1.htm
MZmine 2	MZmine 2	http://mzmine.github.io/download.html
MetaboAnalyst	MetaboAnalyst 3.0	http://www.metaboanalyst.ca/
MSFACTs	NOBLE	http://www.noble.org/plantbio/sumner/msfacts/
COMSPARI	COMSPARI	http://www.biomechanic.org/comspari/
SpectConnect	SpectConnect.	http://spectconnect.mit.edu/

characterization of unknown compounds and is applied for the analysis of metabolites in biological fluid and cell extracts. However, there are some limitations of using NMR analysis due to its longer running time (several hours) per sample and low-sensitivity issue, and the equipment costs are much higher compared to MS-based techniques. The additional advantages of using MS over NMR are its high throughput in combination to identify the unknown metabolites. Moreover, NMR has advantages on the possibility to combine with chromatograph separation techniques (e.g., CE, GC, or LC) expanding the capability for metabolite profiles of the complex biological samples. Undeniably, there are several available tool applications as also seen in Table 5.1 (e.g., AMDIS, MASCOT, MZmine2, XCMS, etc.) for visualizing and searching against the metabolomics library and available databases which can helpfully facilitate the peak identification and functional interpretation of mass spectrometric data.

The recent introduction of ultra-performance liquid chromatography–mass spectrometry (UPLC-MS) has greatly enhanced chromatographic performance, increasing the sensitivity and throughput of liquid chromatography–mass spectrometry (LC-MS) measurements. UPLC-MS routinely detects thousands of features representative of hundreds to thousands of metabolites in biological mixtures. Although the determinations of the exact number of metabolites can be measured by untargeted UPLC-MS, it is complicated to be analyzed due to hard sample preparation, chemical diversity of the metabolite matrix, and the sophisticated isotopes and fragments and also adduct ions. While these analytical chemistry techniques boast high sensitivity and reproducibility and are capable of untargeted detection of a vast number of diverse metabolites, the collection and comparison of a large number of mass spectra or NMR pose great challenge in data analysis regardless of whether NMR, gas chromatography–mass spectrometry (GC-MS), or LC-MS is applied. The spectra generated by these instruments require similar preprocessing steps before comparative analysis even though the capabilities, specificity, and sensitivity are quite different. The signatures of thousands of metabolites generated by UPLC-MS require extensive preprocessing before comparative statistical analysis. A key step in the analysis of UPLC-MS datasets is the transformation of ion intensities. This process is based on the elution time into a matrix of features of each sample (m/z and retention time) which can be applied for peak detection, alignment, and area extraction algorithms. Subsequent statistical and data mining analysis tools can be operated on this matrix of ion intensities and are later mentioned in the following analysis of the metabolomics data section.

5.3.1.2 Metabolomics Data Preprocessing

The most important purpose of metabolomics data preprocessing is to convert different metabolomics data into data matrix suitable and comparable for varieties of statistical analyses. Therefore, the proper metabolomics preprocessing methods must be selected and performed before the metabolomics data analysis. Two major approaches are focused on the quantitative or chemometric (screening/profiling) for data preprocessing. However, some of the preparing steps are regularly required for the assessment of the metabolomics data quality, e.g., deconvolution of overlapping peaks, peak picking, integration, alignment, data cleanup, normalization, as well as metabolite identification [5, 36]. In the following, normalization and metabolite identification are briefly described for data preprocessing.

Normalization

"Normalization" is the most common step for preparing step for data preprocessing. The goal of normalization step of datasets is to allow the direct comparison for metabolome profiling. To reach that point, a representative set of peaks have to be picked from each dataset. The peak sets of key metabolites can be further aligned instead of using all the peaks from the datasets. Parameters of a time shift function based on the mathematical function to model the retention time or migration time shifts between samples are needed to be analyzed. To obtain those parameters, the combination of global optimization and dynamic programming is commonly applied. With the normalization procedure, the optimal parameters can be reliably found, even if the peak sets contain a small number of corresponding peaks.

Commonly, the MS data has some usual problems needed to be considered, i.e., baseline drift, retention time shifts, noise, and artifacts. In general, the targeted metabolomics data quality can be assessed by selecting a number of representative compounds for each metabolite category and calculating their concentration relative to the proper chosen internal standards. On the other hand, the preprocessing of the raw NMR data is usually performed by machine vendor tools which can provide the phase and baseline correction, removal of water and urea resonance, and spectral binning or bucketing [53]. For both MS- and NMR-based techniques, the normalization process can be performed based on the sum or total peak area, a reference compound (i.e., creatinine, internal standard), a reference sample which is also known as "probabilistic quotient normalization," dry mass, volume, etc. [25].

Metabolite Identification

To further achieve the goals in metabolomics investigations, it is necessary to produce a comprehensive metabolome profiling from biological samples. Identification of metabolites is one of the necessary steps in metabolomics studies, and the conclusion drawn from such studies depends on how exactly the metabolites are identified. The experimental identification of unknown metabolites is an essential process. Extensive works have thus been established to identify accurate information on hundreds to thousands of metabolites. In general, the metabolite identification process is largely based on tandem mass spectrometry (MS/MS) spectra generated separately for peaks of interest identified from previous MS runs. Such a delayed and labor-intensive procedure creates a barrier to automation. Further, information embedded in MS data has not been used to its full extent for metabolite identification. Multimers, adducts, multiply charged ions, and fragments of given metabolites occupy a substantial proportion (40–80 %) of the peaks of a quantitative result [21, 76]. However, extensive information on these derivatives, especially fragments, may facilitate metabolite identification.

Beyond this, a procedure with automation capability to group and annotate peaks associated with the same metabolite in the quantitative results of opposite modes and to integrate this information for metabolite identification is proposed. In addition to the conventional mass and isotope ratio matches, the annotated fragments with low-energy MS/MS spectra in public databases have been matched.

Additionally, NMR is one of the most selective analytical techniques, which gives unambiguous structural information of metabolites. Due to complex biological sample matrix, metabolite identification needs the application of advanced NMR techniques and analytical strategies for better accuracy. The major problems that arise in metabolite identification using NMR in biological samples are high spectral crowding, the presence of macromolecule, molecular interaction, dynamic range, enormous solvent concentration, sensitivity, etc. Therefore, several useful techniques are developed and applied to the metabolite identification, such as spiking experiments, standard spectra, NMR data banks and libraries, and literature support. For example, in human body fluids and pathological fluid samples, there is plenty of literature available related to assignments of metabolites [7, 8, 27, 32, 51, 52, 69, 78, 91]. Increasing applications of metabolomics encourage researchers to develop databases and tools for automatic and fast data preprocessing. As listed in Table 5.2, there are several databases, which can be used for metabolite identification and available for MS- and NMR-based techniques.

Table 5.2 List of metabolite identification databases

Databases			URL
NMR	SDBS		http://sdbs.db.aist.go.jp
	BioMagResBank		http://www.bmrb.wisc.edu/metabolomics/
	NMRShiftDB 2		http://nmrshiftdb.nmr.uni-koeln.de/
	MMCD		http://mmcd.nmrfam.wisc.edu/
MS	GMD		http://gmd.mpimp-golm.mpg.de/
	METLIN		https://metlin.scripps.edu/index.php
	MassBank		http://www.massbank.jp/
	NIST		http://webbook.nist.gov/chemistry/name-ser.html
Compound DBs	LIPIDMAPS		http://www.lipidmaps.org/
	KNApSAcK		http://kanaya.naist.jp/KNApSAcK/KNApSAcK-v1200/KNApSAcK.php
	ARM		http://metabolomics.jp/wiki/Main_Page
	ChEBI		https://www.ebi.ac.uk/chebi/init.do
	PubChem		https://pubchem.ncbi.nlm.nih.gov/
	ChemSpider		http://www.chemspider.com/
	Ligand Expo		http://ligand-expo.rcsb.org/
	3DMet		http://www.3dmet.dna.affrc.go.jp/
	MyCompoundID		http://www.mycompoundid.org/

5.3.2 *Analysis of the Metabolomics Data*

After completed acquisition and preprocessing of the metabolomics data, analysis techniques of metabolomics data play an important role helping to extract useful information. Machine learning techniques are incorporated prominently in this area, relating the design and development of algorithms that allow computers to learn based on empirical data [35, 93]. Machine learning can be classified into two commonly known categories, which are the unsupervised and supervised learning methods. Unsupervised learning requires data without labels and determines patterns of the data naturally. On the other hand, supervised learning conducts data with its labels to learn a model to be able to predict new unlabeled data. Besides, feature selection is another component, playing a key role in both supervised and unsupervised learning. This feature selection technique aims to select a subset of relevant features (i.e., metabolites) for constructing robust learning models. In biomedical area, the selected features can be considered as biomarkers for diagnosis and prognosis of diseases or the prediction of therapeutic success. Furthermore, understanding of functional and biological processes of the candidate features/metabolites is also important. Candidate metabolite biomarkers are therefore needed to be interpreted by performing pathway mapping and visualization or enrichment analysis. Here we explain unsupervised and supervised learning and feature selection that are applied to metabolomics data, respectively. For the functional interpretation, it is continuously described in the next section.

5.3.2.1 Unsupervised Learning: Viewing Patterns and Grouping the Metabolomics Data

Unsupervised learning is a machine learning algorithm that determines how the unlabeled data are organized. It helps to find patterns of the data and discover new classes. In the following, we describe concepts of some common unsupervised learning methods, namely, principal component analysis (PCA), clustering, and self-organizing map (SOM), that are currently used for metabolomics studies. In addition, we provide example studies applying these methods for the discovery of metabolomics patterns of biomedical data.

Principal Component Analysis (PCA)

Principal component analysis (PCA) [71] has been used intensively in metabolomics data analysis [16, 61, 63]. The method is used for dimension reduction and provides a visualization of the data. As mentioned previously, the metabolomics data is a high-dimensional data containing a high number of variables. PCA can be used to project the data into a lower dimension, e.g., two or three dimensions, that can be seen and understood by human. The concept of PCA is to find a one-dimensional subspace that captures the most variance of the data, referred to as principal component (PC). From the first PC, the second PC is created by considering the variance that is not captured by the first PC and maximizes the remained variance. The same principle of finding a direction maximizes the variance of the data, which can be repeated for the next PC and so on. Usually, the best two (or three) subspaces are plotted, meaning that the projection of the first two (or three) PCs is used [6]. By using the PCA, samples can be visualized to assess similarities and/or differences among them. The natural grouping of samples can be discovered. This could be used to investigate known classes of the studied samples (e.g., by plotting samples with different colors of their groups) or to discover new subclasses (e.g., investigating the grouping patterns). However, by conducting the PCA, one should keep in mind that PCA reflects the data with a lot lower dimension, and some contents of the data are lost during the dimension reduction. A user should keep in mind that the visualization does not represent all contents of the data. Several studies have applied PCA to analyze metabolomics data for biomedicine.

Odunsi et al. utilized PCA to show a clear separation between serum specimens from 38 patients with epithelial ovarian cancer (EOC), 21 premenopausal normal samples, and patients with the benign ovarian disease by using proton nuclear magnetic resonance (^{1}H-NMR)-based metabolomics [63]. Another study presented by Chen et al. showed that PCA was performed to visualize the metabolic alterations and showed a trend of separation between hepatocellular carcinoma (HCC) patients and healthy controls using serum and urine metabolite profiles [15].

Clustering

Clustering is a method aiming to divide data into clusters according to their properties [93]. The objects (e.g., samples or metabolites) that are grouped together or in the same group are more similar to each other than the objects in other groups. Clustering methods show a natural grouping of the data and help to visualize and reveal patterns of the data. Here we discuss two well-known clustering algorithms which are k-means and hierarchical clustering.

K-means clustering tries to group "n" objects into "k" groups. This technique requires a user to define the number of groups/clusters for dividing the data. The method is useful when the number of the clusters is known and one needs to investigate patterns of the data under each subgroup. However, there are some extensions to the algorithm that provide a computational way to calculate k. For example, an application of fuzzy k-means clustering was shown to classify metabolomics data of NMR spectra of breast cancer cell line and type 2 diabetes patients and animal models [23].

Hierarchical clustering, on the other hand, does not need a number of clusters, but builds a hierarchy of the clusters. The method provides a dendrogram visualizing on how the objects are grouped. Hierarchical clustering has been widely applied in different omics studies including metabolomics, as it displays data grouping, which is easy for investigation and interpretation. For instance, a hierarchical clustering was performed to display the subtype of pancreatic ductal adenocarcinoma (PDAC) through metabolite profiles of cell lines [24]. Three subtypes were identified, which showed proliferating lines, glycolytic lines, and lipogenic lines. An output dendrogram showed metabolic patterns changing in different subtypes.

Self-Organizing Map (SOM)

Self-organizing map (SOM) [45] has been applied to provide visualization of high-dimensional data in low-dimensional space (typically two dimensions). It is a type of neural network. While PCA uses a linear projection to reduce the data dimension, SOM applies more complex of nonlinear pattern recognition. It aims to find a low-dimensional representation of the input data. Samples, which are similar to each other, are placed in a similar region [6]. In metabolomics, for example, SOM was carried out to visualize metabolic changes in breast cancer tissue [4] from normal to different tumor grades. SOM has an advantage of providing a nonlinear mapping for data visualization. However, the algorithm is like a "black box". The key metabolites used for the separation are hidden, which is not useful for functional interpretation.

5.3.2.2 Supervised Learning: Building a Model for Classifying the Metabolomics Data

Supervised learning aims to learn a model from a set of given labeled examples to recognize new examples. For a general procedure of the classification step, firstly a classifier/model is trained using labeled training samples. After training and testing the model, the trained model is used to classify unlabeled subjects or samples. For example, one can use metabolite profiles to create a model to classify between healthy and cancer. A classification algorithm could be combined with a feature selection method to select features/metabolites leading a model to give the best distinguishing between two groups. The selected features could be potential candidate biomarkers for diagnosis, prognosis, or prediction of therapeutic success (see Feature Selection: An Approach to Identify Candidate Metabolite Biomarkers). In addition, to avoid model overfitting, a test set for testing a model should be separated from a training dataset for that is used to construct a model. As described below, we mention some common classification algorithms including partial least squares (PLS), support vector machine (SVM), and random forest (RF) for metabolome analysis and show their example cases in biomedical area.

Partial Least Squares (PLS)

The partial least squares (PLS) [94] method is widely used in metabolomics [6, 15, 61]. The constructed model can be used for classification, e.g., to predict the class of unknown samples. For the high-dimensional data like metabolome, the method facilitates dimensional reduction, and the data can also be projected for visualization into a low-dimensional space. PLS is a regression-based method, which finds a linear regression model by projecting the dependent variables Y and the independent variables X to a new space. The algorithm tries to maximize the variance of the dependent variables that are explained by the independent variables. PLS is similar to PCA in a way that both algorithms focus on finding a subspace capturing the most variance of the data. In PLS, the generated components are called orthogonal vectors, while the outputs from PCA are known as principal components. However, PLS is a supervised learning method. The classes of data or samples are necessary for building a model.

Partial least squares discriminant analysis (PLS-DA) is a version of PLS when Y is categorical. It has been applied in several metabolomics studies [15, 61]. For instance, Nishiumi et al. applied PLS-DA to discriminate stages of pancreatic cancer by using serum metabolome [61]. They suggested candidate metabolites as biomarkers allowing early detection of pancreatic cancer. Besides, by utilizing orthogonal partial least squares discriminant analysis (OPLS-DA) – an extended version of PLS-DA – Chen et al. showed plots by using PLS-DA separating clearly between healthy and HCC patients based on serum data, and the model provided a good performance for diagnostic markers of HCC [15]. PLS and its extended algorithms are popular in metabolomics since it provides both a classification model and visualization of the data, which is easy for data interpretation. As described above, similar to PCA, PLS displays the data overview into low-dimensional space. It can visualize how data is categorized into labeled classes. The output is quite easy to determine by human eyes how the studied classes are separated. Also, the features, which are used to construct a model, could be considered as candidate biomarkers [6, 15, 61].

Support Vector Machine (SVM)

Support vector machine (SVM) [85] is known as a classifier providing good generalization in high-dimensional and noisy data [54, 62]. SVM has been widely and successfully applied to high-throughput data including metabolome [33, 54]. The concept of SVM is that it builds an optimal linear hyperplane separating two classes in a feature space. It seeks the so-called maximum-margin hyperplane, which gives the greatest separation between the classes. The margin is the largest distance between the hyperplane and the data points on either side. The points that are closest to the maximum-margin hyperplane and lie on the margin are called support vectors [62].

As mentioned above, with the concept of finding a hyperplane, SVM is a good generalization classifier and has also showed good performance in metabolomics. For example, Mahadevan et al. showed that SVM gives better predictive model compared to PLS-DA [54]. They demonstrated the comparison by using NMR spectra of urine samples from healthy and pneumonia patients as well as a more ambiguous case, between male and female. Guan et al. employed SVM for a diagnostic purpose to build a classifier that distinguishes between ovarian cancer patients and controls using metabolic data (LC/TOF MS) of serum samples [33]. Classification in a test set was shown with more than 90 % accuracy.

Random Forest (RF)

Random forest (RF) [9] has become popular in bioinformatics as it has been reported to provide high performance on high-dimensional data, such as transcriptomics and metabolomics data [77, 95]. RF is an ensemble-based machine learning method. It relies on aggregated results of several individual decision trees. With its characteristic of the ensemble-based method, it is robust against overfitting [67]. The predicted class of an unknown sample is assumed to be a class that obtained the majority vote from all decision trees. The voting among many decision trees provides better model performance than one decision tree alone [9, 67].

In metabolomics, RF has been demonstrated with a good performance for classification. For instance, Chen et al. showed that RF outperforms the other classifiers, which are PLS, SVM, and linear discriminant analysis (LDA), clinical metabolic data [18]. The performances were evaluated in metabolomics study of GC-MS platform between healthy subjects and patients diagnosed with colorectal cancer. It can be applied for classification and biomarker selection [18].

5.3.2.3 Feature Selection: An Approach to Identify Candidate Metabolite Biomarkers

Rather than building a classification system, many people are more interested to discover candidate biological markers (i.e., metabolites), which give the best distinguishing between two classes of interest, e.g., normal vs. cancer. In the medical area, the candidate biomarkers can then be verified and applied for diagnosis, the prognosis of diseases, or prediction of therapeutic success. A feature selection method can be used as a combination with classification algorithm [14]. An optimal set of features or metabolites that give the best separation between two groups is selected as candidate markers. In addition, feature selection methods can also be applied with unsupervised learning, helping to reduce noises or irrelevant features. Depending on a medical purpose, different types of biomarkers can be classified. Mainly there are three types of biomarkers, which are diagnostic, prognostic, and predictive biomarkers. A diagnostic biomarker helps in diagnosing or distinguishing between a healthy person and patient. For example, prostate-specific antigen (PSA) is used as a biomarker for diagnosis of prostate cancer [64]. A prognostic biomarker provides information on the likely course of the disease, for example, it is used to predict recurrence of a disease. A predictive biomarker can be used to identify subpopulations of patients who are most likely to respond to a given therapy.

In data mining area, the methods for feature selection can be categorized into three main types, which are filter, wrapper, and embedded methods [73]. Filter methods are based on a quality merit of a feature, taking into account its ability to distinguish between predefined classes. In most cases, a feature score is calculated. Features are then ranked according to their scores and k best-ranked features which are commonly selected for supervised or unsupervised learning methods. Wrapper methods use estimations of discrimination performance provided by machine learning approaches to evaluate feature subsets. An optimal subset of features that maximizes the performance of selected machine learning is obtained by using search strategies. Embedded methods are similar to wrapper methods but taking into account searching strategies that require less computational power.

Although wrapper methods can select a feature set with high accuracy, they require high computational power for searching strategies as the dataset has thousands of features. There can be thousands of metabolites in one dataset of metabolic data. Embedded methods need less computational power. Nevertheless, both wrapper and embedded methods depend on an applied machine learning algorithm or classifiers. These classifiers affect the evaluation of a feature set. On the other hand, filter methods are independent of the classification algorithm. In addition, ranked features or metabolites are easier for researchers/biologists for interpretation. Metabolites are ranked or prioritized according to their score and can be identified as candidate biomarkers [48].

As in transcriptomics, filter methods are also commonly used in metabolomics, for example, student's *t*-test or Mann–Whitney *U* test nonparametric testing [95]. Wrapper methods were applied in some studies, for example, MS patterns (with selected metabolites) in serum were utilized to distinguish between cancer and noncancer in ovarian cancer [66]. Embedded methods seem to be alternative methods for feature selection in metabolomics [29, 68, 101]. In the following, we describe some common feature selection methods including statistical hypothesis testing and support vector machine–recursive feature elimination (SVM-RFE).

Statistical Hypothesis Testing

Student's *t*-test [79] is a popular statistical parametric hypothesis testing. It is a univariate filter method for the two-class problem. The *t*-test assesses whether the means of two independent groups are statistically different from each other. The test value or its p-value can be served as a score for a feature. The test is based on an assumption that the data is normally distributed. The variances of the two groups are also expected to be identical. If the data variances are unequal, the Welch test [88] can be employed. Alternatively, for nonparametric tests, Mann–Whitney *U* test or Wilcoxon rank-sum test [55] is commonly applied. These tests are appropriate for two-class problems. The studied metabolites can be ranked according to its score or significant score (p-value). A number of top-ranked metabolites can be considered as candidate biomarkers.

Support Vector Machine–Recursive Feature Elimination (SVM-RFE)

Support vector machine–recursive feature elimination (SVM-RFE) [34] is an embedded selection method. It uses the weights of SVM to rank features and discards features with small weights. It was firstly developed for the discovery of diagnostic markers in gene expression data and was tested in leukemia and colon cancer datasets [34]. Variations of SVM-RFE have been broadly applied for MS data. For example, biomarkers for early stroke diagnosis by using MS data were purposed [68]. Only 13 features were suggested as potential biomarkers providing excellent sensitivity, specificity, and model stability. The purpose of a recursive-support vector machine (R-SVM) [101], which is another variation to SVM-RFE, was to identify biomarkers in noisy high-throughput MS and microarray data. Furthermore, an adapted SVM-RFE was also used for tandem MS quality assessment [26]. The use of feature selection with SVM appears to be a good alternative for biomarker identification in metabolomics; however, it is noted that SVM-RFE is used for a two-class problem only.

5.3.3 *Functional Interpretation of the Metabolomics Data*

Another challenging step is the functional interpretation of the metabolomics data. Gaining only a list of identified candidate metabolites does not explain insight biological process of the studied phenotypes. Notably, the candidate metabolites are often put into biological context, e.g., annotated metabolites, metabolic pathway, or metabolic networks, to gain the biological understanding and meaningful data interpretation. Two main approaches are widely used toward the purpose, which are pathway mapping and visualization and enrichment analysis [13] as described in the following.

5.3.3.1 Pathway Mapping and Visualization

Typically, pathway analysis is performed by mapping candidate metabolites onto metabolic pathways. Several pathway databases and visualization tools are publicly available as provided in Table 5.3. Examples of visualization tools are provided in pathway databases (e.g., Kyoto Encyclopedia of Genes and Genomes (KEGG) [41], Reactome [22], etc.) or intentionally developed by incorporating existing databases (e.g., interactive Pathways Explorer (iPath) [100], MetaMapp [3], etc.). Mostly tools that are provided in pathway databases do not contain many features for visualization as the ones that are intentionally created for only visualization. However, the later one might not contain updated pathway data. Furthermore, some tools also provide an integration and visualization of other omics data [31, 49, 59, 81, 82].

A number of databases contain pathways of various organisms, for example, KEGG [41], while various databases are dedicated for human metabolic pathways. For instance, Reactome [22] is a curated pathway database focusing on the biological pathway of human. HumanCyc [72], as a part of BioCyc [11], is specific to human metabolic pathways and provides a visualization of the human metabolic map. The Small Molecule Pathway Database (SMPDB) [39] is a more specific database containing small molecule pathways that are found in human. It links to the Human Metabolome Database (HMDB) [90], a database containing small molecule metabolites in the human body with their detailed information. In addition, the Urine Metabolome Database [7] is also integrated to HMDB, which contains metabolites particular in human urine.

As mentioned above, many tools are provided aiming for the visualization of metabolic pathways by incorporating data from existing databases. For example, iPath [100] provides an interactive visualization of pathways maps as a web-based

Table 5.3 List of pathway databases and visualization tools

Databases/tools		URL
KaPPA-View		http://kpv2.kazusa.or.jp/kpv4
KEGG		http://www.genome.jp/kegg
HumanCyc		http://humancyc.org
HMDB		http://www.hmdb.ca/
iPath		http://pathways.embl.de
MapMan		http://mapman.gabipd.org/web/guest/mapman
MetaMapp		http://metamapp.fiehnlab.ucdavis.edu/homepage
MetPA		http://metpa.metabolomics.ca/MetPA/faces/Home.jsp
MetScape		http://metscape.ncibi.org
Paintomics		http://www.paintomics.org/cgi-bin/main2.cgi
Pathos		http://motif.gla.ac.uk/Pathos
PathVisio		http://www.pathvisio.org
ProMeTra		http://omictools.com/prometra-s11541.html
Reactome		http://www.reactome.org
SMPDB		http://smpdb.ca

tool. It includes data from KEGG, which are metabolic and regulatory pathways, and biosynthesis of secondary metabolites. MetaMapp [3] is an approach mapping metabolites from metabolomics data into network graph helping to identify metabolic modularity by using Cytoscape [75]. Metabolomics pathway analysis (MetPA) [96] is a web application designed for pathway analysis and visualization of quantitative metabolomics data. Pathos [50] allows displaying metabolites identified by mass spectrometry in the context of the metabolic pathways using data from KEGG [41] and MetaCyc [11].

Furthermore, tools providing an integrated visualization of other omics data have been developed. MapMan [81] displays genomics datasets onto diagrams of metabolic pathways. KaPPA-View [82] allows mapping of experiment data to the metabolic pathway. MetScape [43], a plug-in for Cytoscape [75], provides a framework in the context of human metabolism for the visualization of metabolomics and expression profiling data in a network form. Paintomics [31] is a web tool for the integration and visualization of transcriptomics and metabolomics data. PathVisio [49] is a tool allowing pathway analysis and drawing and visualization of multi-omics data. ProMeTra [59] provides visualization methods for multi-omics datasets, such as genomics, transcriptomics, and metabolomics.

5.3.3.2 Enrichment Analysis

Another approach to facilitate functional interpretation is an enrichment analysis. It helps to determine enrichment of a set of metabolites in predefined groups of annotated functionally related metabolites or pathways. A number of tools are implemented for metabolite enrichment analysis as provided in Table 5.4. Two main approaches have been implemented and applied to these tools, which are overrepresentation analysis (ORA) and set enrichment analysis (SEA) [13]. ORA applies a statistical test to determine whether a set of input metabolites is enriched in a particular annotation compared to a background set. Users could specify the input metabolites, for example, a set of input metabolites that are statistically different between two phenotypes of the study (e.g., healthy and cancer). A weak point of this approach is that the users have to determine, e.g., a cutoff for selected metabolite sets. The results of ORA could be different according to the metabolites used in the analysis. The concept of SEA does not require a cutoff as in ORA. The SEA approach has been widely applied in gene expression analysis, referred to as gene set enrichment analysis (GSEA) [80]. It uses the whole ranked list of differentially expressed genes and evaluates the distribution genes in a particular gene set whether they tend to be in a top or bottom of a ranked list (more upregulated or downregulated). A statistical method is utilized to provide such a score for enrichment evaluation. The same principle can be applied to metabolic data. Instead of evaluating the distribution of genes, distribution of metabolites in a predefined metabolic set can be considered.

Integrated Molecular Pathway-Level Analysis (IMPaLA) [40] applies ORA and SEA and allows integration to other omics data, transcriptomics and proteomics. It takes the concept that genes and metabolites are linked through biochemical

Table 5.4 List of metabolite enrichment analysis tools

Tools		URL
IMPaLA	IMPaLA	http://impala.molgen.mpg.de
MBRole	MBROLE	http://csbg.cnb.csic.es/mbrole
MPEA	MPEA	http://ekhidna.biocenter.helsinki.fi/poxo/mpea
MSEA	MSEA	http://www.msea.ca/MSEA/faces/Home.jsp

reactions and are contained in many pathways. This tool is provided via a web interface.

Metabolite Biological Role (MBRole) [12] performs ORA. The annotations are biological or chemical annotations from several public databases, for instance, KEGG, SMPDB, HMDB, ChEBI, and PubChem. These include a biological pathway; enzyme interaction; disease; tissue, biofluid, and cellular location; pharmacological action; biological role; chemical taxonomy; and chemical group.

Metabolite pathway enrichment analysis (MPEA) [42] is designed for functional analysis and biological interpretation of metabolite profile data, particularly from GC-MS. It applies SEA testing whether metabolites in a particular set (e.g., pathway) appear toward the top or bottom of a ranked compound list.

Metabolite set enrichment analysis (MSEA) [97] has been developed for enrichment analysis and provided as a web-based tool (http://www.msea.ca). It is designed to identify and interpret patterns of metabolite concentration changes in a context for human and mammalian study. It provides three types of enrichment analyses, which are ORA, quantitative enrichment analysis (QEA) – SEA using metabolite concentration – and single sample profiling (SSP). The following method determines whether metabolite concentrations are significantly higher or lower than their normal values. In addition, MSEA can be used as a part of MetaboAnalyst [98], which is a web-based server that integrated several tools for metabolomics data analysis, visualization, and interpretation.

5.4 Perspectives

Advancements in instrumental, methodological developments in pre-metabolomics era toward informatics techniques in post-metabolomics era have been adequate so far. The subsequent challenges associated with metabolomics are the real application to the translational research and biomedical informatics, particularly relevant in metabolite biomarker discovery and key metabolic pathway identification in complex human disease. Conceivably, systems biology is likely to lead to a better understanding of a metabolite biomarker's role in disease phenotype, especially if combined with other omics data to reveal a global representation of the system. As known that the metabolic fluxes and metabolite concentrations can originate from multiple metabolic routes, such that alterations observed in the metabolic phenotype of a biological system can be ambiguous with respect to the origin. Therefore, even although it can reveal disease biomarkers, it may not provide definitive information about the underlying biological processes. Using an integration of high-throughput techniques through multilevel omics data to assess different levels, such as gene expression and regulation, as well as protein synthesis and expression, can provide a way to elucidate bioprocesses that control the metabolome and further identify certain metabolites that appear to be disease biomarkers. The future direction of integration of metabolomics and systems biology for translational biomedical research can thus lead to reveal new biomarkers or potential metabolic targets for future disease therapy.

Acknowledgment We would like to thank the Preproposal Research Fund (grant nos.PRF4/2558 and PRF-PII/59), Faculty of Science, Kasetsart University.

References

1. Baier MC, Barsch A, Kuster H, Hohnjec N. Antisense repression of the Medicago truncatula nodule-enhanced sucrose synthase leads to a handicapped nitrogen fixation mirrored by specific alterations in the symbiotic transcriptome and metabolome. Plant Physiol. 2007;145(4):1600–18.
2. Bais P, Moon-Quanbeck SM, Nikolau BJ, Dickerson JA. Plantmetabolomics.org: mass spectrometry-based Arabidopsis metabolomics-database and tools update. Nucleic Acids Res. 2012;40(Database issue):D1216–20.
3. Barupal DK, Haldiya PK, Wohlgemuth G, Kind T, Kothari SL, Pinkerton KE, Fiehn O. MetaMapp: mapping and visualizing metabolomic data by integrating information from biochemical pathways and chemical and mass spectral similarity. BMC Bioinf. 2012;13:99.
4. Beckonert O, Monnerjahn J, Bonk U, Leibfritz D. Visualizing metabolic changes in breast-cancer tissue using 1H-NMR spectroscopy and self-organizing maps. NMR Biomed. 2003;16 (1):1–11.
5. Bijlsma S, Bobeldijk I, Verheij ER, Ramaker R, Kochhar S, Macdonald IA, van Ommen B, Smilde AK. Large-scale human metabolomics studies: a strategy for data (pre-) processing and validation. Anal Chem. 2006;78(2):567–74.
6. Blekherman G, Laubenbacher R, Cortes DF, Mendes P, Torti FM, Akman S, Torti SV, Shulaev V. Bioinformatics tools for cancer metabolomics. Metabolomics. 2011;7(3):329–43.
7. Bouatra S, Aziat F, Mandal R, Guo AC, Wilson MR, Knox C, Bjorndahl TC, Krishnamurthy R, Saleem F, Liu P, et al. The human urine metabolome. PLoS One. 2013;8 (9):e73076.
8. Boudah S, Olivier MF, Aros-Calt S, Oliveira L, Fenaille F, Tabet JC, Junot C. Annotation of the human serum metabolome by coupling three liquid chromatography methods to high-resolution mass spectrometry. J Chromatogr B Analyt Technol Biomed Life Sci. 2014;966:34–47.
9. Breiman L. Random forests. Mach Learn. 2001;45(1):5–32.
10. Canelas AB, Harrison N, Fazio A, Zhang J, Pitkanen J-P, van den Brink J, Bakker BM, Bogner L, Bouwman J, Castrillo JI, et al. Integrated multilaboratory systems biology reveals differences in protein metabolism between two reference yeast strains. Nat Commun. 2010;1:145.
11. Caspi R, Billington R, Ferrer L, Foerster H, Fulcher CA, Keseler IM, Kothari A, Krummenacker M, Latendresse M, Mueller LA, et al. The MetaCyc database of metabolic pathways and enzymes and the BioCyc collection of pathway/genome databases. Nucleic Acids Res. 2016;44(D1):D471–80.
12. Chagoyen M, Pazos F. MBRole: enrichment analysis of metabolomic data. Bioinformatics. 2011;27(5):730–1.
13. Chagoyen M, Pazos F. Tools for the functional interpretation of metabolomic experiments. Brief Bioinform. 2013;14(6):737–44.
14. Charles EDJ. Optimal algorithm for metabolomics classification and feature selection varies by dataset. Int J Biol. 2015;7(1):100.

15. Chen T, Xie G, Wang X, Fan J, Qiu Y, Zheng X, Qi X, Cao Y, Su M, Wang X, et al. Serum and urine metabolite profiling reveals potential biomarkers of human hepatocellular carcinoma. Mol Cell Proteomics. 2011;10(7):M110 004945.
16. Chen WP, Yang XY, Harms GL, Gray WM, Hegeman AD, Cohen JD. An automated growth enclosure for metabolic labeling of Arabidopsis thaliana with 13C-carbon dioxide – an in vivo labeling system for proteomics and metabolomics research. Proteome Sci. 2011;9 (1):9.
17. Chen YZ, Pang QY, He Y, Zhu N, Branstrom I, Yan XF, Chen S. Proteomics and metabolomics of Arabidopsis responses to perturbation of glucosinolate biosynthesis. Mol Plant. 2012;5(5):1138–50.
18. Chen T, Cao Y, Zhang Y, Liu J, Bao Y, Wang C, Jia W, Zhao A. Random forest in clinical metabolomics for phenotypic discrimination and biomarker selection. Evid Based Complement Alternat Med. 2013;2013:298183.
19. Chumnanpuen P, Zhang J, Nookaew I, Nielsen J. Integrated analysis of transcriptome and lipid profiling reveals the co-influences of inositol-choline and Snf1 in controlling lipid biosynthesis in yeast. Mol Genet Genomics. 2012;287(7):541–54.
20. Chumnanpuen P, Nookaew I, Nielsen J. Integrated analysis, transcriptome-lipidome, reveals the effects of INO-level (INO2 and INO4) on lipid metabolism in yeast. BMC Syst Biol. 2013;7(3):1–14.
21. Chumnanpuen P, Hansen MAE, Smedsgaard J, Nielsen J. Dynamic metabolic footprinting reveals the key components of metabolic network in yeast Saccharomyces cerevisiae. Int J Genomics. 2014;2014:14.
22. Croft D, O'Kelly G, Wu G, Haw R, Gillespie M, Matthews L, Caudy M, Garapati P, Gopinath G, Jassal B, et al. Reactome: a database of reactions, pathways and biological processes. Nucleic Acids Res. 2011;39(Database issue):D691–7.
23. Cuperlovic-Culf M, Belacel N, Culf AS, Chute IC, Ouellette RJ, Burton IW, Karakach TK, Walter JA. NMR metabolic analysis of samples using fuzzy K-means clustering. Magn Reson Chem. 2009;47 Suppl 1:S96–104.
24. Daemen A, Peterson D, Sahu N, McCord R, Du X, Liu B, Kowanetz K, Hong R, Moffat J, Gao M, et al. Metabolite profiling stratifies pancreatic ductal adenocarcinomas into subtypes with distinct sensitivities to metabolic inhibitors. Proc Natl Acad Sci U S A. 2015;112(32): E4410–7.
25. Dieterle F, Ross A, Schlotterbeck G, Senn H. Probabilistic quotient normalization as robust method to account for dilution of complex biological mixtures. Application in 1H NMR metabonomics. Anal Chem. 2006;78(13):4281–90.
26. Ding J, Shi J, Wu FX. SVM-RFE based feature selection for tandem mass spectrum quality assessment. Int J Data Min Bioinform. 2011;5(1):73–88.
27. Edmands WM, Ferrari P, Rothwell JA, Rinaldi S, Slimani N, Barupal DK, Biessy C, Jenab M, Clavel-Chapelon F, Fagherazzi G, et al. Polyphenol metabolome in human urine and its association with intake of polyphenol-rich foods across European countries. Am J Clin Nutr. 2015;102(4):905–13.
28. Eknoyan G. Santorio Sanctorius (1561–1636) – Founding father of metabolic balance studies. Am J Nephrol. 1999;19(2):226–33.
29. Enot DP, Beckmann M, Overy D, Draper J. Predicting interpretability of metabolome models based on behavior, putative identity, and biological relevance of explanatory signals. Proc Natl Acad Sci U S A. 2006;103(40):14865–70.
30. Farag MA, Huhman DV, Dixon RA, Sumner LW. Metabolomics reveals novel pathways and differential mechanistic and elicitor-specific responses in phenylpropanoid and isoflavonoid biosynthesis in Medicago truncatula cell cultures. Plant Physiol. 2008;146(2):387–402.
31. Garcia-Alcalde F, Garcia-Lopez F, Dopazo J, Conesa A. Paintomics: a web based tool for the joint visualization of transcriptomics and metabolomics data. Bioinformatics. 2011;27 (1):137–9.

32. Giskeodegard GF, Davies SK, Revell VL, Keun H, Skene DJ. Diurnal rhythms in the human urine metabolome during sleep and total sleep deprivation. Sci Rep. 2015;5:14843.
33. Guan W, Zhou M, Hampton CY, Benigno BB, Walker LD, Gray A, McDonald JF, Fernandez FM. Ovarian cancer detection from metabolomic liquid chromatography/mass spectrometry data by support vector machines. BMC Bioinf. 2009;10:259.
34. Guyon I, Weston J, Barnhill S, Vapnik V. Gene selection for cancer classification using support vector machines. Mach Learn. 2002;46:389–422.
35. Hastie T, Tibshirani R, Friedman J. The elements of statistical learning: data mining, inference, and prediction, Springer series in statistics. New York: Springer; 2009.
36. Hendriks MMWB, van Eeuwijk FA, Jellema RH, Westerhuis JA, Reijmers TH, Hoefsloot HCJ, Smilde AK. Data-processing strategies for metabolomics studies. TrAC Trends Anal Chem. 2011;30(10):1685–98.
37. Hoult DI, Busby SJW, Gadian DG, Radda GK, Richards RE, Seeley PJ. Observation of tissue metabolites using 31P nuclear magnetic resonance. Nature. 1974;252(5481):285–7.
38. Hu C, Xu G. Mass-spectrometry-based metabolomics analysis for foodomics. TrAC Trends Anal Chem. 2013;52:36–46.
39. Jewison T, Su Y, Disfany FM, Liang Y, Knox C, Maciejewski A, Poelzer J, Huynh J, Zhou Y, Arndt D, et al. SMPDB 2.0: big improvements to the small molecule pathway database. Nucleic Acids Res. 2014;42(Database issue):D478–84.
40. Kamburov A, Cavill R, Ebbels TM, Herwig R, Keun HC. Integrated pathway-level analysis of transcriptomics and metabolomics data with IMPaLA. Bioinformatics. 2011;27 (20):2917–8.
41. Kanehisa M, Goto S. KEGG: kyoto encyclopedia of genes and genomes. Nucleic Acids Res. 2000;28(1):27–30.
42. Kankainen M, Gopalacharyulu P, Holm L, Oresic M. MPEA-metabolite pathway enrichment analysis. Bioinformatics. 2011;27(13):1878–9.
43. Karnovsky A, Weymouth T, Hull T, Tarcea VG, Scardoni G, Laudanna C, Sartor MA, Stringer KA, Jagadish HV, Burant C, et al. Metscape 2 bioinformatics tool for the analysis and visualization of metabolomics and gene expression data. Bioinformatics. 2012;28 (3):373–80.
44. Kikuchi J, Shinozaki K, Hirayama T. Stable isotope labeling of Arabidopsis thaliana for an NMR-based metabolomics approach. Plant Cell Physiol. 2004;45(8):1099–104.
45. Kohonen T, Schroeder MR, Huang TS. Self-organizing maps. New York: Springer; 2001.
46. Krueger S, Steinhauser D, Lisec J, Giavalisco P. Analysis of subcellular metabolite distributions within Arabidopsis thaliana leaf tissue: a primer for subcellular metabolomics. Methods Mol Biol. 2014;1062:575–96.
47. Kusano M, Tohge T, Fukushima A, Kobayashi M, Hayashi N, Otsuki H, Kondou Y, Goto H, Kawashima M, Matsuda F, et al. Metabolomics reveals comprehensive reprogramming involving two independent metabolic responses of Arabidopsis to UV-B light. Plant J. 2011;67(2):354–69.
48. Kusonmano K. Systematic investigation of supervised machine learning strategies and algorithms in biomedical research for functional genomic data. Doctor in Natural Science, Leopold-Franzens-University of Innsbruck. 2011.
49. Kutmon M, van Iersel MP, Bohler A, Kelder T, Nunes N, Pico AR, Evelo CT. PathVisio 3: an extendable pathway analysis toolbox. PLoS Comput Biol. 2015;11(2):e1004085.
50. Leader DP, Burgess K, Creek D, Barrett MP. Pathos: a web facility that uses metabolic maps to display experimental changes in metabolites identified by mass spectrometry. Rapid Commun Mass Spectrom. 2011;25(22):3422–6.
51. Liu R, Li Q, Ma R, Lin X, Xu H, Bi K. Determination of polyamine metabolome in plasma and urine by ultrahigh performance liquid chromatography-tandem mass spectrometry method: application to identify potential markers for human hepatic cancer. Anal Chim Acta. 2013;791:36–45.

52. Llorach-Asuncion R, Jauregui O, Urpi-Sarda M, Andres-Lacueva C. Methodological aspects for metabolome visualization and characterization: a metabolomic evaluation of the 24 h evolution of human urine after cocoa powder consumption. J Pharm Biomed Anal. 2010;51 (2):373–81.
53. Loo RL, Coen M, Ebbels T, Cloarec O, Maibaum E, Bictash M, Yap I, Elliott P, Stamler J, Nicholson JK, et al. Metabolic profiling and population screening of analgesic usage in nuclear magnetic resonance spectroscopy-based large-scale epidemiologic studies. Anal Chem. 2009;81(13):5119–29.
54. Mahadevan S, Shah SL, Marrie TJ, Slupsky CM. Analysis of metabolomic data using support vector machines. Anal Chem. 2008;80(19):7562–70.
55. Mann HB, Whitney DR. On a test of whether one of two random variables is stochastically larger than the other. Ann Math Stat. 1947;18(1):50–60.
56. Misra P, Pandey A, Tiwari M, Chandrashekar K, Sidhu OP, Asif MH, Chakrabarty D, Singh PK, Trivedi PK, Nath P, et al. Modulation of transcriptome and metabolome of tobacco by Arabidopsis transcription factor, AtMYB12, leads to insect resistance. Plant Physiol. 2010;152(4):2258–68.
57. Nakabayashi R, Kusano M, Kobayashi M, Tohge T, Yonekura-Sakakibara K, Kogure N, Yamazaki M, Kitajima M, Saito K, Takayama H. Metabolomics-oriented isolation and structure elucidation of 37 compounds including two anthocyanins from Arabidopsis thaliana. Phytochemistry. 2009;70(8):1017–29.
58. Nakamura Y, Kimura A, Saga H, Oikawa A, Shinbo Y, Kai K, Sakurai N, Suzuki H, Kitayama M, Shibata D, et al. Differential metabolomics unraveling light/dark regulation of metabolic activities in Arabidopsis cell culture. Planta. 2007;227(1):57–66.
59. Neuweger H, Persicke M, Albaum SP, Bekel T, Dondrup M, Huser AT, Winnebald J, Schneider J, Kalinowski J, Goesmann A. Visualizing post genomics data-sets on customized pathway maps by ProMeTra-aeration-dependent gene expression and metabolism of Corynebacterium glutamicum as an example. BMC Syst Biol. 2009;3:82.
60. Nicholson JK, Lindon JC. Systems biology: metabonomics. Nature. 2008;455(7216):1054–6.
61. Nishiumi S, Shinohara M, Ikeda A, Yoshie T, Hatano N, Kakuyama S, Mizuno S, Sanuki T, Kutsumi H, Fukusaki E, et al. Serum metabolomics as a novel diagnostic approach for pancreatic cancer. Metabolomics. 2010;6(4):518–28.
62. Noble WS. What is a support vector machine? Nat Biotechnol. 2006;24:1565–7.
63. Odunsi K, Wollman RM, Ambrosone CB, Hutson A, McCann SE, Tammela J, Geisler JP, Miller G, Sellers T, Cliby W, et al. Detection of epithelial ovarian cancer using 1H-NMR-based metabonomics. Int J Cancer. 2005;113(5):782–8.
64. Oesterling JE. Prostate specific antigen: a critical assessment of the most useful tumor marker for adenocarcinoma of the prostate. J Urol. 1991;145(5):907–23.
65. Pauling L, Robinson AB, Teranishi R, Cary P. Quantitative analysis of urine vapor and breath by gas–liquid partition chromatography. Proc Natl Acad Sci U S A. 1971;68(10):2374–6.
66. Petricoin EF, Ardekani AM, Hitt BA, Levine PJ, Fusaro VA, Steinberg SM, Mills GB, Simone C, Fishman DA, Kohn EC, et al. Use of proteomic patterns in serum to identify ovarian cancer. Lancet. 2002;359:572–7.
67. Polikar R. Ensemble based systems in decision making. IEEE Circuits Syst Mag. 2006;6 (3):21–45.
68. Prados J, Kalousis A, Sanchez JC, Allard L, Carrette O, Hilario M. Mining mass spectra for diagnosis and biomarker discovery of cerebral accidents. Proteomics. 2004;4(8):2320–32.
69. Psychogios N, Hau DD, Peng J, Guo AC, Mandal R, Bouatra S, Sinelnikov I, Krishnamurthy R, Eisner R, Gautam B, et al. The human serum metabolome. PLoS One. 2011;6(2):e16957.
70. Quanbeck SM, Brachova L, Campbell AA, Guan X, Perera A, He K, Rhee SY, Bais P, Dickerson JA, Dixon P, et al. Metabolomics as a hypothesis-generating functional genomics tool for the annotation of Arabidopsis thaliana genes of "Unknown Function". Front Plant Sci. 2012;3:15.

71. Ringnér M. What is principal component analysis? Nat Biotechnol. 2008;26:303–4.
72. Romero P, Wagg J, Green ML, Kaiser D, Krummenacker M, Karp PD. Computational prediction of human metabolic pathways from the complete human genome. Genome Biol. 2005;6(1):R2.
73. Saeys Y, Inza I, Larranaga P. A review of feature selection techniques in bioinformatics. Bioinformatics. 2007;23(19):2507–17.
74. Saito K. Plant metabolomics: a basis for plant functional genomics and biotechnology. New Biotechnol. 2009;25:S317–8.
75. Shannon P, Markiel A, Ozier O, Baliga NS, Wang JT, Ramage D, Amin N, Schwikowski B, Ideker T. Cytoscape: a software environment for integrated models of biomolecular interaction networks. Genome Res. 2003;13(11):2498–504.
76. Smedsgaard J, Nielsen J. Metabolite profiling of fungi and yeast: from phenotype to metabolome by MS and informatics. J Exp Bot. 2005;56(410):273–86.
77. Statnikov A, Wang L, Aliferis CF. A comprehensive comparison of random forests and support vector machines for microarray-based cancer classification. BMC Bioinf. 2008;9:319.
78. Stringer KA, Younger JG, McHugh C, Yeomans L, Finkel MA, Puskarich MA, Jones AE, Trexel J, Karnovsky A. Whole blood reveals more metabolic detail of the human metabolome than serum as measured by 1H-NMR spectroscopy: implications for sepsis metabolomics. Shock. 2015;44(3):200–8.
79. Student. Probable Error Mean Biometrika. 1908;5(6):1–25.
80. Subramanian A, Tamayo P, Mootha VK, Mukherjee S, Ebert BL, Gillette MA, Paulovich A, Pomeroy SL, Golub TR, Lander ES, et al. Gene set enrichment analysis: a knowledge-based approach for interpreting genome-wide expression profiles. Proc Natl Acad Sci U S A. 2005;102(43):15545–50.
81. Thimm O, Blasing O, Gibon Y, Nagel A, Meyer S, Kruger P, Selbig J, Muller LA, Rhee SY, Stitt M. MAPMAN: a user-driven tool to display genomics data sets onto diagrams of metabolic pathways and other biological processes. Plant J. 2004;37(6):914–39.
82. Tokimatsu T, Sakurai N, Suzuki H, Ohta H, Nishitani K, Koyama T, Umezawa T, Misawa N, Saito K, Shibata D. KaPPA-view: a web-based analysis tool for integration of transcript and metabolite data on plant metabolic pathway maps. Plant Physiol. 2005;138(3):1289–300.
83. Trethewey RN, Krotzky AJ, Willmitzert L. Metabolic profiling: a rosetta stone for genomics? Curr Opin Plant Biol. 1999;2(2):83–5.
84. van der Greef J, Smilde AK. Symbiosis of chemometrics and metabolomics: past, present, and future. J Chemom. 2005;19(5–7):376–86.
85. Vapnik VN. Statistical learning theory. New York: Wiley; 1998.
86. Watson BS, Bedair MF, Urbanczyk-Wochniak E, Huhman DV, Yang DS, Allen SN, Li W, Tang Y, Sumner LW. Integrated metabolomics and transcriptomics reveal enhanced specialized metabolism in Medicago truncatula root border cells. Plant Physiol. 2015;167 (4):1699–716.
87. Weingart GJF, Lawo NC, Forneck A, Krska R, Schuhmacher R. Study of the volatile metabolome in plant–insect interactions. In: The handbook of plant metabolomics. Weinheim: Wiley; 2013. p. 125–53.
88. Welch BL. The generalisation of student's problems when several different population variances are involved. Biometrika. 1947;34(1–2):28–35.
89. Wishart DS. Proteomics and the human metabolome project. Expert Rev Proteomics. 2007;4 (3):333–5.
90. Wishart DS, Tzur D, Knox C, Eisner R, Guo AC, Young N, Cheng D, Jewell K, Arndt D, Sawhney S, et al. HMDB: the human metabolome database. Nucleic Acids Res. 2007;35: D521–6.
91. Wishart DS, Lewis MJ, Morrissey JA, Flegel MD, Jeroncic K, Xiong Y, Cheng D, Eisner R, Gautam B, Tzur D, et al. The human cerebrospinal fluid metabolome. J Chromatogr B. 2008;871(2):164–73.

92. Wishart DS, Knox C, Guo AC, Eisner R, Young N, Gautam B, Hau DD, Psychogios N, Dong E, Bouatra S, et al. HMDB: a knowledgebase for the human metabolome. Nucleic Acids Res. 2009;37:D603–10.
93. Witten IH, Eibe F, Hall MA. Data mining: practical machine learning tools and techniques. Amsterdam/Boston: Morgan Kaufmann; 2011.
94. Wold H. Path models with latent variables: the NIPALS approach. New York: Acad Press; 1975.
95. Wu B, Abbott T, Fishman D, McMurray W, Mor G, Stone K, Ward D, Williams K, Zhao H. Comparison of statistical methods for classification of ovarian cancer using mass spectrometry data. Bioinformatics. 2003;19(13):1636–43.
96. Xia J, Wishart DS. MetPA: a web-based metabolomics tool for pathway analysis and visualization. Bioinformatics. 2010;26(18):2342–4.
97. Xia J, Wishart DS. MSEA: a web-based tool to identify biologically meaningful patterns in quantitative metabolomic data. Nucleic Acids Res. 2010;38(Web Server issue):W71–7.
98. Xia J, Sinelnikov IV, Han B, Wishart DS. MetaboAnalyst 3.0-making metabolomics more meaningful. Nucleic Acids Res. 2015;43(W1):W251–7.
99. Xu YJ, Luo F, Gao Q, Shang Y, Wang C. Metabolomics reveals insect metabolic responses associated with fungal infection. Anal Bioanal Chem. 2015;407(16):4815–21.
100. Yamada T, Letunic I, Okuda S, Kanehisa M, Bork P. iPath2.0: interactive pathway explorer. Nucleic Acids Res. 2011;39(Web Server issue):W412–5.
101. Zhang X, Lu X, Shi Q, Xu XQ, Leung HC, Harris LN, Iglehart JD, Miron A, Liu JS, Wong WH. Recursive SVM feature selection and sample classification for mass-spectrometry and microarray data. BMC Bioinf. 2006;7:197.
102. Zhang J, Vaga S, Chumnanpuen P, Kumar R, Vemuri GN, Aebersold R, Nielsen J. Mapping the interaction of Snf1 with TORC1 in Saccharomyces cerevisiae. Mol Syst Biol. 2011;7:545.

Chapter 6
Metagenomics and Single-Cell Omics Data Analysis for Human Microbiome Research

Maozhen Han*, **Pengshuo Yang***, **Hao Zhou, Hongjun Li, and Kang Ning**

Abstract Microbes are ubiquitous on our planet, and it is well known that the total number of microbial cells on earth is huge. These organisms usually live in communities, and each of these communities has a different taxonomical structure. As such, microbial communities would serve as the largest reservoir of genes and genetic functions for a vast number of applications in "bio"-related disciplines, especially in biomedicine. Human microbiome is the area in which the relationships between ourselves as hosts and our microbiomes have been examined.

In this chapter, we have first reviewed the researches in microbes on community, population and single-cell levels in general. Then we have focused on the effects of recent metagenomics and single-cell advances on human microbiome research, as well as their effects on translational biomedical research. We have also foreseen that with the advancement of big-data analysis techniques, deeper understanding of human microbiome, as well as its broader applications, could be realized.

Keywords Metagenomics • Single-cell • Omics • Human Microbiome

6.1 Introduction

Microbes are ubiquitous on our planet, and it is well known that the total number of microbial cells on earth is huge [89]. These organisms usually live in communities, and each of these communities has a different taxonomical structure. As such, microbial communities would serve as the largest reservoir of genes and genetic functions for a vast number of applications in "bio"-related disciplines, including biomedicine, bioenergy, bioremediation, and biodefense [36].

---

*These authors equally contributed to this chapter.

M. Han • P. Yang • H. Zhou • H. Li • K. Ning (✉)
Key Laboratory of Molecular Biophysics of the Ministry of Education,
College of Life Science and Technology, Huazhong University of Science and Technology,
Wuhan, Hubei 430074, China
e-mail: ningkang@hust.edu.cn

B. Shen et al. (eds.), *Translational Biomedical Informatics*, Advances in
Experimental Medicine and Biology 939, DOI 10.1007/978-981-10-1503-8_6

In most cases, a microbe itself is a unicellular cell, but different species of microbes usually live in communities, meaning that microbes of different species live and interact within a physically close space. Therefore, to understand these microbes in their native status, it's essential and indispensable to perform research on microbes at the community level, including the taxonomical structure, functional profiling, regulation network, etc. By analyzing the taxonomical structure and the dynamic changes of the target microbial community, we could obtain hints about new important microbial functional groups, which in turn could provide rich information that could guide us toward community function regulation.

Human microbiome research (HMR) has mainly focused on human microbiome. Human microbiome refers to the total genetic material in a microbial community that live on or in the human body. Human microbial communities could have different habitats, and they possessed rich resources for functional genomic studies and applications [19, 43, 93, 116]. For example, the gut bacteria in total have genomes larger than human genome in size, and they are in dynamic changes [1]. Gut bacteria are strictly anaerobiotic, and they obtain energy through fermentation process while performing reductive reactions such as methanogenesis, acetogenesis, nitrate reduction, and sulfate reduction to influence human body [107]. As gut bacteria have played important roles in the digestive system, they could profoundly influence our health status.

One of the most famous projects in HMR is the Human Microbiome Project (HMP), which has been considered as the second Human Genome Project. The HMP is funded by the National Institutes of Health, launched officially in year 2007. It planned to conduct 900 human microbial whole-genome sequencing in 5 years at a cost of 150 million US dollars [18]. The aim was to explore the feasibility of the research on human microbiome, to study the relationship between the changes of the human microbiome for healthy and diseased hosts, and to provide informatics and technical support for other scientific research areas. After the completion of HMP, it has been clear that a new chapter in the human exploration has been opened for the analysis of the relationship between human themselves and the microbes; thus, it is a milestone for medical research.

6.1.1 The Research in Microbes on Community, Population, and Single-Cell Levels

The complete understanding of the community genotypes and phenotypes depends on the understanding of its functions and activities at different levels, namely, the levels of community, population and single cells (Fig. 6.1). Firstly, at the community level, taxonomical structure is the basic character of microbial community and is also the foundation for analysis of community's function. Interactions between species are indispensable for the formation of microbial communities as well. Only by understanding the interactions of species could we understand the role of

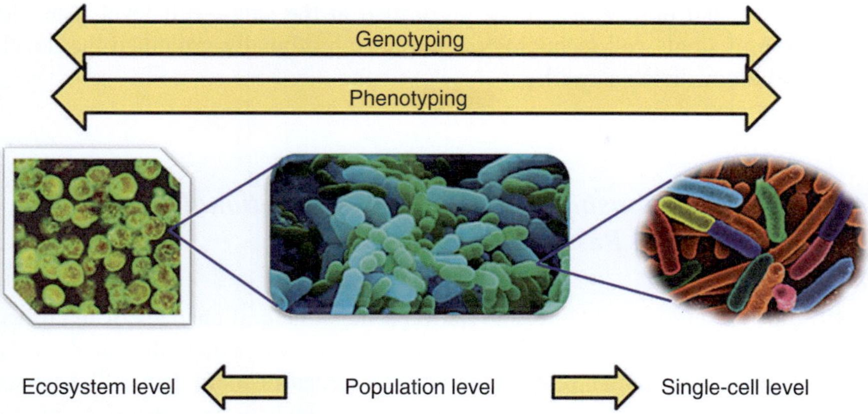

Fig. 6.1 Analysis of the genotypes and phenotypes of microbial communities at the levels of community, population, and single cells

individual species in the community. Secondly, at the population level, the analysis of genomic variations and gene expressions within a specific species could reveal the functional characters of this species in the community. Thirdly, even in the same population of a species, growth rates and gene expression profiles of different single cells are different. Therefore, researches at the single-cell level could provide researchers with better understanding of the phenotypical and functional heterogeneities among individuals in the community and are also a method to analyze uncultured and low-abundant microbes in the community.

6.1.2 Metagenomics and Single-Cell Omics

Metagenome refers to the set of all genetic materials in the whole microbial community. The metagenomic approach is different from the traditional method for microbiome research, as for metagenomics, the genetic material for all microorganisms, including those that cannot be cultured [48], would be analyzed as a whole in one run. Among the methods designed to gain access to the physiology and genetics of uncultured organism, metagenomics has emerged as a powerful solution. Apparently, the research in metagenomics would have broad impact in various application areas such as biomedicine and microbial ecosystem.

As for single-cell omics, the sequencing of a single microorganism enables the discovery and research of unknown microorganisms, from human microbiome to the microbes living in deep sea, which sometimes would not be possible to sequence as the traditional ways would need enough number of cells [83]. It heralds a new era that allows "omics" analysis, including genomics, transcriptomics, epigenomics, and proteomics, to be performed at the single-cell level. And the application of single-cell technology could have profoundly promoted the development of biomedicine as well.

6.1.3 The Relationship of HMR with Translational Biomedical Research

HMR aims to study the relationship between the structural changes of microbial flora on the surface and inside human bodies, by using the method of metagenomics. The human body is the host for a total of more than 10^{14} bacteria [96, 117]; therefore, it can be said that a human body is an ecosystem that is composed of bacteria and its own cells. In general, human microbial communities would live in harmony with their host, but when dysbiosis occurs, they may cause many diseases including respiratory diseases, nervous system diseases, cardiovascular diseases, connective tissue diseases, immune diseases, endocrine diseases and cancer [10, 54, 87]. Understanding the taxonomical structure of microbial communities in their stable status and their dysbiosis events would lead to the better healthcare for many microbial community-related diseases such as inflammatory bowel disease (IBD) and colorectal cancer (CRC). The former is an unexplained intestinal tract disease, which occurs at random location of the intestinal tract. The onset is accompanied by blood in the stool symptoms, fever, etc., recurrent and difficult to effect a radical cure. Recent studies on IBD provided a feasible treatment [7, 100]: in comparison to healthy adults, the intestinal microbial samples of patients were found to have obvious disorders and deletions. And then through the treatment of fecal microbiota transplantation (FMT), transplanting the intestinal microbiota of healthy adults into intestinal tracts of patients, there are no signs of recurrence in the treated patients. For the latter, the structural variation of gut microbiota has been found to be responsible for the progression of the CRC [11]. As exemplified by these, metagenomics provides a brand new perspective for research on pathogenic mechanism of some complex diseases.

6.2 Metagenomics for HMR

6.2.1 Metagenomics and Its Context

Since over 90 % of strains in a microbial community could not be isolated or cultured [49], metagenomic methods have become popular for analysis of a microbial community as a whole. Such an approach enables the exploration of the relationships among microbes, their communities and habitats at the most fundamental genomic level. Understanding the taxonomical structure of a microbial community (alpha diversity) and the differences in taxa among microbial communities (beta diversity) have become two of the most important problems in metagenomic research [57, 67, 68], in which understanding beta diversity is especially critical for studying microbial communities' heterogeneity ecology. For example, the Human Microbiome Project [108] and related efforts to study microbial communities occupying various human body habitats have shown a surprising amount of diversity among individuals in the skin [73, 109], gut [73] and mouth ecosystems [45, 77]. Furthermore, even microbial communities from similar types of environment might differ significantly [76].

Next-generation sequencing (NGS) techniques have enabled fast profiling of large volumes of metagenomic samples. As a result, a rapidly increasing number of metagenomic profiles (and datasets) of microbial communities have been archived in public repositories and research labs around the world. Therefore, it is becoming more and more important to perform the in-depth analysis for the valuable biological information that is hidden in large number of samples. Hence, a system that provides functionalities for data analysis would be of significant value to a worldwide user base from multiple disciplines.

6.2.2 NGS Techniques for Metagenomic Researches

The increased throughput and decreased cost of sequencing have made NGS more and more popular and widely used in various fields. The advantages of NGS have also allowed an explosion in sequencing of microbial communities and opened the gate for a revolution in microbial community sequencing and analysis [60]. Based on metagenomic approach, NGS can sequence all microorganisms in a microbial community in one run without the need to separate the microorganisms or to establish a gene library [48]. Currently, both 16S rRNA profiling (amplicon sequencing) and whole-genome sequencing (WGS) were commonly used to describe and interpret the taxonomical structure and functional profile of microbial communities. Sequencing analysis of 16s rRNA amplification is mainly used for the analysis of microbial community's taxonomical structure and relative abundance of taxon in the community. For the application of WGS on microbial communities, the main purpose would be to understand the functional profiles of the communities. Based on WGS data, standard functional genomic analyses including genome assembly, gene prediction and annotation, as well as regulation network reconstruction would be included. Phylogenetic marker gene in the genome is used to characterize species diversity as well as genetic diversity, based on which the distribution and function of species in the community can be investigated [99].

6.2.3 Computational Techniques for Metagenomic Researches

As for the metagenomic technology, new algorithms and tools have been developed, which could provide more powerful technological support for metagenomic researches. From the preprocessing of sequences, to genome assembly and gene prediction, to the calculation of diversity of species, to metagenomic sample comparison, to functional profiling and other processes, more and more excellent tools will be invented, and these tools will provide powerful help for metagenomic data analysis.

Based on these researches about functional genomes and metagenomes, some large sequence databases with different signature sequence, such as Greengenes [23], SILVA [90], RDP [16], etc., and a series of large datasets have been established in this field now. Current databases include MG_RAST [73] (http://metagenomics.anl.gov/, Metagenome database), CAMERA [52] (http://camera.calit2.net/, Metagenome database, now renamed to iMicrobe), and universal databases like NCBI (http://www.ncbi.nlm.nih.gov/). These databases are playing a significant role for metagenomic research. At present, the number of the publicly available metagenomic projects in NCBI, MG_RAST, and CAMERA2 sums up to more than 10,000, and the data volume is up to several hundred TB. However, these databases serve mainly as data repositories, with neither comprehensive tools for comparative analysis nor capabilities for extensive comparison and search.

Currently, the tools for 16s rRNA profiling are more mature than metagenome analysis tools for metagenomic data. MEGAN [45] is a metagenomic analysis tool for taxonomical comparisons [77] and statistical analyses [52], which can only compare single pairs of metagenomic samples based on taxonomical annotations, as is also the case with STAMP [85]. ShotgunFunctionalizeR [52], Mothur [85], and METAREP [32] identify the differences between samples using standard statistical tests (mainly *t*-tests with some modifications). UniFrac [61] and Fast UniFrac [35] examine the similarities among species based on their overlaps in phylogenetic tree to discover ecological patterns.

Because of the lacking of reference database, the process flow of the whole-genome analysis (based on WGS data) for metagenomic data is not very mature at present. Every laboratory might have its own method to deal with the whole-genome sequencing data for microbial communities. STAMP [108] and MEGAN5 [28] provide some convenience for the whole-genome sequencing data. MEGAN based on the JAVA mainly uses LCA algorithm to analyze BLAST result. Except for analyzing species abundance and diversity, MEGAN also can analyze functional genetic diversity and abundance. STAMP is also an open-source platform analysis tool and can operate under Linux and Windows. Friendly interface and simple graphic analysis could provide researchers with various statistical models to evaluate either a small or a large number of samples and could provide them with high-quality graphical results.

The integrated tools have also been developed recently. These tools mainly include QIIME [12], Mothur [98], and Meta-Mesh system [103]. QIIME is an open-source process based on Python and developed for specially analyzing high-throughput sequencing data of microbe PCR products by Rob Knight et al. in 2009, which collects many tool packages and conducts charts directly during the progress of analysis. The analysis process includes three steps: firstly, conducting the data preprocessing including possible assembly; secondly, producing OTU, classifying OTU, and conducting statistics of OTU abundance by known sequence of character sequence database; and thirdly, calculating α- and β-diversity and drawing several graphs including the PCoA graph. In 2009, Mothur was also issued by Patrick Schloss (University of Michigan, USA), which is popular and widely used in the bioinformatics analysis field now. Although it can't draw graph directly from data, ideal analysis graphs can be gotten with the help of R programming language. At present, Mothur can already support sequencing data produced by various paltforms including Sanger, PacBio, IonTorrent, 454, and Illumina (MiSeq/HiSeq).

6.2.4 Metagenomics: Milestone Works for Biomedical Research

Current emphases of metagenomics researches in applications are mostly on biomedicine, biofuel, environmental monitoring and agriculture [22, 25, 41, 48], in which several milestone works have been conducted.

Firstly, researches have outlined the global profile of human microbiome and have showed that many diseases have close relations with human microbiome. In 2010, the completion of the first human microbiome profile has provided a deep insight for the research of human metagenome [17]. There are thousands of billions of bacteria cells that live on the surface of the human body and inside of the human body; in other words, the human body is an organic whole that consists of bacteria and somatic cells. The maintaining of demic normal microbe is significant to the health of the human body, mainly reflected in the following four sides: (1) synthesizing vitamin and bacteriocin, such as *Lactobacillus acidophilus* that can synthesize vitamin K; (2) promoting the metabolism of the host, such as *Lactobacillus casei* that can promote the progress of digestion and reduce the content of cholesterol in the blood; (3) promoting demic immunity, such as *L. rhamnosus* that can increase the number and activity of the immune globulin and macrophage; and (4) microbial antagonism, like *Lactobacillus* that can restrain the growth of *Staphylococcus aureus*.

Secondly, current studies on microbial community in human digestive system have revealed previously unseen connections between gut microbial communities and human health. For example, some researches have been done to identify that oral microbiota has effects on dental caries [86]. In addition, in the research of obesity, two major flora *Bacteroidetes* and *Firmicutes* have been found, which have different abundances between healthy people and those with obesity [29]. In type II diabetes study, researchers found an obvious decrease of lactic acid bacteria in the gut microbial community for those with type II diabetes [110]. Furthermore, oral microbiome is antibiotic resistant in their native environment [24]. Gut microbial communities even can influence our brain by interaction [70]. The relationship between premature birth and maternal bacterial infection indicated that related translational medicine treatment might be conducted through mediation of microbial communities [26]. Maternal bacterial and different delivery way would influence the baby's intestinal microbiota [5, 66]. This may explain the continuity of the microorganisms in human body.

Thirdly, microbial community studies might lead to some novel findings for the research on "agelessness." Agelessness is our human's long-cherished wish since our civilization, for which we have made unremitting efforts and countless attempts. In the old times, the first emperor developed alchemy, and pharaohs made mummy to achieve this goal. In modern ages, many are still devoted in researches on cell reconstruction, chromosome telomere, with aim to extend human life and finally achieve the longevity goal. And microbes might possess solutions for agelessness: it has been reported that several kinds of bacteria can produce a defensive chemical called rapamycin [55], which can make these species live longer. On the permafrost of Siberia and Canada and in the South Pole, several bacteria which have survived for half a million years have been studied [88]. These studies have provided a new avenue on which we might use the metagenome method to screen chemical substances about disease treatment and antiaging from environment [65]. Since there are several close ties between microbes and animals on different levels, metagenomic approach for study on microbes might discover novel solutions for agelessness treatment as well.

6.3 Single-Cell Omics for HMR

6.3.1 Single-Cell Genotyping and Phenotyping

The monitoring of microbial cells during the time course is a very effective method to analyze the adaptation of a cell population to changing conditions, such as nutrient supply and stress exposure. Notwithstanding culminating evidences for adaptation diversities among individual population members, such endeavors have only been undertaken recently due to enormous technical challenges that are faced. Regardless of these obstacles, such studies hold great promise to provide substantially new insight into fundamental physiological processes in microorganisms as well as to accelerate the development of superior strains for industrial biotechnology.

Single-cell technologies, like the classical FACS (fluorescence-activated cell sorting) analysis, possess the capabilities to detect population heterogeneities by observing distinct phenotypic parameters. On this level, single-cell-based FACS techniques would be highly appropriate at resolving the dynamics of cells at individual level by recording and comparing whole cell phenotypes. Additionally, it is rational to combine FACS with subsequent omics analysis, as described in recent publications to (a) detect population heterogeneities with FACS [95] and to (b) thereupon perform omics characterization on selected (sub-) populations [6, 33]. However, such novel investigations of microbial cell population dynamics would require the development of improved microbe FACS profiling and increased specificity and sensitivity in microbial data acquisition and analysis [38].

6.3.2 NGS Techniques for Single-Cell Omics Researches

When it comes to single-cell omics, the first problem that we will face is about single-cell DNA extraction, for which novel methods and techniques are needed. One popular technique is making use of microfluidics to extract single-cell DNA: single-cell nuclei are immobilized in a micro-channel, and each nucleus will release the chromosomal DNA when exposed to protease solution [114]. Another single-cell DNA extraction method is based on using N-lauroylsarcosine salt solution [106], which is efficient when compared to using other chemicals, and also with simplified manipulation process.

There are currently several DNA whole-genome amplification (WGA) techniques for single-cell sequencing, among which the most mature one is Multiple Displacement Amplification (MDA). The rolling circle amplification mechanism in circular DNA such as plasmid and virus DNA provides the inspiration for the MDA technique [20]. MDA makes use of annealing random primers to denatured DNA and then conducts strand displacement synthesis under a constant temperature, which is catalyzed by φ29 DNA polymerase or the large fragment of Bst (*Bacillus stearothermophilus*) DNA polymerase [21]. Thus, MDA has a more uniform coverage of the single-cell genome and facilitates the analysis of genomes from a great number of uncultivable microbial species [58, 92].

Multiple annealing and looping-based amplification cycles (MALBAC), which combine MDA with PCR, use quasi-linear pre-amplification to reduce the bias associated with nonlinear amplification [53, 120]. As for the primers, random sequence and a common sequence tag are included. And primers are annealed to template DNA. Then in the isothermal strand displacement reaction, primers are extended by the large fragment of Bst DNA polymerase. The strand displacement synthesis produces some partial amplicons, which are later denatured at the template of 94 °C. In a quasi-linear amplification stage, there are five cycles of annealing, and extension and denaturation are performed. And during this stage, initial priming events are more evenly distributed over the course of multiple cycles, which is in order to limit the reaction rate. In addition, when the newly made amplicons are finished, for which they have a sequence common to each primer, closed loops can be formed. The advantage is that closed loops can prevent them from being copied again. Consequently, the amplification remains linear, subsequently, making use of PCR amplified on the preamplified DNA to generate adequate amounts of DNA for sequencing [58]. Based on these reasons, MALBAC is better than other WGA methods when it comes to amplification bias on amplifying the single-cell genome [120].

Recently, there are more developments in third-generation sequencing technologies, including nanopore sequencing, single-molecule real-time (SMRT) sequencing, and direct imaging [97], many might revolutionize single-cell sequencing. Third-generation sequencing technologies can improve sequencing length and facilitate the analysis of long repetitive sequences and alternative splicing events. Most importantly, there is no need to convert RNA into cDNA in third-generation sequencing [82]. As for the RNA amplification and other RNA sequencing, many new methods have been developed to improve the global transcriptome sequencing, like PCR-based methods, IVT (in vitro transcription)-based methods, and phi29 DNA polymerase-based methods (rolling cycle amplification (RCA)) [59, 84]. And further improvement in the breadth of single-cell mRNA analysis has been achieved recently using mRNA sequencing (mRNA-Seq) [104] and quantitative PCR (qPCR) [105].

6.3.3 Computational Techniques for Single-Cell Omics Researches

Compared with traditional multi-cell samples that contain multiple cells, single-cell analysis needs amplification of the whole genome (namely, WGA or the whole transcriptome) before sequencing. Currently, both classical method such as MDA [14] and new methods such as MALBAC [120] are used for the amplification of whole-genome DNA. However, these technologies have limitations embodied in the following two aspects: (1) amplification is difficult for the whole genome, rendering some certain regions in genome not amplified (and thus not sequenced) at all, and (2) amplification process might result in bias (uneven distribution of reads), which means that within amplified regions, some might be amplified more than other regions. Both of these two problems will create certain challenges for the subsequent bioinformatics analyses.

For example, for single-cell genome sequencing, the copy number variation (CNV) and single nucleotide polymorphism (SNP) are two important topics at present [80]. For SNP, the WGA of single-cell genome will bring the following problems: (1) As the genome coverage is low, for the uncovered fragments of amplification, SNP information in these areas would be unknown. And at the same time, the incomplete diploid genome amplification is likely to have allele dropout. (2) As there might have errors in the process of amplification (the DNA polymerase used in amplification has certain error rate), some new "SNP" would be introduced. In addition, the cost of single-cell sequencing is still very high, rendering deep sequencing for a large number of single-cell impractical. Thus, how to accurately call SNPs from low-coverage sequencing data is desired but still difficult. For CNV, bias in the process of amplification would also affect the accuracy and resolution of CNV identification.

6.3.4 Single-Cell Omics for Human Microbiome: Milestone Works for Biomedical Research

There are huge numbers of microbes in our body; most of them are in the intestinal tract. The highly diverse intestinal microbiota forms a structured community, which is engaged in constant communication with itself and its host and is characterized by extensive ecological interactions [101]. A main benefit is that the microbiota can protect its host from infecting in a process termed colonization resistance (CR), which remains insufficiently understood [101]. Now, we can make full use of single-cell omics to research human microbiome and related diseases, such as colorectal adenoma carcinoma [31]. In addition, single-cell omics for human microbiome can be used for diagnosing: Preimplantation genetic diagnosis (PGD), for example, is the analysis of a single cell from a biopsy of an embryo after in vitro fertilization. PGD is used to test for genetic diseases and chromosome aneuploidies [113].

6.4 Combining Metagenomics and Single-Cell Omics Data Toward Better Interpretation of Human Microbiome

6.4.1 Problems About Single-Cell Heterogeneity

Cellular heterogeneity within an isogenic cell population is a widespread phenomenon [34, 46]. Cellular heterogeneity that arises from stochastic expression of genes, proteins, and metabolites is a fundamental principle of cell biology [113]. However, it is difficult to probe single-cell heterogeneity only relying on metagenomics approach. On the contrary, probing single-cell heterogeneity without considering the physical and chemical properties of the whole community and the interactions of microbes in the community would tend to be biased. Moreover, it would be important to probe the physical and chemical indicators of microenvironment and the reaction of single cells on both single-cell and community levels [15]. Therefore, it would be advantageous to combine metagenomics and single-cell omics data as well as related techniques for more in-depth analysis human microbiome.

6.4.2 Integrated Data Analysis

As the integration of single-cell omics and phenotype data would provide in-depth understanding of microbial communities, the collection and organization of these data from various sources and update of integrated databases would be of paramount importance.

Taking single nucleotide polymorphism (SNP) for example, a microbial community contains many different species, thus containing many different genomes. Yet each of the single cells in the community has a single genome. Therefore, we can easily establish a one-to-one relationship by comparing the single-cell genome and microbial community total genomes. However, single cells have variations even within the same species, so it will lead to an incomplete match between single-cell genome and metagenome. Through the ways like SNP detection, we will identify the differences of genomes between single-cell and microbial community and thus discover the SNP pattern for the community (Fig. 6.2).

And the integration of metagenomics and single-cell omics data would not only be beneficial for SNP analyses. Rather, the combination of biological techniques and computational techniques for both metagenomics and single-cell samples would result in more powerful tools for microbial community analysis. For example, the "mini-metagenome" approach has been proposed recently, which could divide a community into several small sets of species by sorting techniques (such as FACS) and then analyze representative sets at single-cell level in more details [80]. This approach has been successfully applied on the discovery of TM6 genome (a Gram-negative organism) from a hospital sink biofilm [50, 72] and has provided some insights for TM6's functions that might affect clinical practices.

6.5 Discussions

6.5.1 The Needs for the Improvement of Big Data Analysis Techniques for HMR

WGS approaches for HMR are still costly; therefore, there is a need for better computational model that could map species to their main functions, especially for large-scale studies [44, 48].

There are generally two approaches for metagenomic functional profiling, one relies on sequence alignment, and the other is based on alignment-free approach [8]. Sequence alignment must search for sequence similarities against the reference database [8]. However, we know a little about the whole microorganism, and the annotation is limited and can't deal with large-scale studies. On the contrary, an alignment-free sequence can use short sub-sequences (*k*-mers) [115]. Alignment-free sequence does not need to use PCR amplification and could be directly used for analysis. It is an appropriate way to analysis metagenome. There already have many

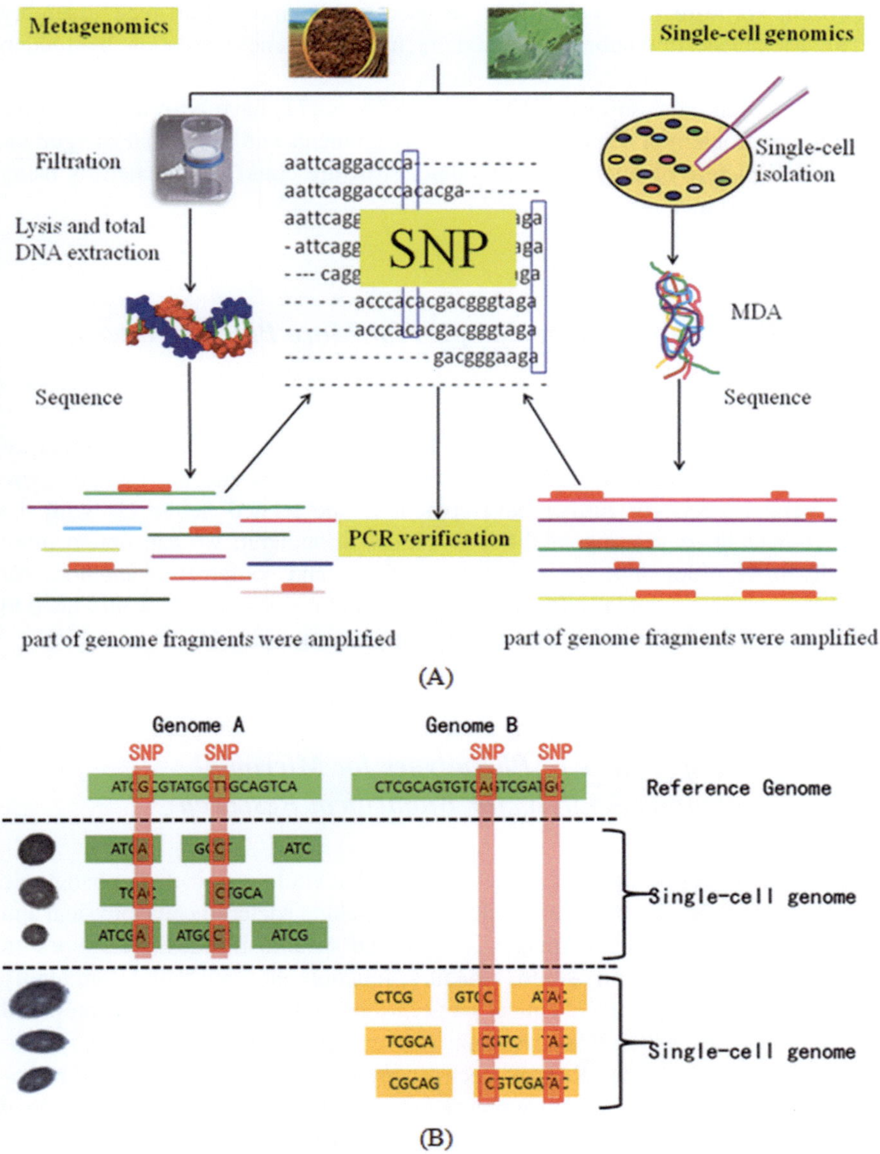

Fig. 6.2 Methods for integrating single-cell and metagenomic data to analyze the differences of genotypes at single-cell and community levels. (**a**) An illustration of using both metagenome and single-cell gel electropheresis to identify SNPs. (**b**) An illustration of using multiple single-cell omics data for SNP analysis

tools for this such as CoMeta [51]. Using next-generation sequencing (NGS) with GPU computing is another way to improve the quality of large-scale studies [102], and the sequencing simulator NESSM can help to evaluate this [47]. For the functional annotation using traditional assembly-gene prediction-annotation approach, software like Velvet can be used for de novo short-read assembly [118]. MetaGeneAnnotator [79] could be used for gene prediction by hidden Markov model (HMM).

As more and more software [102] and databases [71] have been developed for solving the problem of sequence to species assignments and functional categorization, a full picture of microbial community structure could be revealed by using more advanced methods.

6.5.2 The Need for More Samples for More Precise Pattern Discovery

Although pattern discovery is desired to be as precise as possible, the amount of samples collected might be the limiting factor [30]. To overcome the obstacle, there are several aspects for further development in the related fields: (1) with the development of sequencing facilities of higher throughput, we can obtain more samples more efficiently, and (2) more modeling and computation methods for pattern discovery, especially for heterogeneous and noisy samples, would need to be developed for data mining based on existing samples [30, 44].

6.5.3 The Importance of Phenotypes for Microbial Community Analysis for Biomedical Research

Omics data are not limited to genetic materials, but could also include phenotypic data [30]. Microbes' phenotypes would usually include bacterial cells' physical and chemical properties, such as images and infrared spectra. Besides phenotypes on unitary level such as composition, structure, function, etc., microbial community also possesses physical and chemical properties on cellular level as it is composed of bacterial cells, which also determine the uniqueness of the community. For example, both infrared and Raman spectroscopy methods have shown good reproducibility and high discriminatory power for biological samples [112]. Thus, they provide the unique advantage of differentiating taxonomical structure at the species or subspecies level of bacterial cells on the basis of variations in the spectral features [62]. Represented by the breakthrough publications in nature about using infrared spectroscopy [78] and Raman spectroscopy [91] to study microorganisms, these two properties have become important phenotypes of a microbial population or community. Other physical and chemical properties of bacterial cells such as images also are more and more involved in research on microbes [3, 81].

For biomedical research, the integration of genotypes and phenotypes would undoubtedly yield novel insights that were not discovered by traditional methods.

For example, such integration would reveal the molecular mechanisms responsible for phenotypic heterogeneity [37] and could provide microbial groups with new functionality [2].

6.5.4 Broader Application Areas for HMR

With the development of metagenomic and single-cell analytical techniques, HMR would undoubtedly expand in its application areas. These new application areas include but not limited to phage therapy for gut bacteria, brain function examination, and drug development.

1. *Human microbial communities that involve in digestive system, phage therapy.* Phage therapy is an old approach to treat bacterial infection: as phage has the specific host, phage treatment has a great treatment for specific bacteria [64]. Phage has a flexible genome, and we can modify their genome to change their capsid protein to bind specific bacteria [4]. Though sometimes phages might infect bacteria and cause little damage to our host cell, their specificity is relatively high. Therefore, they would be useful for treating specific bacterial infection [13].
2. *Human microbial communities that involve brain research.* It is a novel finding that our gut microbiota could interact with our brain [69, 94]: first, they influence intestinal permeability and immune function, then activity in the enteric nervous system, the HPA axis, pain modulation systems, and the brain [70]. Additionally, research has found that the diet and diet-related changes could change our gut microbiota and then influence the gut-brain axis and in turn influence your behaviors [63]. Furthermore, the gut microbiome will influence our temperament during our early childhood [74]. Autism spectrum disorder (ASD) is reported to have a relation with our disordered gut bacteria. A mouse model which displays features of ASD, maternal immune activation (MIA) model shows a link between ASD and gut bacteria [42]. Most recently, the human satiety has been linked with gut microbial community [9]. All of these studies have shown that there is a potential link between gut microbiota and brain function.
3. *Human microbial communities that involve in drug development.* Human microbial communities have been found to be very useful for drug production and drug scanning in recent years [56, 111]. For example, human microbial communities could produce bioactive flavonoids by biotransformation like dehydroxylation, O-methylation, O-demethylation, hydrogenation, etc. [39]. They can also produce antibiotic: lactocillin, a big community member in vaginal microbiota, can produce thiopeptide antibiotic [27]. A kind of N-acyl-homoserinelactones-producing bacteria was also discovered, which could prevent plant-originated infections with human pathogens [40]. Furthermore, researchers have already mined Human Microbiome Project data to identify 3118 small molecule

biosynthetic gene clusters (BGCs) in genomes of human-associated bacteria, and they have found a new thiopeptide antibiotic from these candidates [27]. Therefore, with the development of metagenomic approach, we could find more drugs from this drug pool and find bacteria which can produce new drug for us [119].

All in all, HMR area is at the stage at which massive amount of sequencing data has been generated, based on which quite a lot of exciting (sometimes astonishing) discoveries have been reported [94, 100]. These findings have already deepen our understanding of human as host and human microbial communities and provided clues for clinical practices [7, 100]. However, the data bonanza in human microbial community researches has also created great obstacles for data analyses, interpretations, and applications [44, 99]. Yet we believe that with the rapid acumination of data and more devotion to the development of analytical methods for big data in HMR, such obstacles could be overcome [48, 75]. And such breakthroughs can lead to even better understanding of human microbiome and thus push related translational medicine applications to a higher level.

References

1. Ackerman J. The ultimate social network. Sci Am. 2012;306(6):36–43.
2. Ackermann M. A functional perspective on phenotypic heterogeneity in microorganisms. Nat Rev Microbiol. 2015;13(8):497–508.
3. Amann R, et al. In situ visualization of high genetic diversity in a natural microbial community. J Bacteriol. 1996;178(12):3496–500.
4. Bakhshinejad B, Sadeghizadeh M. Bacteriophages and their applications in the diagnosis and treatment of hepatitis B virus infection. World J Gastroenterol. 2014;20(33):11671–83.
5. Biasucci G, et al. Cesarean delivery may affect the early biodiversity of intestinal bacteria. J Nutr. 2008;138(9):1796s–800.
6. Blainey PC. The future is now: single-cell genomics of bacteria and archaea. FEMS Microbiol Rev. 2013;37(3):407–27.
7. Borody TJ, Finlayson S, Paramsothy S. Is Crohn's disease ready for fecal microbiota transplantation? J Clin Gastroenterol. 2014;48(7):582–3.
8. Borozan I, Watt S, Ferretti V. Integrating alignment-based and alignment-free sequence similarity measures for biological sequence classification. Bioinformatics. 2015;31(9):1396–404.
9. Breton J, et al. Gut commensal E. coli proteins activate host satiety pathways following nutrient-induced bacterial growth. Cell Metab. 2015;23(2):324–34.
10. Campbell-Valois FX, Sansonetti PJ. Tracking bacterial pathogens with genetically-encoded reporters. FEBS Lett. 2014;588(15):2428–36.
11. Candela M, et al. Inflammation and colorectal cancer, when microbiota-host mutualism breaks. World J Gastroenterol. 2014;20(4):908–22.
12. Caporaso JG, et al. QIIME allows analysis of high-throughput community sequencing data. Nat Methods. 2010;7(5):335–6.
13. Chan BK, Abedon ST, Loc-Carrillo C. Phage cocktails and the future of phage therapy. Future Microbiol. 2013;8(6):769–83.
14. Chen M, et al. Comparison of multiple displacement amplification (MDA) and multiple annealing and looping-based amplification cycles (MALBAC) in single-cell sequencing. PLoS One. 2014;9(12):e114520.

15. Chen M, et al. Correction: comparison of multiple displacement amplification (MDA) and multiple annealing and looping-based amplification cycles (MALBAC) in single-cell sequencing. PLoS One. 2015;10(4):e0124990.
16. Cole JR, et al. The ribosomal database project: improved alignments and new tools for rRNA analysis. Nucleic Acids Res. 2009;37 suppl 1:D141–5.
17. Consortium HMJRS. A catalog of reference genomes from the human microbiome. Science. 2010;328(5981):994–9.
18. Consortium HMP. A framework for human microbiome research. Nature. 2012;486 (7402):215–21.
19. Costello EK, et al. The application of ecological theory toward an understanding of the human microbiome. Science. 2012;336(6086):1255–62.
20. Dean FB, et al. Rapid amplification of plasmid and phage DNA using Phi 29 DNA polymerase and multiply-primed rolling circle amplification. Genome Res. 2001;11(6):1095–9.
21. Dean FB, et al. Comprehensive human genome amplification using multiple displacement amplification. Proc Natl Acad Sci U S A. 2002;99(8):5261–6.
22. DeLong EF. Microbial community genomics in the ocean. Nat Rev Microbiol. 2005;3 (6):459–69.
23. DeSantis TZ, et al. Greengenes, a chimera-checked 16S rRNA gene database and workbench compatible with ARB. Appl Environ Microbiol. 2006;72(7):5069–72.
24. Diaz-Torres ML, et al. Determining the antibiotic resistance potential of the indigenous oral microbiota of humans using a metagenomic approach. FEMS Microbiol Lett. 2006;258 (2):257–62.
25. Diene SM, et al. Bacterial genomics and metagenomics: clinical applications and medical relevance. Rev Med Suisse. 2014;10(450):2155–61.
26. Dominguez-Bello MG, et al. Delivery mode shapes the acquisition and structure of the initial microbiota across multiple body habitats in newborns. Proc Natl Acad Sci U S A. 2010;107 (26):11971–5.
27. Donia MS, et al. A systematic analysis of biosynthetic gene clusters in the human microbiome reveals a common family of antibiotics. Cell. 2014;158(6):1402–14.
28. El Hadidi M, Ruscheweyh HJ, Huson D. Improved metagenome analysis using MEGAN5. In: Joint 21st annual international conference on Intelligent Systems for Molecular Biology (ISMB) and 12th European Conference on Computational Biology (ECCB), 2013.
29. Everard A, et al. Microbiome of prebiotic-treated mice reveals novel targets involved in host response during obesity. ISME J. 2014;8(10):2116–30.
30. Falony G, Vieira-Silva S, Raes J. Microbiology Meets Big Data: the case of gut microbiota-derived trimethylamine. Annu Rev Microbiol. 2015;69:305–21.
31. Feng Q, Liang S, Jia H. Gut microbiome development along the colorectal adenoma-carcinoma sequence. Nat Commun. 2015;6:6528.
32. Goll J, et al. METAREP: JCVI metagenomics reports—an open source tool for high-performance comparative metagenomics. Bioinformatics. 2010;26(20):2631–2.
33. Gomariz M, et al. From community approaches to single-cell genomics: the discovery of ubiquitous hyperhalophilic Bacteroidetes generalists. ISME J. 2015;9(1):16–31.
34. Graf T, Stadtfeld M. Heterogeneity of embryonic and adult stem cells. Cell Stem Cell. 2008;3 (5):480–3.
35. Hamady M, Lozupone C, Knight R. Fast UniFrac: facilitating high-throughput phylogenetic analyses of microbial communities including analysis of pyrosequencing and PhyloChip data. ISME J. 2010;4(1):17–27.
36. Handelsman J et al. The new science of metagenomics: revealing the secrets of our microbial planet. Nat Res Council Report. 2007. http://www.ncbi.nlm.nih.gov/pubmed/?term=The+new+science+of+metagenomics%3A+revealing+the+secrets+of+our+microbial+planet.
37. Hatzenpichler R, et al. In situ visualization of newly synthesized proteins in environmental microbes using amino acid tagging and click chemistry. Environ Microbiol. 2014;16 (8):2568–90.

38. Hedlund BP, et al. Impact of single-cell genomics and metagenomics on the emerging view of extremophile “microbial dark matter”. Extremophiles. 2014;18(5):865–75.
39. Hegazy ME, et al. Microbial biotransformation as a tool for drug development based on natural products from mevalonic acid pathway: a review. J Adv Res. 2015;6(1):17–33.
40. Hernandez-Reyes C, et al. N-acyl-homoserine lactones-producing bacteria protect plants against plant and human pathogens. Microb Biotechnol. 2014;7(6):580–8.
41. Hess M, et al. Metagenomic discovery of biomass-degrading genes and genomes from cow rumen. Science. 2011;331(6016):463–7.
42. Hsiao EY, et al. Microbiota modulate behavioral and physiological abnormalities associated with neurodevelopmental disorders. Cell. 2013;155(7):1451–63.
43. Human Microbiome Project, C. Structure, function and diversity of the healthy human microbiome. Nature. 2012;486(7402):207–14.
44. Hunter CI, et al. Metagenomic analysis: the challenge of the data bonanza. Brief Bioinform. 2012;13(6):743–6.
45. Huson DH, et al. MEGAN analysis of metagenomic data. Genome Res. 2007;17(3):377–86.
46. Irish JM, Kotecha N, Nolan GP. Mapping normal and cancer cell signalling networks: towards single-cell proteomics. Nat Rev Cancer. 2006;6(2):146–55.
47. Jia B, et al. NeSSM: a next-generation sequencing simulator for metagenomics. Plos One. 2013;8(10):108–10.
48. Jiahuan C, et al. Research in metagenomics and its applications in translational medicine. Yi Chuan. 2015;37(7):645–54.
49. Jurkowski A, Reid AH, Labov JB. Metagenomics: a call for bringing a new science into the classroom (while it’s still new). CBE-Life Sci Educ. 2007;6(4):260–5.
50. Kashtan N, et al. Single-cell genomics reveals hundreds of coexisting subpopulations in wild Prochlorococcus. Science. 2014;344(6182):416–20.
51. Kawulok J, Deorowicz S. CoMeta: classification of metagenomes using k-mers. PLoS One. 2015;10(4):e0121453.
52. Kristiansson E, Hugenholtz P, Dalevi D. ShotgunFunctionalizeR: an R-package for functional comparison of metagenomes. Bioinformatics. 2009;25(20):2737–8.
53. Lasken RS. Single-cell sequencing in its prime. Nat Biotechnol. 2013;31(3):211–2.
54. Laufer AS, et al. Microbial communities of the upper respiratory tract and otitis media in children. MBio. 2011;2(1):e00245–10.
55. Law BK. Rapamycin: an anti-cancer immunosuppressant? Crit Rev Oncol Hematol. 2005;56 (1):47–60.
56. Leslie M. MICROBIOME. Microbes aid cancer drugs. Science. 2015;350(6261):614–5.
57. Ley RE, et al. Worlds within worlds: evolution of the vertebrate gut microbiota. Nat Rev Microbiol. 2008;6(10):776–88.
58. Liang J, Cai W, Sun Z. Single-cell sequencing technologies: current and future. J Genet Genomics. 2014;41(10):513–28.
59. Liu N, Liu L, Pan X. Single-cell analysis of the transcriptome and its application in the characterization of stem cells and early embryos. Cell Mol Life Sci. 2014;71(14):2707–15.
60. Loman NJ, et al. High-throughput bacterial genome sequencing: an embarrassment of choice, a world of opportunity. Nat Rev Microbiol. 2012;10(9):599–606.
61. Lozupone C, Knight R. UniFrac: a new phylogenetic method for comparing microbial communities. Appl Environ Microbiol. 2005;71(12):8228–35.
62. Lu X, et al. Application of mid-infrared and Raman spectroscopy to the study of bacteria. Food Bioprocess Technol. 2011;4(6):919–35.
63. Luna RA, Foster JA. Gut brain axis: diet microbiota interactions and implications for modulation of anxiety and depression. Curr Opin Biotechnol. 2015;32:35–41.
64. Lusiak-Szelachowska M, et al. Phage neutralization by sera of patients receiving phage therapy. Viral Immunol. 2014;27(6):295–304.
65. Mackelprang R, et al. Metagenomic analysis of a permafrost microbial community reveals a rapid response to thaw. Nature. 2011;480(7377):368–71.

66. Mackie RI, Sghir A, Gaskins HR. Developmental microbial ecology of the neonatal gastrointestinal tract. Am J Clin Nutr. 1999;69(5):1035s–45.
67. Magurran AE. Measuring biological diversity. Afr J Aquat Sci. 2004;29(2):285–6.
68. Magurran AE. Measuring biological diversity. 1st ed. Malden: Wiley; 2013.
69. Mayer EA. Gut feelings: the emerging biology of gut-brain communication. Nat Rev Neurosci. 2011;12(8):453–66.
70. Mayer EA, Savidge T, Shulman RJ. Brain-gut microbiome interactions and functional bowel disorders. Gastroenterology. 2014;146(6):1500–12.
71. McDonald D, et al. An improved Greengenes taxonomy with explicit ranks for ecological and evolutionary analyses of bacteria and archaea. ISME J. 2012;6(3):610–8.
72. McLean JS, et al. Candidate phylum TM6 genome recovered from a hospital sink biofilm provides genomic insights into this uncultivated phylum. Proc Natl Acad Sci U S A. 2013;110 (26):E2390–9.
73. Meyer F, et al. The metagenomics RAST server–a public resource for the automatic phylogenetic and functional analysis of metagenomes. BMC Bioinf. 2008;9(1):386.
74. Minot S, et al. The human gut virome: inter-individual variation and dynamic response to diet. Genome Res. 2011;21(10):1616–25.
75. Mitchell A et al. EBI metagenomics in 2016 – an expanding and evolving resource for the analysis and archiving of metagenomic data. Nucleic Acids Res. 2015;44(D1):D595–603.
76. Mitra S, Klar B, Huson DH. Visual and statistical comparison of metagenomes. Bioinformatics. 2009;25(15):1849–55.
77. Mitra S, et al. Comparison of multiple metagenomes using phylogenetic networks based on ecological indices. ISME J. 2010;4(10):1236–42.
78. Naumann D, Helm D, Labischinski H. Microbiological characterizations by FT-IR spectroscopy. Nature. 1991;351(6321):81–2.
79. Noguchi H, Taniguchi T, Itoh T. MetaGeneAnnotator: detecting species-specific patterns of ribosomal binding site for precise gene prediction in anonymous prokaryotic and phage genomes. DNA Res. 2008;15(6):387–96.
80. Nurk S, et al. Assembling single-cell genomes and mini-metagenomes from chimeric MDA products. J Comput Biol. 2013;20(10):714–37.
81. O'Donnell AG, et al. Visualization, modelling and prediction in soil microbiology. Nat Rev Microbiol. 2007;5(9):689–99.
82. Ozsolak F, et al. Direct RNA sequencing. Nature. 2009;461(7265):814–8.
83. Pan X. Single cell analysis: from technology to biology and medicine. Single cell Biol. 2014;3(1):106.
84. Pan X. Single cell analysis: from technology to biology and medicine. Single Cell Biol. 2014;3(1). pii: 106. http://www.ncbi.nlm.nih.gov/pubmed/25177539.
85. Parks DH, Beiko RG. Identifying biologically relevant differences between metagenomic communities. Bioinformatics. 2010;26(6):715–21.
86. Peterson SN, et al. Functional expression of dental plaque microbiota. Front Cell Infect Microbiol. 2014;4:108.
87. Pettigrew MM, et al. Upper respiratory tract microbial communities, acute otitis media pathogens, and antibiotic use in healthy and sick children. Appl Environ Microbiol. 2012;78(17):6262–70.
88. Price PB. Microbial life in glacial ice and implications for a cold origin of life. FEMS Microbiol Ecol. 2007;59(2):217–31.
89. Proctor GN. Mathematics of microbial plasmid instability and subsequent differential growth of plasmid-free and plasmid-containing cells, relevant to the analysis of experimental colony number data. Plasmid. 1994;32(2):101–30.
90. Pruesse E, et al. SILVA: a comprehensive online resource for quality checked and aligned ribosomal RNA sequence data compatible with ARB. Nucleic Acids Res. 2007;35 (21):7188–96.

91. Puppels GJ, et al. Studying single living cells and chromosomes by confocal Raman microspectroscopy. Nature. 1990;347(6290):301–3.
92. Raghunathan A, et al. Genomic DNA amplification from a single bacterium. Appl Environ Microbiol. 2005;71(6):3342–7.
93. Reid G, et al. Microbiota restoration: natural and supplemented recovery of human microbial communities. Nat Rev Microbiol. 2011;9(1):27–38.
94. Rhee SH, Pothoulakis C, Mayer EA. Principles and clinical implications of the brain–gut–enteric microbiota axis. Nat Rev Gastroenterol Hepatol. 2009;6(5):306–14.
95. Rinke C, et al. Obtaining genomes from uncultivated environmental microorganisms using FACS-based single-cell genomics. Nat Protoc. 2014;9(5):1038–48.
96. Savage DC. Microbial ecology of the gastrointestinal tract. Annu Rev Microbiol. 1977;31 (31):107–33.
97. Schadt EE, Turner S, Kasarskis A. A window into third-generation sequencing. Hum Mol Genet. 2010;19(R2):R227–40.
98. Schloss PD, et al. Introducing mothur: open-source, platform-independent, community-supported software for describing and comparing microbial communities. Appl Environ Microbiol. 2009;75(23):7537–41.
99. Sharpton TJ. An introduction to the analysis of shotgun metagenomic data. Front Plant Sci. 2014;5:209.
100. Smits LP, et al. Therapeutic potential of fecal microbiota transplantation. Gastroenterology. 2013;145(5):946–53.
101. Stecher B, Berry D, Loy A. Colonization resistance and microbial ecophysiology: using gnotobiotic mouse models and single-cell technology to explore the intestinal jungle. FEMS Microbiol Rev. 2013;37(5):793–829.
102. Su X, et al. Parallel-META 2.0: enhanced metagenomic data analysis with functional annotation, high performance computing and advanced visualization. PLoS One. 2014;9 (3):e89323.
103. Su X, et al. Application of Meta-Mesh on the analysis of microbial communities from human associated-habitats. Quant Biol. 2015;3(1):4–18.
104. Tang F, et al. mRNA-Seq whole-transcriptome analysis of a single cell. Nat Methods. 2009;6 (5):377–82.
105. Taniguchi K, Kajiyama T, Kambara H. Quantitative analysis of gene expression in a single cell by qPCR. Nat Methods. 2009;6(7):503–6.
106. Tsuchiya S, et al. The "spanning protocol": a new DNA extraction method for efficient single-cell genetic diagnosis. J Assist Reprod Genet. 2005;22(11–12):407–14.
107. Tuohy KM, et al. Metabolism of Maillard reaction products by the human gut microbiota-implications for health. Mol Nutr Food Res. 2006;50(9):847–57.
108. Turnbaugh PJ, et al. An obesity-associated gut microbiome with increased capacity for energy harvest. Nature. 2006;444(7122):1027–131.
109. Turnbaugh PJ, et al. A core gut microbiome in obese and lean twins. Nature. 2009;457 (7228):480–4.
110. Udayappan SD, et al. Intestinal microbiota and faecal transplantation as treatment modality for insulin resistance and type 2 diabetes mellitus. Clin Exp Immunol. 2014;177(1):24–9.
111. Vetizou M, et al. Anticancer immunotherapy by CTLA-4 blockade relies on the gut microbiota. Science. 2015;350(6264):1079–84.
112. Wagner M. Single-cell ecophysiology of microbes as revealed by Raman microspectroscopy or secondary ion mass spectrometry imaging. Annu Rev Microbiol. 2009;63:411–29.
113. Wang D, Bodovitz S. Single cell analysis: the new frontier in 'omics'. Trends Biotechnol. 2010;28(6):281–90.
114. Wang X, et al. Microfluidic extraction and stretching of chromosomal DNA from single cell nuclei for DNA fluorescence in situ hybridization. Biomed Microdevices. 2012;14 (3):443–51.

115. Wen J, et al. K-mer natural vector and its application to the phylogenetic analysis of genetic sequences. Gene. 2014;546(1):25–34.
116. Yatsunenko T, et al. Human gut microbiome viewed across age and geography. Nature. 2012;486(7402):222–7.
117. Zeglin LH, et al. Landscape distribution of microbial activity in the McMurdo dry valleys: linked biotic processes, hydrology, and geochemistry in a cold desert ecosystem. Ecosystems. 2009;12(4):562–73.
118. Zerbino DR, Birney E. Velvet: algorithms for de novo short read assembly using de Bruijn graphs. Genome Res. 2008;18(5):821–9.
119. Zhang YJ, et al. Impacts of gut bacteria on human health and diseases. Int J Mol Sci. 2015;16 (4):7493–519.
120. Zong C, et al. Genome-wide detection of single-nucleotide and copy-number variations of a single human cell. Science. 2012;338(6114):1622–6.

Chapter 7
Text Mining for Precision Medicine: *Bringing Structure to EHRs and Biomedical Literature to Understand Genes and Health*

Michael Simmons, Ayush Singhal, and Zhiyong Lu

Abstract The key question of precision medicine is whether it is possible to find clinically actionable granularity in diagnosing disease and classifying patient risk. The advent of next-generation sequencing and the widespread adoption of electronic health records (EHRs) have provided clinicians and researchers a wealth of data and made possible the precise characterization of individual patient genotypes and phenotypes. Unstructured text—found in biomedical publications and clinical notes—is an important component of genotype and phenotype knowledge. Publications in the biomedical literature provide essential information for interpreting genetic data. Likewise, clinical notes contain the richest source of phenotype information in EHRs. Text mining can render these texts computationally accessible and support information extraction and hypothesis generation. This chapter reviews the mechanics of text mining in precision medicine and discusses several specific use cases, including database curation for personalized cancer medicine, patient outcome prediction from EHR-derived cohorts, and pharmacogenomic research. Taken as a whole, these use cases demonstrate how text mining enables effective utilization of existing knowledge sources and thus promotes increased value for patients and healthcare systems. Text mining is an indispensable tool for translating genotype-phenotype data into effective clinical care that will undoubtedly play an important role in the eventual realization of precision medicine.

Keywords Precision medicine • Text mining • Genotype • Phenotype • EHR • NLP • Database curation • Cancer • Outcome prediction • Pharmacogenomics • Biomedical literature

---

M. Simmons • A. Singhal • Z. Lu (✉)
National Center for Biotechnology Information (NCBI), National Library of Medicine (NLM), 8600 Rockville Pike, Bldg 38A, 10N1003A, Bethesda, MD 20894, USA
e-mail: zhiyong.lu@nih.gov

B. Shen et al. (eds.), *Translational Biomedical Informatics*, Advances in Experimental Medicine and Biology 939, DOI 10.1007/978-981-10-1503-8_7

7.1 Introduction

The precision medicine ideal is that data about genes, environment, and lifestyle can enable optimal patient care by allowing physicians to customize each person's treatment to reflect these unique health determinants. Governments and healthcare organizations around the globe have taken interest in this ideal, with the recent notable instance of the US Precision Medicine Initiative (PMI) announced by President Barack Obama in January 2015 [12]. Prior to President Obama's announcement, many countries, including China, the UK, Iceland, Japan, Canada and others, had established infrastructures for precision medicine research through the development of biobanks (repositories of patient DNA with accompanying databases that link the medical history and lifestyle information of donors to their biologic samples) [90]. Precision medicine is thus a global hope founded in the belief that it is possible to harness big data in healthcare and biology to promote health and relieve suffering.

The core challenge of precision medicine (PM) is that of classification: is it possible to discern differences between individuals in a heterogeneous population that can guide treatment decisions and support improved care? Which people, for example, are going to develop cancer? What medications will treat their cancers most effectively? What is it about patient A that makes her fundamentally distinct from patient B, and how should doctors tailor the care of these patients to reflect these distinctions? These questions have always been relevant to clinical practice, but a trade-off has always existed between increasing information and decreasing clinical utility of that information. Precision medicine is a relevant concept now because technology has advanced in key areas of medicine to such a degree that it is possible that precise classification of individuals may indeed enable clinically effective, personalized treatment.

The last decade witnessed the birth of two key sources of data with great promise to enable the precise classification of individuals for medical care: the sequencing of the human genome with accompanying improvements in sequencing technology and the widespread adoption of the electronic health record (EHR). For both these data sources, much of the information required to conduct precision medicine is contained within unstructured, written texts such as the biomedical literature and clinical notes. In its current state, this information is not computable; hence, "unlocking" this information via natural language processing (NLP) is an essential and truly exciting area of study.

This chapter is about text mining for precision medicine (TM for PM). Text mining is a subfield of NLP dedicated to enabling computational analysis of text-locked data. The text mining workflow generally involves identification of specific entities in surface text such as diseases, genes, or relational terms and the deep normalization of these entities to standardized ontologies. Data thus processed become the input values for a variety of computations. There are two core functions of text mining: (1) information extraction and (2) hypothesis generation via relationship extraction.

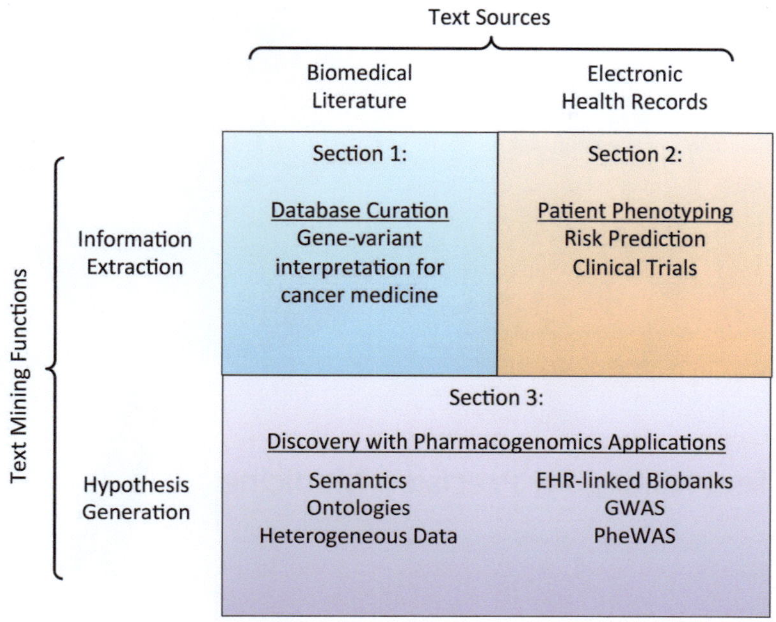

Fig. 7.1 The structure of this chapter reflects the two core functions of text mining and the two foremost text sources related to precision medicine. Section 7.1 discusses how text mining published literature can facilitate curation of genotype-phenotype databases for support of personalized cancer medicine. Section 7.2 discusses how text mining is useful in defining patient phenotypes from EHRs. Section 7.3 is about using text mining of both text sources for hypothesis generation in pharmacogenomics

In this chapter, we devote two sections to the information extraction functionality of text mining. The first section addresses mining biomedical literature for the purpose of assisting database curation in personalized cancer medicine. The second addresses mining EHRs for the purpose of cohort identification. The third section of this chapter explores the role of TM for PM as a vehicle for hypothesis generation and ties the previous two sections together with a discussion of methods of using biomedical literature and EHR texts to conduct pharmacogenomic research (Fig. 7.1).

Two key terms to any discussion of precision medicine are the terms "genotype" and "phenotype." These terms can be confusing to people who are new to genetics research because the meaning of both terms is contextual. *Genotype*, for example, can refer to the entirety of an individual's unique assortment of genes, or it can refer to a specific variant of a single gene that distinguishes an individual from others. Likewise the term *phenotype* can be defined as broadly or as narrowly as context demands (a specific disease, such as age-related macular degeneration (AMD), could be considered a phenotype, but within a group of people with AMD, the presence or absence of specific findings such as aberrant blood vessel growth could also be considered a phenotype). In this chapter, we discuss the biomedical

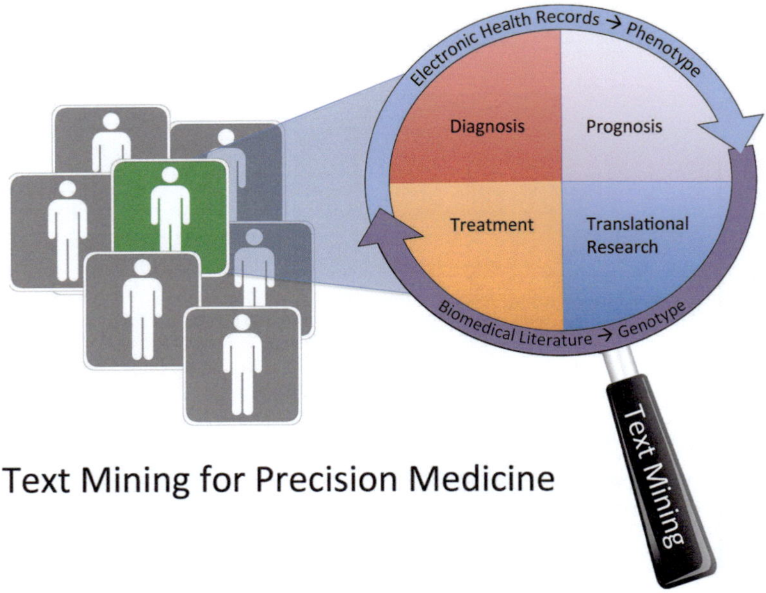

Fig. 7.2 Text mining brings unstructured information into focus to characterize genotypes and phenotypes in precision medicine

literature as an authoritative source of genotype information, and we discuss electronic health records as a dynamic resource of human phenotypes (Fig. 7.2).

7.2 TM for PM: Personalized Cancer Medicine and Database Curation

One area where PM has already demonstrated value and great promise is the field of cancer medicine. This is why the near-term focus of the US Precision Medicine Initiative is cancer diagnosis and treatment [12].

Cancer is a collection of diseases, all of which involve the development of a population of cells in the body that gain the potential to replicate infinitely. Cancers are typically named after the location of the cells that have undergone this transformation (e.g., "breast cancer" if the altered cells were initially breast cells). Cancer of any form is a disease of the genome [25, 26]. The changes that lead to cancer development occur within the DNA sequence and result in the removal of physiologic protections against cancer (loss of function mutations) and the production of new stimuli that promote cellular growth (gain of function mutations). Doctors are hopeful that genomics-driven precision medicine will be particularly effective in treating cancer because of the genetic nature of the disease process

[12, 21] and because of the demonstrated effectiveness of therapies directed precisely at the genomic alterations that cause cancer [76].

Text mining has an intuitive place in the conceptual framework for the implementation of personalized cancer medicine, which involves (1) characterization of the "driver" mutations in a given patient's tumor and (2) identification of the drugs that will best counteract the effects of those driver mutations [21]. Both of these steps are information extraction tasks, which is a key function of text mining. Additionally, much of the information needed for performing these two tasks is contained within the biomedical literature. This section will discuss the current issues and science behind text mining for assisting database curation in personalized cancer medicine.

7.2.1 Considerations for Text Mining in Database Curation

The biomedical literature helps clinicians and researchers interpret genetic information, and many databases in the cancer domain include literature references. Some prominent databases with literature curations include the Catalogue of Somatic Mutations in Cancer (COSMIC), Online Mendelian Inheritance in Man (OMIM), ClinVar, ClinGen (the proposed manually curated component of ClinVar), Swiss-Prot, Human Gene Mutation Database (HGMD), Pharmacogenomics Database (PharmGKB), and Comparative Toxicogenomics Database (CTD). All the above databases are examples of genotype-phenotype databases [8]. The gold standard of quality in the curation of literature references for these databases is manual expert curation, but there is an indisputable need for text mining tools to assist in the curation process. Baumgartner et al. elegantly illustrated this need by applying a found/fixed graph to examine the curation completeness of two databases – Entrez Gene and Swiss-Prot. Ignoring the pace of new publications, they instead examined the numbers of missing entities within these two databases and compared the rate of generation of missing data annotations over time to the rate of resolution of these missing data points. They concluded that neither database would ever "catch up" to the pace of generation of information without changes to their curation processes [3]. The rate of biomedical discovery exceeds the curation capacity of these comprehensive resources.

Although it is true that the pace of article publication exceeds human curation capabilities, it is a fallacy to conclude that assimilation of *all* new information is *necessary*. In our experience, text mining applications are most likely to be adopted by domain experts such as clinicians, researchers, and curation teams when the applications correctly *limit* the amount of information they return. Curators of databases may recognize a need to increase the pace and breadth of their curation efforts, but their intent is not to curate *all* new articles but rather to curate only the articles that further their institutional goals [48]. As an example, compare two high-quality genotype-phenotype databases, Swiss-Prot and ClinVar. Both databases contain information about diseases associated with protein/gene sequence variations, but their institutional scope is different. The chief aim of Swiss-Prot is to identify variants that alter protein function, while the chief aim of ClinVar is to provide evidence of the relationship between human genetic variants and phenotypes. Because of this difference, an article demonstrating a causative association between a variant and a given disease would likely rank higher in priority for curation in ClinVar (or its manually curated partner, ClinGen) than in Swiss-Prot. To be useful to domain experts and database curators, text mining tools must balance (1) comprehensive analysis of the literature with (2) filtering tools for ranking and identifying the most useful literature.

7.2.2 Text Mining in Database Curation

The curation workflow for genotype-phenotype databases involves three important steps where text mining can play a crucial role [41, 100]: (1) information retrieval and document triage, (2) named entity recognition and normalization, and (3) relation extraction. Fig. 7.3 provides a schematic overview of this process. For an excellent treatment of the entire workflow, we direct readers to the review by Hirschman et al. [28]. In the remainder of this section, we will discuss the latter two aspects of the workflow.

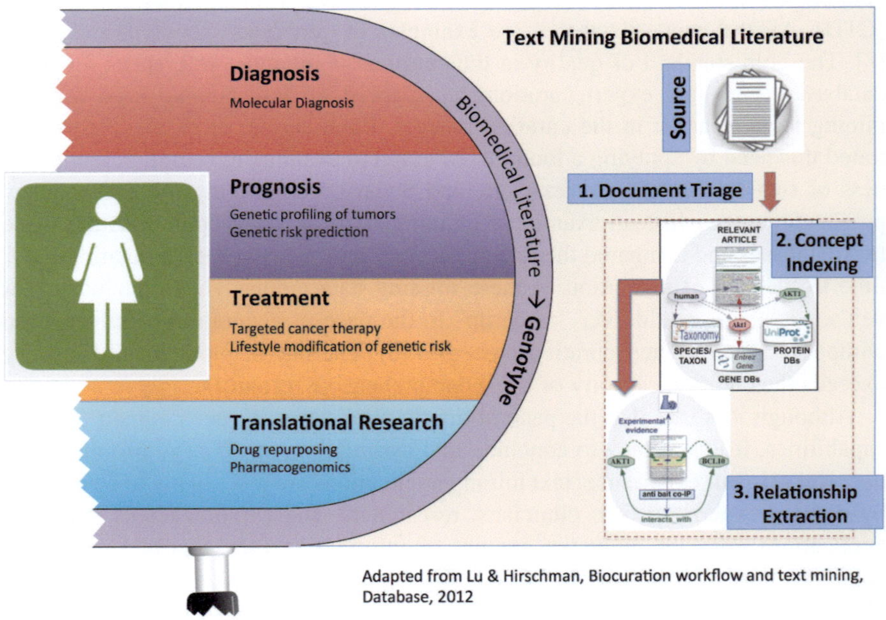

Fig. 7.3 Genotype data permits incredibly deep classification of individuals. The biomedical literature contains a wealth of information regarding how to clinically interpret genetic knowledge. Text mining can facilitate expert curation of this information into genotype-phenotype databases

An important step in curating genotype-phenotype databases is identifying relevant entities within text. A wide variety of entities are appropriate for text mining in precision medicine, including genes, gene variants, chemical/drug names, species, cohort types, diseases, or even other biological concepts such as analysis techniques or evidence levels. Tagging these entities is called named entity recognition (NER), and mapping tagged entities to standard vocabularies is called normalization. NER and normalization constitute the second step of the biocuration workflow. Mining named entities from text is challenging because of the variability of natural language expressions. We will discuss identification of three core entities for genotype-phenotype database curation: genes, variants, and diseases.

Genes Several forms of natural language variation complicate gene NER: orthographical variations (e.g., "*ESR1*" vs "*ESR-1*"), morphological variations (e.g., "GHF-1 transcription*al* factor" vs "GHF-1 transcription factor"), and abbreviations (e.g., "estrogen receptor alpha (ER)"). Ambiguity also arises in texts that discuss multiple species, since separate species frequently share genes of the same name with distinct sequences (e.g., *ERBB2* can be either a human gene or mouse gene name). It is also possible for different genes to share the same name. For example, "*AP-1*" can refer to either the "jun proto-oncogene" (Entrez Gene: 3725) or the "FBJ murine osteosarcoma viral oncogene homolog" (Entrez Gene: 2353). GNormPlus is a state-of-the-art, open-source text mining system for gene NER and normalization [102]. GNormPlus utilizes multiple sophisticated text mining techniques and achieves a balanced performance in recall and precision (please see the text box, "Text Mining Performance Metrics," in the last section of the chapter for more information about precision and recall).

Variants Gene variants and mutations are not uniformly reported with standard nomenclature in biomedical texts so identification and normalization of variant mentions is challenging. Variant/mutation normalization is also complicated by the fact that nomenclature standards have evolved over time as researchers have gained additional insights into genetic complexities. The state-of-the-art tool for variant extraction is tmVar [100].

Diseases Literature mentions of diseases involve frequent use of abbreviations, synonyms, morphological or orthographical variations, and inconsistent word ordering. DNorm is an open-source tool for disease NER and normalization that maps concepts to the MEDIC vocabulary using a pairwise learning to rank machine learning algorithm [41].

One of the most recent advancements in curation support is a tool produced by NCBI called PubTator [101]. This web-based application incorporates multiple state-of-the-art tools, including all three NER tools discussed above, to support three curation tasks: document triage, NER and normalization, and relationship annotation. PubTator combines an intuitive interface with comprehensive access to all references in PubMed and is truly an excellent multipurpose tool for curation (Fig. 7.4).

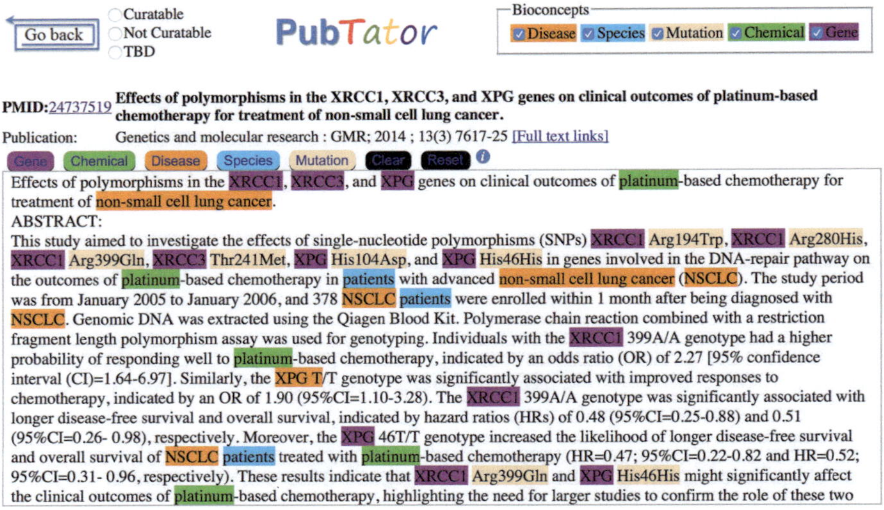

Fig. 7.4 This abstract includes examples of each of the five bio-entities that PubTator identifies. Note the correct identification of mentions to non-small cell lung cancer regardless of whether the text uses the full term or its abbreviation, NSCLC. Likewise, PubTator correctly interprets the term "patients" as a reference to a species, *Homo sapiens*. Although this abstract uses protein-level nomenclature to describe gene variants (e.g., "XRCC1 Arg399Gln"), the authors distinguish genotypes with nucleotides rather than amino acids (e.g., "the XRCC1 399A/A genotype"). This variability is an example of the challenges inherent to named entity recognition of gene mutations

The identification of semantic relationships between entities is called relationship extraction and is considered the third step in the curation workflow. Conventionally, most relation extraction techniques have used co-occurrence metrics to relate two entities within a text (e.g., co-occurrence metrics make the assumption that if gene A and variant B are both in the same abstract, then variant B must be a variant of gene A). However, co-occurrence approaches ignore contextual content and result in many errors. For example, in variant-disease relationship extraction, co-occurrence methods will wrongly interpret negative results as support for an association. The high rate of false positives associated with co-occurrence methods significantly lowers the utility of these methods for genotype-phenotype database curation workflows. In response to these challenges Singhal et al. [82, 83] developed a machine learning and text mining approach to extract disease-variant relations from biomedical text. Their machine learning approach learns patterns from the text to decide whether two entities co-occurring within the text have any stated relationship or association. The patterns are learned on six predefined features that capture both in-sentence and cross-sentence relation mentions. Negative findings within the text are taken into account using numeric sentiment descriptors. They demonstrate that machine learning delivers significantly higher performance than the previous co-occurrence-based state of the art [18].

7.2.3 Applications of Text Mining in Personalized Cancer Medicine

Text mining plays an important role in the curation workflow of many cancer-related genotype-phenotype databases. For example, curators of Swiss-Prot use a number of text mining resources for document triage in their curation workflow, including TextPresso and iHOP [94]. In the Pharmacogenomics Database (PharmGKB), curators use an adaptation of the TextPresso tool, Pharmspresso, for information retrieval and document triage [22, 94]. The miRCancer database, which catalogs literature mentions of microRNA expression in cancer cells, uses a rule-based text mining approach to identify miRNA-cancer mentions for manual curation [107]. Two groups have used text mining to develop gene methylation databases for cancer research: MeInfoText [19] and PubMeth [61]. Still other groups have created cancer-specific databases using text mining. Examples of such databases include the Osteosarcoma Database [65] and the Dragon Database of Genes associated with Prostate Cancer (DDPC) [52].

One of the most prominent databases to use text mining to curate information related to precision cancer medicine is the Comparative Toxicogenomics Database (CTD), a publicly available database containing manually curated relationships between chemicals, genes, proteins, and diseases. CTD employs text mining to triage documents and identify entities in text for curation. They developed a metric called a document relevancy score to quantify how important a given literature reference might be to their curation goals, and they found that text mining reliably identifies articles that are most likely to provide the highest yield of new, relevant, and biologically interpretable information [14]. CTD has also featured prominently in several BioCreative community challenges[1]. Track I of the BioCreative-2012 Workshop involved developing a document triage process with a web interface [105] for curation in CTD. Likewise, Track 3 of BioCreative IV involved developing web tools for performing named entity recognition on text passages using CTD's controlled vocabulary [1]. More recently, the 2015 BioCreative V challenge included a chemical-disease relation (CDR) task involving extraction of chemically induced diseases. The best systems in this task achieved an F1 score of 57 % (Wei et al. [103]).

The need for using text mining in database curation is extremely strong. The interdisciplinary nature of precision cancer medicine and the volume of information relevant to customizing patient care necessitate the use of databases to integrate information and create broad access to it. The biomedical literature constitutes a relevant and authoritative source of information for such databases, and text mining can structure and summarize this information for rapid assimilation. As the science

[1] BioCreative is one of a number of community-wide competitions and collaborative efforts designed to assess and advance the state of the art in text mining for biomedicine. Past challenges have addressed many issues related to TM for PM. For more information regarding this unique aspect of the text mining community, please see the review by Huang and Lu [31].

of text mining advances and tools become more robust and accurate, the immediate relevance of TM for PM will only increase.

7.3 TM for PM: EHRs and Phenotype Determination

The previous section discussed how text mining biomedical literature supports database curation and thus informs clinicians and researchers as they look *deeply* into individuals' *genotypes*. In this next section, we consider a completely separate perspective of TM for PM—how text mining electronic health record (EHR) data can enable physicians to look *broadly* at an entire population by classifying patient *phenotypes*. Such patient phenotypes may be the most accurate means of representing the interplay of all the health determinants of precision medicine: genes, environment, and lifestyle.

Consider the case of an actual clinical dilemma that occurred involving a teenage girl who was hospitalized for an autoimmune disorder called systemic lupus erythematosus (SLE or lupus) [20]. Several factors complicated this girl's condition and predisposed her to forming blood clots. Although blood-thinning medications could protect against these clots, her providers were also concerned about paradoxical bleeding if they prescribed these medications. The key clinical question in this situation—whether to prescribe a blood thinner—was not readily answerable from published research studies or guidelines, so her provider turned to her institution's EHR and used its research informatics platform [47] to identify an "electronic cohort" of pediatric patients who had in previous years experienced similar lupus exacerbations. From this cohort, they identified a trend of complications from blood clots, which convinced them to administer anticoagulants.

This story illustrates the potential of using EHR data to direct personalized care. The physicians in this case used EHR data to identify a group of similar patients with known outcomes. The outcome data then enabled them to estimate their patient's risk and intervene to modify that risk. Although this analysis did not yield the statistical confidence of a formal clinical trial, the patient's physicians felt that this EHR cohort analysis provided superior information than the alternatives—pooled opinion and physician recollection [20]. In a large healthcare system, EHR-derived cohorts can reflect the interplay of genes, environment, and lifestyle in the health outcomes of a specific group of patients. Many exciting applications of cohort identification from EHR data exist. This section will discuss two use cases: patient outcome prediction via patient similarity analytics and cohort identification for clinical trials. Text mining is integral to the development of these applications because in many cases the richest and most diverse patient information in EHRs is contained in free, unstructured notes written by healthcare providers [15] (Fig. 7.5).

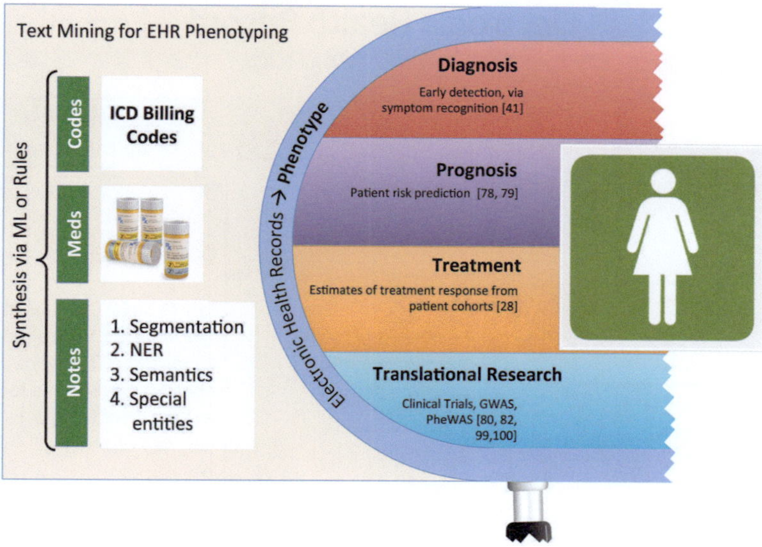

Fig. 7.5 EHRs are rich sources of phenotype information. Algorithms to extract phenotypes commonly incorporate text mining of clinical notes as well as billing codes and medications. In contrast to the deeply individual nature of genotype information, phenotype algorithms generate clinical insights by first looking broadly at aggregated populations of people with similar conditions and known health outcomes

7.3.1 EHR Phenotype Determination

Identifying populations of people with shared health characteristics amounts to defining a phenotype. EHR data has been shown to be an effective source for comprehensive measurement of the phenotypic characteristics of populations [38]. Simply defined, EHRs are information systems that contain electronic records of all the key clinical and administrative data relevant to the medical history of each patient in a specific healthcare institution [10]. EHRs consist of both structured and unstructured data fields (Fig. 7.6). Some of the key structured data fields include billing data, laboratory and vital signs, and medication records. Unstructured data is largely present in notes, of which there are two main types: clinical notes (e.g., history and physical notes or discharge summaries) and test results (e.g., radiology reports or pathology reports) [15]. EHR data is a promising source of phenotype information because

1. information in EHRs is relatively inexpensive to obtain since EHR data is generated as a by-product of the healthcare process;
2. the scope of EHR data is vaster than the scope of any organized study, both in terms of the variety of pathology and in terms of longitudinal coverage; and

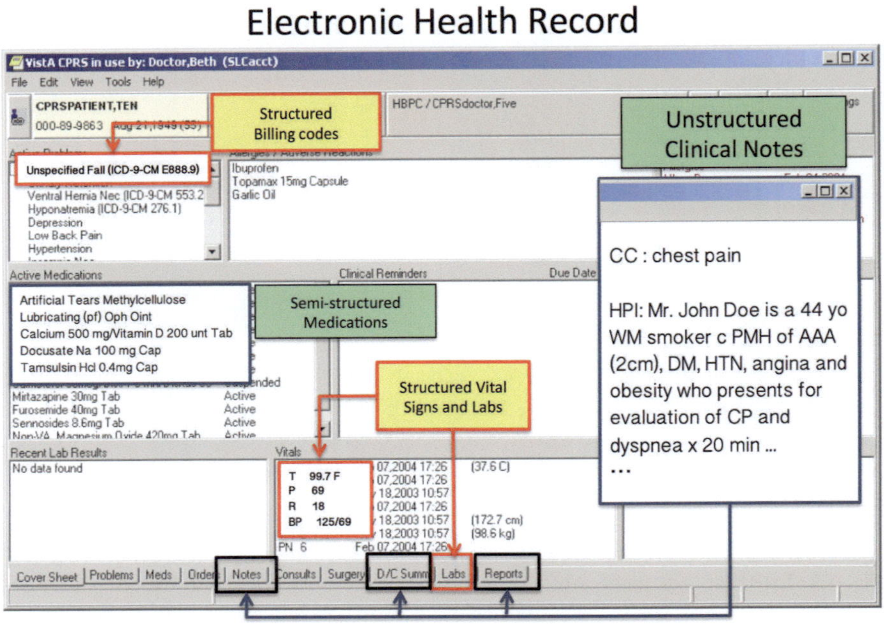

Fig. 7.6 The Veterans Information Systems and Technology Architecture (VISTA) is the most widely used EHR in the United States. Like most EHRs, it contains structured data and unstructured text

3. the resolution of EHR data continually improves over time because additional encounters for any given individual lead to increased certainty regarding the presence or absence of a given diagnosis.

There are several noteworthy challenges inherent in using EHR data for precision medicine [29]. (1) EHR data consists of sensitive and highly confidential information with extensive legal protections [70] and real ethical pitfalls related to privacy [81]. (2) EHR data is incomplete for many reasons. One reason is that patients often utilize multiple healthcare systems (e.g., a given patient may see specialists at different, unrelated hospitals), but separate systems do not share EHR data so information within EHRs can be fragmented [15]. (3) EHR data is complicated by multiple biases. The highest quality manner of collecting population phenotype data is through a prospective observational cohort, like the Framingham Heart Study (FHS) [46]. In comparison with such studies, EHR data mining ranks much lower in an evidence hierarchy [27] and should be suspected of significant biases such as multiple confounders, selection bias (EHRs represent sick people more than healthy people), and measurement bias (e.g., documented physical exam findings may differ in reliability between physicians) [78].

In addition to the above challenges inherent to EHR data, the process of identifying patients with a specific diagnosis (i.e., defining a phenotype from an

EHR) has its own attendant challenges. Sun et al. illustrated these challenges in their work to identify people with the condition of high blood pressure (hypertension) from their institution's EHR [86]. Physicians diagnose hypertension by observing blood pressures that consistently exceed a certain threshold over time. Even though many EHRs contain blood pressure measurements in structured formats, using blood pressure measurements alone to identify patients with the condition of hypertension is surprisingly inaccurate. This is because blood pressure measurements are inherently variable [33] (e.g., blood pressures may rise in response to pain or anxiety) and are modifiable through treatment (i.e., the use of antihypertensive medication by people with the condition will result in normal blood pressure levels). Thus, in the approach utilized by Sun et al. to detect changes in HTN control status, detection models incorporating only aggregated blood pressure measurements identified many more false positives and false negatives than models that incorporated multiple features from the EHR.

Because of the complexity of determining phenotypes from EHR data, a common convention is to use billing data such as International Classification of Disease (ICD) codes and Current Procedural Terminology (CPT) codes for representing phenotypes [15]. These billing codes are universal between healthcare systems and are available in structured formats. Yet billing data alone is also insufficient for accurate representation of disease phenotypes [106]. The best performing approaches to identifying EHR phenotypes incorporate multiple data fields, including text mining [16]. Wei et al. demonstrated the benefit of text mining in phenotype development in a study where they examined the advantages and utilities of billing codes, clinical notes (using text mining), and medications from EHRs in detection of ten separate disease phenotypes. They found that information collected from clinical notes through text mining offered the best average sensitivity (0.77) out of all individual components, whereas billing code data had a sensitivity of 0.67. They also found that the relevance of using text mining of clinical notes in identifying phenotypes varied by disease (e.g., 84 % of all information in the EHR necessary for identifying rheumatoid arthritis was contained in clinical notes, whereas only 2 % of information needed for identifying atrial fibrillation was contained in notes) (Wei et al. [104]). Ultimately, the highest F1 scores resulted from combinations of two or more separate EHR components, such as billing data and text mining. In general, incorporating text mining in phenotype algorithms results in improvements in both the precision and recall [15].

7.3.2 Phenotype Extraction from EHR Text

In this section, we present a classification of the text mining approaches used in information extraction for EHR phenotype development, and we provide a brief overview of phenotyping. For a more comprehensive treatment of the entire approach for identifying patient phenotype cohorts from EHRs, please see the review by Shivade et al. [80].

Text in clinical notes differs from text in biomedical literature in several ways. The foundational difference is that biomedical literature contains formal, peer-reviewed writing, whereas clinical notes are comparatively informal. Physicians write clinical notes with the goal of maximizing time efficiency, so clinical notes contain heavy use of abbreviations, often only decipherable from contextual cues. For legal reasons, modification of existing clinical notes is not permitted, yet due to time constraints, physicians often submit notes with only cursory proofreading. Consequently, spelling errors and unconventional sentence structures are common. Lastly, clinical notes frequently contain copied and pasted sections of values or text such as lab findings or vital signs, which complicate parsing of notes [56].

The extraction of relevant information from EHR texts involves four steps: (1) text segmentation, (2) entity identification and normalization, (3) evaluation for semantic modifiers such as negation and possibility phrases, and (4) extraction of special entities from text.

1. Text segmentation. Free text from clinical notes needs to be segmented into fundamental units called "tokens" before any further processing takes place. Text segmentation is done using NLP parsers and tokenizers. cTAKES and the GATE framework are examples of open-source NLP parsers designed for handling clinical text [73]. Several commercial options are also available.
2. NER and normalization. The most common approach for entity identification is to map tokens from segmented text into target dictionaries such as SNOMED-CT, ICD, UMLS, etc. Tools for mapping concepts in biomedical literature such as MetaMap can also process tokenized clinical texts [2, 4], but other clinically customized tools exist [43, 55] such as the HITex system [109] and the Knowledge Map Concept Identifier [9, 109].
3. Evaluation for semantic modifiers. Words that modify the semantic meaning of sentences are important for accurate phenotyping. For example, consider the significance of the word "no" in the following list that might appear in a provider's note for a patient with chest pain: "no history of heart palpitations, dizziness/fainting, or tobacco use." In a similar sense, identifying possibility phrases or status keywords such as "confirmed," "possible," "probable," etc. is very helpful.
4. Special entity extraction. Other important entities in free text include special keywords such as numerical measurements and units [23], dates and time-related words (e.g., "before," "during," or "after") [84], and phenotype-specific keywords [85].

7.3.3 Phenotype Algorithm Development

The entities extracted from clinical texts serve as inputs along with other structured EHR data such as billing codes and medications for phenotype algorithms. It is also worth acknowledging that several studies have looked at external sources such as imaging data [5, 50], drug characterization databases [45], and scientific articles from biomedical literature [110] to extract EHR phenotypes. Regardless of the data types used, phenotype algorithm development for EHRs involves identifying relevant features and then synthesizing those features—either through the application of expert-derived rules or through machine learning (ML).

Rule Application For some conditions, clear clinical guidelines exist to guide phenotype algorithm development. The breathing disorder asthma is one such condition [49]. One way of defining a phenotype algorithm is to incorporate these clinical guidelines into a set of rules that can guide identification of patients with a given condition. Wu et al. took this approach in modifying the asthma diagnostic criteria and developing a set of rules to identify patients with *definite* and *probable* asthma [106]. Other conditions also lend to rule-based algorithms. Nguyen et al. [59] used a set of rules to build a classification system to identify lung cancer stages from free text data from pathology reports. Schmiedeskamp et al. [74] used a set of empirical rules to combine ICD-9-CM codes, labs, and medication data to classify patients with nosocomial *Clostridium difficile* infection. A few studies such as those by Kho et al. [34], Klompas et al. [37], Trick et al. [95], and Mathias et al. [53] have used guidelines published by organizations such as the American Diabetes Association, the Centers for Disease Control and Prevention, the American Cancer Society, and other trusted organizations to develop rule-based algorithms.

Machine Learning As opposed to rule-based techniques where expert knowledge determines parameter significance, ML techniques identify patterns (not necessarily rules) from data. ML techniques are ideal for extracting phenotypes when rules are either not available or not comprehensive. There are two steps to ML phenotype algorithm development: feature selection and model building. "Feature selection" refers to the identification of parameters to use in ML algorithms. Establishing a sufficiently robust feature set prior to model building is important for achieving the best performance. We encourage the readers to read Bishop [6] for this topic.

Researchers have experimented with several ML models including probability [32, 77, 110], decision tree [50, 51, 97], discriminant [9, 77] and other types of ML models [35, 43, 92] to build phenotype categorization systems. In each case, the models essentially approach phenotyping as a classification or categorization problem with a positive class (the target phenotype) and a negative class (everything else). No consensus exists about which ML model is best for phenotype algorithm development. Wu et al., who developed the rule-based algorithm for identifying patients with asthma that we previously discussed, also developed an algorithm for asthma detection using ML. They chose a decision tree model and compared the results of their ML and rule-based algorithms. Both algorithms substantially outperformed a phenotyping approach using ICD codes alone, and the ML algorithm performed slightly better than the rule-based algorithm across all performance metrics [106].

Phenotype extraction from EHRs is a challenging task that has become the rate-limiting step for applications of EHR phenotypes; nevertheless, broad, flexible phenotyping for point-of-care uses like patient outcome prediction is achievable. One factor that may aid in broad phenotype determination is the transferability of phenotype algorithms from one institute to another. The eMERGE Network—a collaboration between healthcare systems with EHR-linked biobanks—demonstrated this transferability in a study where separate healthcare systems measured the performance of phenotype algorithms developed by a primary site at other sites within the network. They found that the majority of algorithms transferred well despite dramatic differences in EHR structures between sites. The majority of algorithms yielded PPV values above 90 % [58]. Many phenotype algorithms are publicly available at the Phenotype Knowledgebase, an online repository of algorithms produced in partnership with the eMERGE Network and the NIH Collaboratory [93]. Other groups have investigated the possibility of automatic, high-throughput phenotype development. Yu et al. employed a penalized logistic regression model to identify phenotypic features automatically and generated algorithms for identifying cases of rheumatoid arthritis and coronary artery disease cases using data obtained through text mining of EHRs. Their approach demonstrated comparable accuracy to algorithms trained with expert-curated features such as those in the previously mentioned study by the eMERGE Network [108].

7.3.4 Patient Outcome Prediction Through Similarity Analytics

Widespread interest exists in utilizing EHR phenotype data to produce point-of-care tools for clinicians [75]. One potential function is patient outcome prediction through similarity analytics. This is the use case of EHR phenotypes that we presented at the beginning of this section. Patient outcome prediction is a relatively new field of study. Consequently, even though text mining is an important component of the phenotype-determination algorithms that such prediction tools require, relatively few studies have examined the role of text mining in end-to-end risk prediction models. Most studies focus on either patient risk prediction or phenotype definition [80]. Regarding the former, Lee et al. showed the potential benefit of phenotype-derived predictions in their work developing a patient similarity metric to quantify the degree of similarity between patients in an intensive care unit for the purpose of predicting 30-day mortality [42]. Their model demonstrated that using data from a relatively small (~100) subset of patients who possessed the greatest degree of similarity delivered the best predictive performance. One notable study has used text mining in an end-to-end fashion with patient outcome prediction: Cole et al. used the NCBO Annotator to process clinical notes related to juvenile idiopathic arthritis (JIA) and employed this information in combination with billing code data to predict risk of developing uveitis (a vision-threatening complication of JIA) [11]. The field of patient similarity analytics using phenotype detection to drive outcome prediction is an exciting field of research where TM for PM promises great results.

7.3.5 *Clinical Trial Recruitment*

Another application of text mining for phenotype definition from EHRs is automated clinical trial eligibility screening. Precise phenotype cohort identification facilitates improvements in both the effectiveness (identifying the best patients) and efficiency (enrolling the most patients) of clinical trial recruitment. The key goals of this process are to (1) identify those populations who meet the inclusion criteria for a study and (2) facilitate the most efficient workflow for enrollment of those patients into the correct trial [39]. Ni et al. showed that a text mining-based screening system could accomplish both goals [60]. They identified patient phenotypes using text mining of notes and billing code data to enable automated patient screening. At the same time, they obtained from ClinicalTrials.gov the narrative text of eligibility criteria of the trials being conducted at their institution and used NLP to extract pattern vectors for each clinical trial. They used these vectors to identify which trials would best fit a given patient. Ultimately, their process reduced workload of physicians in screening patients for clinical trials by 85 %. Many other studies have shown similar benefits [39].

Another exciting application of precision EHR phenotyping related to trial recruitment is that of automated point-of-care clinical trial generation using EHRs [75]. Conducting randomized interventional studies using the existing infrastructure of the EHR involves building point-of-care tools into the EHR that will activate an enrollment process when a clinician is faced with a clinical decision where medical knowledge is insufficient to guide care [13]—for example, consider hypothetically the case study of the teen with lupus at the beginning of this section. In her situation, if an EHR trial was in place regarding the use of anticoagulation in a lupus exacerbation, and if the patient and her physician were truly ambivalent about what the right choice was, after she provided consent, the EHR would randomize the patient to one intervention or the other (i.e., give anticoagulants or withhold them). All subsequent follow-up would be purely observational through the data recorded in the course of the patient's care. The randomized intervention in this type of study resolves some of the issues of confounding associated with observational data, and as such, this type of randomized observational study falls between observational studies and randomized trials in an evidence hierarchy. Vickers and Scardino proposed that this model might be applied in four areas: comparison of surgical techniques, "me too" drugs, rare diseases, and lifestyle interventions [98]. Point-of-care clinical trials are an application of EHR phenotyping that might be the only cost-effective way to effectively study a large number of clinical questions related to precision medicine.

7.4 TM for PM: Hypothesis Generation

Text Mining Performance Metrics
The three most common evaluation metrics for text mining tools are precision, recall, and F1 score [30]. These metrics apply equally to tools for processing clinical text and published literature. Precision and recall are also common metrics for evaluating clinical diagnostic tests. We describe each below.

Precision is the text mining equivalent of the clinical metric of *positive predictive value* and is equal to the ratio of true positives to all positive values from a test. In lay terms, recall answers the question, "How likely is it that the results of this tool actually reflect what I was searching for?" Tools with high precision scores have low numbers of false positives and can thus be trusted to produce only correct results.

$$\begin{aligned}\text{Precision} &= \text{Positive Predictive Value}\\ &= \text{True Positives}/(\text{True Positives} + \text{False Positives})\end{aligned}$$

Recall is the text mining equivalent of the clinical metric of *sensitivity* and is equal to the proportion of all true positives that a test detects. In lay terms, recall answers the question,

"Can I rely on this tool to identify all relevant entities?" Tools with high recall scores have low numbers of false negatives and are thus most reliable.

$$\begin{aligned}\text{Recall} &= \text{Sensitivity}\\ &= \text{True Positives}/(\text{True Positives} + \text{False Negatives})\end{aligned}$$

A trade-off exists between precision and recall such that it is possible to improve the precision of any tool at the expense of recall and vice versa. For this reason, it is common in text mining evaluations to use a composite metric called *F1 score*, which is the harmonic mean of precision and recall.

Text mining is useful for identifying genotypes and phenotypes from biomedical literature and EHRs. Yet information extraction is not the only function of TM for PM. Text mining tools can also generate hypotheses and by so doing support precision medicine research. In this section, we will address hypothesis generation using biomedical literature and EHR data related to one area of precision medicine—pharmacogenomics discovery.

Pharmacogenomics is the study of how genes affect drug response. In the context of this chapter, it may be helpful to view pharmacogenomics as a particular kind of genotype-phenotype relationship (i.e., consider response to a drug as a phenotype). Two areas of applied pharmacogenomics research where text mining tools for hypothesis generation have proved useful are drug repurposing (identifying new indications for existing, approved drugs) and drug adverse effect prediction.

Text mining tools for hypothesis generation universally function through identification of relationships between entities. These relationships can be semantically defined relationships, formal relationships in structured ontologies, or relationships across heterogeneous data sources [24]. Text mining tools that synthesize such relationships and successfully identify new information can increase research productivity and provide substantial savings in terms of opportunity cost. Hypothesis-generating tools are particularly important in precision medicine because the large data sources associated with precision medicine encourage execution of multiple tests, which results in increased statistical penalties [8]. For example, although genetics researchers now have the capability of sequencing thousands of genes to identify genetic determinants of response to a particular drug, the analysis of each of these genes results in thousands of tests, each of which carries a specific probability of returning a false positive. Thus, for every test that researchers perform, they must increase the value of their significance threshold, which in turn creates a bias preventing detection of rare variants and variants with small effect sizes. Well-formulated and supported hypotheses derived in silico from data such as biomedical literature grant researchers the ability to find support for a potential gene association before committing resources to test that association experimentally. As researchers are enabled to test fewer genes, the significance thresholds for discovery grow smaller, and the likelihood of discovering true associations increases.

7.4.1 Pharmacogenomic Hypothesis Generation from Text Mining Biomedical Literature

The world's biomedical literature, when accessed via text mining, can become an incredibly rich database of multidimensional relationships, including core pharmacogenomic relationships such as those between genes, proteins, chemicals, and diseases. Text mining makes such relationships computationally mappable and enables discovery of "hidden" relationships that are not explicitly described in published literature and, indeed, are not yet known. The validity of this conceptual approach to hypothesis generation was demonstrated in an early application of text mining that explored disease-chemical relationships to predict a benefit for using fish oil in Raynaud's syndrome [87] and for using magnesium to treat migraines [88]. The discoveries hypothesized in these studies identified a straightforward form of relationship: if drug A is related to phenotype B in one body of literature, and disease C is related to phenotype B in a separate body of literature, drug A and disease C might be related to each other [89]. Reflecting an understanding of the complexity of biological disease processes, more recent approaches using text mining to generate hypotheses for precision medicine have explored drug-gene relationships, drug-protein interaction networks, and gene pathway relationships.

Regarding drug-gene relationships, Percha et al. hypothesized that drug-drug interactions (DDIs) occur when different drugs interact with the same gene product. To prove this hypothesis, they extracted a network of gene-drug relationships from Medline (the indexed component of PubMed). Their work is notable for their extraction of the type of relationship between drugs and genes (e.g., "inhibit," "metabolize," etc.) as well as the gene and drug entities themselves. They verified their approach by predicting a number of known DDIs as well as several DDIs that were not reported in the literature [64]. In another work, Percha and Altman approached mapping the rich networks of drug-gene relationships in published literature by explicitly defining the ways in which gene-drug interactions are described in literature. They employed a novel algorithm, termed ensemble biclustering for classification, which they validated against manually curated sets from PharmGKB and DrugBank. Finally they applied it to Medline, creating a map of all drug-gene relationships in Medline with at least five mentions. This map contained 2898 pairs of drug-gene interactions that were novel to PharmGKB and DrugBank [63].

Drug-protein interactions (DPIs) are another form of relationship from which to generate pharmacogenomic hypotheses. Li et al. used DPI data to create disease-specific drug-protein connectivity maps as a tool for drug repurposing. Their approach involved establishing connectivity maps between text-mined disease-drug relationships and an outside database of protein-drug interactions. They demonstrated the utility of this approach by applying their work to Alzheimer disease, where they used these maps to generate the hypotheses that prazosin, diltiazem, and quinidine might have therapeutic effects in AD patients. By searching ClinicalTrials.gov, they discovered that one of these drugs, prazosin, was already under investigation as a therapy for agitation and aggression in AD patients [44].

Biological pathways are sequences of interactions between biological entities such as genes, chemicals, and proteins that combine to exert a change in a cell. Because these pathways are essentially complex networks of relationships, they have great potential for hypothesis generation, yet pathways are necessarily high in order. As we discussed in the biocuration section of this chapter, mining high-order entities from text remains challenging. Tari et al. made significant progress in this domain with an approach to construction of pharmacogenomic pathways [91]. They produced pharmacokinetic pathways for 20 separate drugs (pharmacokinetic pathways describe how the body processes a drug). They extracted molecular drug interaction facts from databases such as DrugBank and PharmGKB and then expanded this information with text-mined data from Medline. The key contribution of their approach is their use of automated reasoning to sort these facts and construct pathways according to logical time points by assigning pre- and post-condition states to entities. A comparison between their automatically constructed pathways and manually curated pathways for the same drugs in PharmGKB revealed that the automated approach achieved 84 % precision for the extraction of enzymes and 82 % for the extraction of metabolites. Their system enabled them to propose an additional 24 extra enzymes and 48 metabolites that were not included in the manually curated resources.

7.4.2 Hypothesis Generation from EHRs

Many applications of TM for PM using EHR text for hypothesis generation require the integration of phenotypic data with genetic data. Although this is uncommon in EHRs, it is possible with EHR-linked biobanks. Examples of such biobanks include the NUgene Project at Northwestern University [62], the Personalized Medicine Research Project of the Marshfield Clinic [54], and BioVu at Vanderbilt [7]. A fitting starting point to a discussion of TM for PM in hypothesis generation using EHR text is the experimental design of a genome-wide association study (GWAS), which detects disease-causing genetic variants [99]. GWA studies are traditionally conducted by enrolling patients and then obtaining their phenotype and genotype through physical exam and gene sequencing; however, these studies can also be conducted using EHR-linked genetic data in conjunction with EHR phenotyping.

In a GWAS, researchers compare genes of people with a disease (the "cases") to genes of people without the disease (the "controls"). Gene variants that are found more commonly among cases than controls are evidence of an association between a variant and a disease if the difference reaches statistical significance. Because of the statistical hazards of multiple testing mentioned in the introduction to this section, significance thresholds in GWAS are often quite stringent [36]. Ritchie et al. demonstrated the feasibility of using EHR-linked genetic data in performing GWAS [69]. Text mining is important in EHR-based GWAS since accurately defining case and control phenotypes is a prerequisite to distinguishing genetic associations. For example, in an EHR-based GWAS regarding cardiac rhythm, Denny et al. employed text mining to detect negated concepts and family history from all physician-generated clinical documents. They linked this text-mined data with electrocardiogram data, billing codes, and labs to define phenotypes for cases and controls. The inclusion of text mining in these phenotype algorithms resulted in substantial improvements in recall while maintaining a high precision. Ultimately, this approach identified a novel gene for an ion channel involved in cardiac conduction [15, 17].

Text mining-enabled GWAS using EHR-linked genetic data are a cost-efficient and flexible avenue of discovery because EHRs can support exploration of an incredibly dynamic array of phenotypes. For example, Kullo et al. used NLP of clinical notes and medication lists to identify case and control phenotypes for an EHR-based GWAS that employed genetic data from a previous study about one phenotype (peripheral artery disease) to perform an EHR-based GWAS about a completely separate phenotype (red blood cell traits) and identified a new variant in a gene previously unknown to be related to RBC function while also successfully replicating results of previous dedicated GWAS about RBC function [40]. Although the biases inherent in EHR data limit the reliability of these findings, this in silico method of discovery demonstrates the utility of text mining in EHR notes to generate hypotheses through GWAS.

One limitation of GWA studies is that the selection of cases and controls permits investigation of only one phenotype at a time and prevents the detection of gene variants that might predispose to multiple diseases. For example, it took two separate studies performed at different times to demonstrate that variants in the *FTO* gene predispose to both diabetes and to obesity. Text mining EHRs can enable discovery of such gene-disease relationships through an experimental modality called a phenome-wide association study (PheWAS), which is essentially the reverse design of a GWAS. In a PheWAS, the cases and controls are people with or without a specific gene variant that is suspected of causing disease. Comparison of a broad array of hundreds of disease phenotypes experienced by people with and without the variant allows the detection of multiple gene-disease associations and suggests etiologic relationships between disease types. The first PheWAS used only billing code data to define phenotypes, but subsequent studies have shown that using text mining of clinical notes in addition to billing data improves the significance of results [15].

PheWAS are a powerful hypothesis-generating application of TM for PM. Moore et al., noting that PheWAS enable discovery of multiple phenotypes associated with single genes, used clinical trial data from the AIDS Clinical Trials Group (ACTG) to explore phenotypes related to drug adverse effects in AIDS therapies. Their first published work established baseline associations between clinical measurements and patient genotypes, replicating 20 known associations and identifying several that were novel in HIV-positive cohorts [57]. In other areas of medicine, Rzhetsky et al. used a PheWAS approach to hypothesize a relationship between autism, bipolar, and schizophrenia [72]. Likewise, Shameer et al. performed PheWAS and demonstrated that gene variants that affect characteristics of platelets also have an association with heart attack and autoimmune diseases [79]. Each of these findings identifies a potential avenue for pharmacogenomic therapy and demonstrates the potential of text mining EHRs for hypothesis generation.

7.5 TM for PM Conclusion: Value in Healthcare

One final concept that merits discussion is that of value. Value in healthcare is defined as health outcomes achieved per dollar spent [66]. Every medical intervention, including precision medicine and TM for PM, should be weighed in terms of this framework. How much will precision medicine benefit patients and at what cost?

In many circumstances, value actually opposes the implementation of precision medicine. For many conditions, increasing the granularity with which we understand our patients may result in benefits, but those benefits may be so slight that the cost of obtaining them renders the technology valueless [71]. In some settings, the current therapies or diagnostics may be so effective and inexpensive that the costs of PM will not merit the marginal gains. Alternatively, even if PM does greatly enhance diagnosis and prediction of disease, if no effective therapies exist for that disease, the overall value of PM will be reduced [67]. It is difficult to predict in the early stages of adoption of PM which diseases and therapies will benefit from PM and which will not.

Value is also the tantalizing target of precision medicine. In the 2016 Precision Medicine Initiative Summit, which took place one year after the announcement of the Precision Medicine Initiative, US President Barack Obama reviewed the status of the initiative and asserted the potential of precision medicine to produce efficient and cost-effective healthcare [68]. Many factors support this assertion. Regarding the numerator of the value equation (healthcare outcomes), it is likely that precision medicine will indeed increase prevention of many diseases and improve therapeutic options for diseases that are detected. Regarding the denominator (cost), two factors may lower the relative costs and favor its adoption: (1) human DNA is largely unvarying (with the exception of cancer) within a single individual throughout the lifespan. Therefore, although genetic sequence analysis may be initially expensive compared to other diagnostic tests, as our understanding of the role of genes in health and disease increases, the repeated utility of sequence data will lower the comparative cost. (2) Data in electronic health records are already widely collected and stored, so use of this data should require only minimal expense.

Text mining is a vehicle to obtain increased utility from existing information resources, and it offers several advantages in the precision medicine value equation. Mining biomedical literature, for example, can help streamline curation and can improve research efficiency through hypothesis generation. Likewise, mining EHR text facilitates the use of this underutilized source of important patient phenotype information and enables a host of useful applications. As far as TM for PM can demonstrate increased value, its merit and ultimate adoption into mainstream medicine is assured.

Acknowledgments This research was supported by the NIH Intramural Research Program, National Library of Medicine, and the NIH Medical Research Scholars Program, a public-private partnership supported jointly by the NIH and generous contributions to the Foundation for the NIH from the Doris Duke Charitable Foundation, the Howard Hughes Medical Institute, the American Association for Dental Research, the Colgate-Palmolive Company, and other private donors. No funds from the Doris Duke Charitable Foundation were used to support research that used animals.

References

1. Arighi CN, Wu CH, Cohen KB, et al. BioCreative-IV virtual issue. Database. 2014. doi:10.1093/database/bau039.
2. Aronson AR. Effective mapping of biomedical text to the UMLS Metathesaurus: the MetaMap program. Proc AMIA Symp. 2001;17–21.

3. Baumgartner Jr WA, Cohen KB, Fox LM, et al. Manual curation is not sufficient for annotation of genomic databases. Bioinformatics. 2007;23:i41–8.
4. Bejan CA, Xia F, Vanderwende L, et al. Pneumonia identification using statistical feature selection. J Am Med Inform Assoc. 2012;19:817–23.
5. Berty HL, Simon M, Chapman BE. A semi-automated quantification of pulmonary artery dimensions in computed tomography angiography images. AMIA Annu Symp Proc. 2012;2012:36–42.
6. Bishop CM. Pattern recognition and machine learning. New York: Springer; 2006.
7. Bowton EA, Collier SP, Wang X, et al. Phenotype-driven plasma biobanking strategies and methods. J Pers Med. 2015;5:140–52.
8. Brookes AJ, Robinson PN. Human genotype-phenotype databases: aims, challenges and opportunities. Nat Rev Genet. 2015;16:702–15.
9. Carroll RJ, Eyler AE, Denny JC. Naïve Electronic Health Record phenotype identification for Rheumatoid arthritis. AMIA Annu Symp Proc. 2011;2011:189–96.
10. CMS.gov – EHR Overview. 2012.
11. Cole TS, Frankovich J, Iyer S, et al. Profiling risk factors for chronic uveitis in juvenile idiopathic arthritis: a new model for EHR-based research. Pediatr Rheumatol Online J. 2013;11:45.
12. Collins FS, Varmus H. A new initiative on precision medicine. N Engl J Med. 2015;372:793–5.
13. D'Avolio L, Ferguson R, Goryachev S, et al. Implementation of the Department of Veterans Affairs' first point-of-care clinical trial. J Am Med Inform Assoc. 2012;19:e170–6.
14. Davis AP, Wiegers TC, Johnson RJ, et al. Text mining effectively scores and ranks the literature for improving chemical-gene-disease curation at the comparative toxicogenomics database. PLoS One. 2013;8:e58201.
15. Denny JC. Chapter 13: mining electronic health records in the genomics era. PLoS Comput Biol. 2012;8:e1002823.
16. Denny JC, Peterson JF, Choma NN, et al. Extracting timing and status descriptors for colonoscopy testing from electronic medical records. J Am Med Inform Assoc. 2010;17:383–8.
17. Denny JC, Ritchie MD, Crawford DC, et al. Identification of genomic predictors of atrioventricular conduction: using electronic medical records as a tool for genome science. Circulation. 2010;122:2016–21.
18. Doughty E, Kertesz-Farkas A, Bodenreider O, et al. Toward an automatic method for extracting cancer- and other disease-related point mutations from the biomedical literature. Bioinformatics. 2011;27:408–15.
19. Fang Y-C, Lai P-T, Dai H-J, Hsu W-L. MeInfoText 2.0: gene methylation and cancer relation extraction from biomedical literature. BMC Bioinf. 2011;12:471.
20. Frankovich J, Longhurst CA, Sutherland SM. Evidence-based medicine in the EMR era. N Engl J Med. 2011;365:1758–9.
21. Garraway LA, Verweij J, Ballman KV. Precision oncology: an overview. J Clin Oncol. 2013;31:1803–5.
22. Garten Y, Altman RB. Pharmspresso: a text mining tool for extraction of pharmacogenomic concepts and relationships from full text. BMC Bioinf. 2009;10 Suppl 2:S6.
23. Garvin JH, DuVall SL, South BR, et al. Automated extraction of ejection fraction for quality measurement using regular expressions in Unstructured Information Management Architecture (UIMA) for heart failure. J Am Med Inform Assoc. 2012;19:859–66.
24. Hahn U, Cohen KB, Garten Y, Shah NH. Mining the pharmacogenomics literature-a survey of the state of the art. Brief Bioinform. 2012;13:460–94.
25. Hanahan D, Weinberg RA. The hallmarks of cancer. Cell. 2000;100:57–70.
26. Hanahan D, Weinberg RA. Hallmarks of cancer: the next generation. Cell. 2011;144:646–74.
27. Harris RP, Helfand M, Woolf SH, et al. Current methods of the US Preventive Services Task Force: a review of the process. Am J Prev Med. 2001;20:21–35.

28. Hirschman L, Burns GAPC, Krallinger M, et al. Text mining for the biocuration workflow. Database. 2012;bas020.
29. Hripcsak G, Albers DJ. Next-generation phenotyping of electronic health records. J Am Med Inform Assoc. 2013;20:117–21.
30. Huang J. Performance measures of machine learning. University of Western Ontario, Ontario. 2006. ISBN: 978-0-494-30363-4.
31. Huang C-C, Lu Z. Community challenges in biomedical text mining over 10 years: success, failure and the future. Brief Bioinform. 2016;17:132–44.
32. Kawaler E, Cobian A, Peissig P, et al. Learning to predict post-hospitalization VTE risk from EHR data. AMIA Annu Symp Proc. 2012;2012:436–45.
33. Kawano Y. Diurnal blood pressure variation and related behavioral factors. Hypertens Res. 2011;34:281–5.
34. Kho AN, Hayes MG, Rasmussen-Torvik L, et al. Use of diverse electronic medical record systems to identify genetic risk for type 2 diabetes within a genome-wide association study. J Am Med Inform Assoc. 2012;19:212–8.
35. Kim D, Shin H, Song YS, Kim JH. Synergistic effect of different levels of genomic data for cancer clinical outcome prediction. J Biomed Inform. 2012;45:1191–8.
36. Klein RJ, Zeiss C, Chew EY, et al. Complement factor H polymorphism in age-related macular degeneration. Science. 2005;308:385–9.
37. Klompas M, Haney G, Church D, et al. Automated identification of acute hepatitis B using electronic medical record data to facilitate public health surveillance. PLoS One. 2008;3: e2626.
38. Kohane IS. Using electronic health records to drive discovery in disease genomics. Nat Rev Genet. 2011;12:417–28.
39. Köpcke F, Prokosch H-U. Employing computers for the recruitment into clinical trials: a comprehensive systematic review. J Med Internet Res. 2014;16:e161.
40. Kullo IJ, Ding K, Jouni H, et al. A genome-wide association study of red blood cell traits using the electronic medical record. PLoS One. 2010. doi:10.1371/journal.pone.0013011.
41. Leaman R, Islamaj Dogan R, Lu Z. DNorm: disease name normalization with pairwise learning to rank. Bioinformatics. 2013;29:2909–17.
42. Lee J, Maslove DM, Dubin JA. Personalized mortality prediction driven by electronic medical data and a patient similarity metric. PLoS One. 2015;10:e0127428.
43. Lehman L-W, Saeed M, Long W, et al. Risk stratification of ICU patients using topic models inferred from unstructured progress notes. AMIA Annu Symp Proc. 2012;2012:505–11.
44. Li J, Zhu X, Chen JY. Building disease-specific drug-protein connectivity maps from molecular interaction networks and PubMed abstracts. PLoS Comput Biol. 2009;5:e1000450.
45. Liu M, Wu Y, Chen Y, et al. Large-scale prediction of adverse drug reactions using chemical, biological, and phenotypic properties of drugs. J Am Med Inform Assoc. 2012;19:e28–35.
46. Long MT, Fox CS. The framingham heart study – 67 years of discovery in metabolic disease. Nat Rev Endocrinol. 2016. doi:10.1038/nrendo.2015.226.
47. Lowe HJ, Ferris TA, Hernandez PM, Weber SC. STRIDE-an integrated standards-based translational research informatics platform. AMIA Annu Symp Proc. 2009;2009:391–5.
48. Lu Z, Hirschman L. Biocuration workflows and text mining: overview of the BioCreative 2012 Workshop Track II. Database 2012:bas043.
49. Lung NH, Institute B, National centre for Biotechnology Information. Expert Panel Report 3 (EPR 3): guidelines for the diagnosis and management of asthma, National Institutes of Health 40. Bethesda: National centre for Biotechnology Information; 2007.
50. Mani S, Chen Y, Arlinghaus LR, et al. Early prediction of the response of breast tumors to neoadjuvant chemotherapy using quantitative MRI and machine learning. AMIA Annu Symp Proc. 2011;2011:868–77.
51. Mani S, Chen Y, Elasy T, et al. Type 2 diabetes risk forecasting from EMR data using machine learning. AMIA Annu Symp Proc. 2012;2012:606–15.

52. Maqungo M, Kaur M, Kwofie SK, et al. DDPC: dragon database of genes associated with prostate cancer. Nucleic Acids Res. 2011;39:D980–5.
53. Mathias JS, Gossett D, Baker DW. Use of electronic health record data to evaluate overuse of cervical cancer screening. J Am Med Inform Assoc. 2012;19:e96–101.
54. McCarty CA, Nair A, Austin DM, Giampietro PF. Informed consent and subject motivation to participate in a large, population-based genomics study: the Marshfield Clinic Personalized Medicine Research Project. Public Health Genomics. 2006;10:2–9.
55. McCowan IA, Moore DC, Nguyen AN, et al. Collection of cancer stage data by classifying free-text medical reports. J Am Med Inform Assoc. 2007;14:736–45.
56. Meystre SM, Savova GK, Kipper-Schuler KC, Hurdle JF. Extracting information from textual documents in the electronic health record: a review of recent research. Yearb Med Inform. 2008;47:128–44.
57. Moore CB, Verma A, Pendergrass S, et al. Phenome-wide association study relating pretreatment laboratory parameters with human genetic variants in AIDS clinical trials group protocols. Open Forum Infect Dis. 2015;2:ofu113.
58. Newton KM, Peissig PL, Kho AN, et al. Validation of electronic medical record-based phenotyping algorithms: results and lessons learned from the eMERGE network. J Am Med Inform Assoc. 2013;20:e147–54.
59. Nguyen AN, Lawley MJ, Hansen DP, et al. Symbolic rule-based classification of lung cancer stages from free-text pathology reports. J Am Med Inform Assoc. 2010;17:440–5.
60. Ni Y, Wright J, Perentesis J, et al. Increasing the efficiency of trial-patient matching: automated clinical trial eligibility pre-screening for pediatric oncology patients. BMC Med Inform Decis Mak. 2015;15:28.
61. Ongenaert M, Van Neste L, De Meyer T, et al. PubMeth: a cancer methylation database combining text-mining and expert annotation. Nucleic Acids Res. 2008;36:D842–6.
62. Ormond KE, Cirino AL, Helenowski IB, et al. Assessing the understanding of biobank participants. Am J Med Genet A. 2009;149A:188–98.
63. Percha B, Altman RB. Learning the structure of biomedical relationships from unstructured text. PLoS Comput Biol. 2015;11:e1004216.
64. Percha B, Garten Y, Altman RB. Discovery and explanation of drug-drug interactions via text mining. Biocomputing. 2012. World Scientific, pp 410–421.
65. Poos K, Smida J, Nathrath M, et al. Structuring osteosarcoma knowledge: an osteosarcoma-gene association database based on literature mining and manual annotation. Database. 2014. doi:10.1093/database/bau042.
66. Porter ME. What is value in health care? N Engl J Med. 2010;363:2477–81.
67. Prasad V, Fojo T, Brada M. Precision oncology: origins, optimism, and potential. Lancet Oncol. 2016;17:e81–6.
68. Remarks by the president in precision medicine panel discussion. In: whitehouse.gov. 2016. https://www.whitehouse.gov/the-press-office/2016/02/25/remarks-president-precision-medicine-panel-discussion. Accessed 2 Mar 2016.
69. Ritchie MD, Denny JC, Crawford DC, et al. Robust replication of genotype-phenotype associations across multiple diseases in an electronic medical record. Am J Hum Genet. 2010;86:560–72.
70. Rosano G, Pelliccia F, Gaudio C, Coats AJ. The challenge of performing effective medical research in the era of healthcare data protection. Int J Cardiol. 2014;177:510–1.
71. Rubin R. Precision medicine: the future or simply politics? JAMA. 2015;313:1089–91.
72. Rzhetsky A, Wajngurt D, Park N, Zheng T. Probing genetic overlap among complex human phenotypes. Proc Natl Acad Sci U S A. 2007;104:11694–9.
73. Savova GK, Masanz JJ, Ogren PV, et al. Mayo clinical Text Analysis and Knowledge Extraction System (cTAKES): architecture, component evaluation and applications. J Am Med Inform Assoc. 2010;17:507–13.

74. Schmiedeskamp M, Harpe S, Polk R, et al. Use of international classification of diseases, ninth revision, clinical modification codes and medication use data to identify nosocomial Clostridium difficile infection. Infect Control Hosp Epidemiol. 2009;30:1070–6.
75. Schneeweiss S. Learning from big health care data. N Engl J Med. 2014;370:2161–3.
76. Schwaederle M, Zhao M, Lee JJ, et al. Impact of precision medicine in diverse cancers: a meta-analysis of phase II clinical trials. J Clin Oncol. 2015;33:3817–25.
77. Sesen MB, Kadir T, Alcantara R-B, et al. Survival prediction and treatment recommendation with Bayesian techniques in lung cancer. AMIA Annu Symp Proc. 2012;2012:838–47.
78. Sessler DI, Imrey PB. Clinical research methodology 2: observational clinical research. Anesth Analg. 2015;121:1043–51.
79. Shameer K, Denny JC, Ding K, et al. A genome- and phenome-wide association study to identify genetic variants influencing platelet count and volume and their pleiotropic effects. Hum Genet. 2014;133:95–109.
80. Shivade C, Raghavan P, Fosler-Lussier E, et al. A review of approaches to identifying patient phenotype cohorts using electronic health records. J Am Med Inform Assoc. 2014;21:221–30.
81. Shoenbill K, Fost N, Tachinardi U, Mendonca EA. Genetic data and electronic health records: a discussion of ethical, logistical and technological considerations. J Am Med Inform Assoc. 2014;21:171–80.
82. Singhal A, Simmons M, Lu Z. Text mining for precision medicine: automating disease mutation relationship extraction from biomedical literature. J Am Med Inform Assoc. 2016;23(4):766–772.
83. Singhal A, Simmons M, Lu Z. Text mining genotype-phenotype relationships from biomedical literature for database curation and precision medicine. PLOS Comput Biol. 2016 (In Press).
84. Sohn S, Savova GK. Mayo clinic smoking status classification system: extensions and improvements. AMIA Annu Symp Proc. 2009;2009:619–23.
85. Sohn S, Kocher J-PA, Chute CG, Savova GK. Drug side effect extraction from clinical narratives of psychiatry and psychology patients. J Am Med Inform Assoc. 2011;18 Suppl 1: i144–9.
86. Sun J, McNaughton CD, Zhang P, et al. Predicting changes in hypertension control using electronic health records from a chronic disease management program. J Am Med Inform Assoc. 2014;21:337–44.
87. Swanson DR. Fish oil, Raynaud's syndrome, and undiscovered public knowledge. Perspect Biol Med. 1986;30:7–18.
88. Swanson DR. Migraine and magnesium: eleven neglected connections. Perspect Biol Med. 1988;31:526–57.
89. Swanson DR. Medical literature as a potential source of new knowledge. Bull Med Libr Assoc. 1990;78:29–37.
90. Swede H, Stone CL, Norwood AR. National population-based biobanks for genetic research. Genet Med. 2007;9:141–9.
91. Tari L, Anwar S, Liang S, et al. Synthesis of pharmacokinetic pathways through knowledge acquisition and automated reasoning. Biocomputing. 2010. World Scientific address = year = 2012 edition=,year = 2012 edition =, pp 465–476.
92. Tatari F, Akbarzadeh-T M-R, Sabahi A. Fuzzy-probabilistic multi agent system for breast cancer risk assessment and insurance premium assignment. J Biomed Inform. 2012;45:1021–34.
93. The Phenotype KnowledgeBase | PheKB. https://phekb.org/. Accessed 1 Mar 2016.
94. Thorn CF, Klein TE, Altman RB. Pharmacogenomics and bioinformatics: PharmGKB. Pharmacogenomics. 2010;11:501–5.
95. Trick WE, Zagorski BM, Tokars JI, et al. Computer algorithms to detect bloodstream infections. Emerg Infect Dis. 2004;10:1612–20.
96. UniProt UniProt: Annotation guidelines.

97. Van den Bulcke T, Vanden Broucke P, Van Hoof V, et al. Data mining methods for classification of medium-chain acyl-CoA dehydrogenase deficiency (MCADD) using non-derivatized tandem MS neonatal screening data. J Biomed Inform. 2011;44:319–25.
98. Vickers AJ, Scardino PT. The clinically-integrated randomized trial: proposed novel method for conducting large trials at low cost. Trials. 2009;10:14.
99. Visscher PM, Brown MA, McCarthy MI, Yang J. Five years of GWAS discovery. Am J Hum Genet. 2012;90:7–24.
100. Wei C-H, Harris BR, Kao H-Y, Lu Z. tmVar: a text mining approach for extracting sequence variants in biomedical literature. Bioinformatics. 2013;29:1433–9.
101. Wei C-H, Kao H-Y, Lu Z. PubTator: a web-based text mining tool for assisting biocuration. Nucleic Acids Res. 2013;41:W518–22.
102. Wei C-H, Kao H-Y, Lu Z. GNormPlus: an integrative approach for tagging genes, gene families, and protein domains. Biomed Res Int. 2015;2015:918710.
103. Wei C-H, Peng Y, Leaman R, et al. Overview of the BioCreative V chemical disease relation (CDR) task. Proceedings of the fifth BioCreative challenge evaluation workshop, Sevilla, Spain. 2015b.
104. Wei W-Q, Teixeira PL, Mo H, et al. Combining billing codes, clinical notes, and medications from electronic health records provides superior phenotyping performance. J Am Med Inform Assoc. 2015. doi:10.1093/jamia/ocv130.
105. Wiegers TC, Davis AP, Mattingly CJ. Collaborative biocuration-text-mining development task for document prioritization for curation. Database. 2012;bas037.
106. Wu ST, Sohn S, Ravikumar KE, et al. Automated chart review for asthma cohort identification using natural language processing: an exploratory study. Ann Allergy Asthma Immunol. 2013;111:364–9.
107. Xie B, Ding Q, Han H, Wu D. miRCancer: a microRNA–cancer association database constructed by text mining on literature. Bioinformatics. 2013;29(5):638–44.
108. Yu S, Liao KP, Shaw SY, et al. Toward high-throughput phenotyping: unbiased automated feature extraction and selection from knowledge sources. J Am Med Inform Assoc. 2015;22:993–1000.
109. Zeng QT, Goryachev S, Weiss S, et al. Extracting principal diagnosis, co-morbidity and smoking status for asthma research: evaluation of a natural language processing system. BMC Med Inform Decis Mak. 2006;6:30.
110. Zhao D, Weng C. Combining PubMed knowledge and EHR data to develop a weighted bayesian network for pancreatic cancer prediction. J Biomed Inform. 2011;44:859–68.

Chapter 8
Medical Imaging Informatics

William Hsu, Suzie El-Saden, and Ricky K. Taira

Abstract Imaging is one of the most important sources of clinically observable evidence that provides broad coverage, can provide insight on low-level scale properties, is noninvasive, has few side effects, and can be performed frequently. Thus, imaging data provides a viable observable that can facilitate the instantiation of a theoretical understanding of a disease for a particular patient context by connecting imaging findings to other biologic parameters in the model (e.g., genetic, molecular, symptoms, and patient survival). These connections can help inform their possible states and/or provide further coherent evidence. The field of radiomics is particularly dedicated to this task and seeks to extract quantifiable measures wherever possible. Example properties of investigation include genotype characterization, histopathology parameters, metabolite concentrations, vascular proliferation, necrosis, cellularity, and oxygenation. Important issues within the field include: signal calibration, spatial calibration, preprocessing methods (e.g., noise suppression, motion correction, and field bias correction), segmentation of target anatomic/pathologic entities, extraction of computed features, and inferencing methods connecting imaging features to biological states.

Keywords Radiomics • Radiogenomics • Magnetic resonance imaging • Glioblastoma multiforme • Quantitative imaging • Imaging standards • Imaging informatics

8.1 Introduction

Diagnostic imaging represents an important component of precision medicine. Clinically, it constitutes a frequent, noninvasive, longitudinal, in vivo approach to gathering objective patient evidence related to a patient's condition. Imaging can capture structural, compositional, and functional information across multiple scales of evidence, including manifestations of disease processes at the molecular, genetic, cellular, tissue, and organ levels [56]. It plays a central role in disease

W. Hsu, Ph.D • S. El-Saden • R.K. Taira (✉)
Department of Radiological Sciences, University of California, Los Angeles, CA, USA
e-mail: whsu@mednet.ucla.edu; rtaira@ucla.edu

B. Shen et al. (eds.), *Translational Biomedical Informatics*, Advances in Experimental Medicine and Biology 939, DOI 10.1007/978-981-10-1503-8_8

screening, disease detection, disease assessment, treatment planning, and prognostic assessment. *Imaging informatics* specifically deals with optimizing those clinical decisions that could be rationalized via radiological, pathological, and/or ophthalmological evidence. As such, it must work in close association with a more comprehensive systems view of disease in order to maximize its clinical benefits.

Using the term *imaging* to indicate a subspecialty of informatics might seem a bit odd. Indeed, imaging informatics shares common methodologies with other subdisciplines of medical informatics. Due to the specific challenges related to the storage, management, distribution, processing, and visualization of these voluminous, high-dimensional datasets, imaging informatics had developed into its own subspecialty [8, 110]. Today, the goals of the field have expanded to include optimizing the use of imaging data across the entire process of patient care. Routine radiological data can now bridge evidence variables from multiple scales, thereby helping to substantiate clinical theories of a patient's diseased condition. Imaging can provide a whole-body perspective of a diseased state such as metastasis and can also assist in detecting tumoral mutations, providing important clues as to the evolving heterogeneity of a patient's cancer. In this chapter, we explore the evolving role of imaging informatics with respect to precision medicine, touching on current views, open computational problems, and developmental approaches. We explore the field from a number of perspectives including theoretical aspects, experimental issues, engineering/computational concerns, community goals, and clinical/patient concerns. We will limit the discussion to radiology imaging, with emphasis on magnetic resonance imaging. Examples will be mainly drawn from the clinical area of oncology due to the concentration of applied work in the field. In this chapter we attempt to crystallize the diversity of tasks using the running example of treatment planning and management for a patient with glioblastoma multiforme (GBM), a type of malignant brain cancer.

8.2 The Big Picture: Imaging and Precision Medicine

There has been a tremendous amount of research and development in the advancement of imaging methods for the human body. Specifically, considerable research has been directed to developing imaging biomarkers, defined as "... anatomic, physiologic, biochemical, or molecular parameters detectable with imaging methods used to establish the presence or severity of disease which offers the prospect of improved early medical product development and pre-clinical testing" [191]. Yet the full utility of image data is not realized, with prevailing interpretation methods almost entirely relying on the conventional subjective analysis of grayscale images. The interdisciplinary field of imaging informatics addresses many issues that currently prevent the systematic, scientific understanding of radiological evidence and how this imaging evidence can inform a biologically inspired model of a disease to improve medical decisions and predictions. The field attempts to operationalize fragments of knowledge from medical imaging, clinical medicine, genomics, systems biology, and cognitive sciences. Briefly, imaging informatics intersects a number of diverse disciplines including:

Medical Imaging Physics An understanding of how exactly a pixel value is grounded to physical properties of the patient is nebulous even to most radiologists. To optimize downstream analysis methods, both the semantics of the imaging data and the variability due to technical/noise factors need to be considered as part of the interpretive process.

Clinical Medicine Imaging examinations are motivated to answer a clinical question. Thus, mappings between what are the optimal imaging protocols (acquisition and analysis methods) needed to best answer a clinical question are crucial. An intelligent order entry system would anticipate relevant propositional entailments of the clinical question in order to rationalize what particular imaging methods should be considered. This factoring of the clinical question could also be utilized to provide a targeted comprehensive report based on the motivations of the study.

Genomics and Systems Medicine It has been observed that patients presenting with similar clinical phenotypes, provided with similar treatments, can have vastly different therapeutic responses and prognostic outcomes [65, 84, 134]. High-throughput technology has shown evidence that the underlying gene expression can differentiate subgroups of tumors, adding to tissue classification efforts in pathology and imaging [211, 216, 222]. Thus, bidirectional association studies to/from low-level biological parameters from/to observable imaging evidence features are an important line of research. Imaging informatics efforts include creating knowledge bases and appropriate representations for these associations such that various queries related to clinical, research, and statistical applications can be performed. Currently, radiomics studies are geared toward the expansion of quantitative imaging phenotypes and the integration of molecular high-throughput data [37, 76]. Together, the intersection of these studies produces the budding field of radiogenomics.

Molecular Biology Imaging scientists and molecular biologists are developing imaging probes (i.e., biomarkers) that are used to chemically bind to biological targets causing imaging signal intensity changes in the area of interest. A number of important applications of this technique have demonstrated its great promise, including characterization of gene expression and detection of molecular properties associated with pre-disease states [73, 220].

Computation Transforming imaging data into a useable form to answer high-level clinical questions is complex, likely involving a number of fragmented models spanning variables over multiple domains. Storage, representation, computational complexity, and computing speed are all critical issues in dealing with this high-dimensional feature space. Issues at this level also include: how to best integrate knowledge reported in the primary literature for clinical trials and/or observational studies, how to address the issue of reproducibility in the context of a state space that is extremely large, and how to address issues of sparse training sets that may be noisy in and of themselves.

Cognitive Science Cognitive frameworks are needed in order to accurately represent and synthesize fragments of knowledge for the purpose of intelligent reasoning about a given clinical problem. A rich ontologic representation of concepts, frames, and processes needs to be maintained so that it can be easily extended and include the necessary knowledge and details to support various precision medicine type queries. For example, the extension of ontologies includes those entities that exist purely in the imaging (e.g., shadows, edges, textures) or computational worlds. The Quantitative Imaging Biomarker Ontology (QIBO) is an example of such a development. In addition, causal and probabilistic inferencing frameworks in large state spaces need to be developed in order to support imaging-based precision medicine queries.

Psychophysics Imaging data from an acquisition device produces a matrix of numerical data. This numerical information is traditionally transformed into some human perceptual form such as brightness and hue. Radiologists have developed pattern-matching skills to targeted findings based on these spatial-temporal light signal patterns. With complex imaging and analysis methods, imaging data can be mapped to a large number of variations; furthermore, these variations can be visualized in a number of forms. Thus, methods for summarizing, highlighting, compressing, and organizing this complex data such that important patterns can be easily perceived are an important research area in imaging.

Operations Research Finally, the field of imaging informatics must consider umbrella issues that bring all these components together into an application for assisting physicians in the interpretation of imaging information. Thus, various end-to-end operational issues are critical for deployment. Issues in this regard include: the role of the patient; workflow issues given that the imaging consultation process may be complicated by advanced acquisition, processing, and reporting steps; quality assurance; payment policies for computationally intensive consultations; and coordination of goals and community group efforts in order to accelerate/stimulate global developments.

Figure 8.1 summarizes the main topics of discussion in this chapter including: the relationship between imaging informatics and precision medicine (Sect. 8.3), the nature of imaging signals and their representation (Sect. 8.4), the compilation of cases for observational research (Sect. 8.5.3.1), the standardization and preprocessing requirements (Sect. 8.5.3.2), the computation of imaging features (Sect. 8.5.3.4), the model building (Sect. 8.5.3.5), the validation of models (Sect. 8.5.3.6), and the clinical implementation issues (Sect. 8.6).

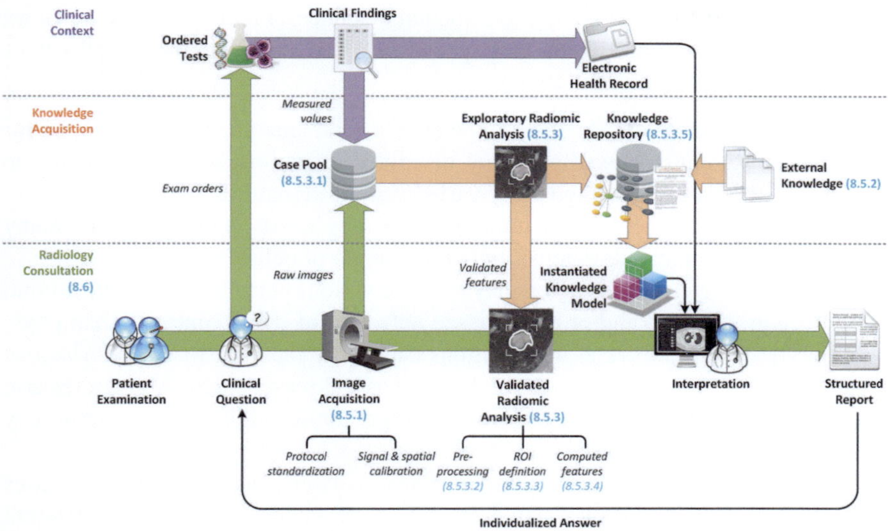

Fig. 8.1 A high-level framework for radiomics research and application. The *blue labels* in *parenthesis* are references to descriptions in the corresponding sections in the text

8.3 Role of Imaging in Instantiating Systems Models of Disease

Diseases are complex phenomena of nature with etiologies and presentations that vary tremendously among different individuals. Often, our first notion of how we perceive a disease is via its name (e.g., breast cancer, non-small cell carcinoma of the lung, or glioblastoma multiforme). This label primes a physician's cognitive system with respect to the patient's prognosis, the expected disease course, and a general bias toward how to treat and monitor the disease. However, current classification schemes for many diseases are far from complete and have marginalized this definition over what actually are a number of distinct "subtypes" of a disease [152]. Thus, most classifications and staging systems are underspecified and based only on partial constraints such as location, histology, or morphological/structural characteristics [182].

Current research in genetics and systems biology, for example, has revealed that breast cancer is better classified into five distinct subtypes [77]; glioblastoma multiforme can be stratified into four distinct subtypes [211]; and the disease we label as "autism" may in fact be on the order of 500 different genetic subtypes [181]. The implications, and indeed the driving motivation, of identifying a patient's precise subtype are many, most important of which include matching the most appropriate targeted therapies for the given subtype based on a theoretical mechanistic model supported by experimental evidence.[1] Precision medicine views of cancer must be adopted by imaging specialists. An overview of the complexity of this situation involves the following:

1. A patient's disease may be distributed over several anatomical sites. Tumors for a cancer patient, for example, may be rooted at various locations within the patient's body, each with possibly varied genetic signatures.
2. Within each diseased site, a tumoral mass may itself be composed of many genetically distinct subregions (mutant colonies) of cells.
3. Each distinct genetic subregion may result in various perturbations of the normal molecular, cellular, and/or organ/tissue networks of our complex human physiology [52, 78, 226]. The malfunctions caused by these network perturbations are then the basis for what we might view as a diseased state. A given disease may have many associated perturbed configurations, adding to the complexity for a disease description.
4. The expressed phenotypes caused by these network perturbations at various biological scales can vary based on a host of factors operating at different biological levels and networks including the states of regulatory systems, epigenetic regulation, cellular microenvironments, bacteriome, diet, etc. Different edge anomalies within a cellular network caused by a given mutant gene set could result in different phenotypes that may be percolated up and observed differently at various biological scales of organization [59]. The abilities of a patient's autoregulation system toward dealing with such perturbed network states can influence disease incidence and manifestation.
5. Furthermore, there are scale-level entanglement interactions. Higher-scale phenotypes (e.g., tumor level local resources) can trigger low-level genetic mutations (i.e., a downward causation effect). Thus, the monitoring of phenotypes can be an important factor in predicting probabilities of tumor cell mutations [3].

What Is the Role Then of Imaging Informatics in the Overall Goals of Precision Medicine? Imaging is one of the most important sources of objective observable evidence that provides broad coverage, can provide insight on low-level scale properties, is noninvasive, has few side effects, and can be performed frequently. Multiple magnetic resonance imaging and x-ray computed tomography studies, for example, are routinely performed on virtually all cancer patients. Thus, imaging data provides a viable observable within an evidence-based precision medicine model. Other variables in such a model (e.g., biological parameters and their joint complex states), which may be vital for therapeutic decisions, may not be practically measurable or cannot be practically monitored through time. For example, it is not possible to obtain a large number of biopsy samples that cover all suspect areas of disease at frequent intervals due to patient discomfort among other

[1]"Endotypes" are subtypes in which an underlying causal mechanism has been identified (see [24]).

"Verotypes" are used to refer to the true population of similar patients for treatment purposes.

factors. Therefore, imaging information facilitates the instantiation of a theoretical understanding of a disease for a particular patient context by connecting imaging findings to other biologic parameters in the model (e.g., genetic, molecular, symptoms, and patient survival). These connections can help inform their possible states and/or provide further coherent evidence.

8.4 Imaging Data in a Nutshell

Modern scanners, from CT, MR, PET, ultrasound, nuclear medicine, etc. gather raw analog signals generated from low-level phenomena (e.g., atomic/nuclear interactions) [34], which are detected, electronically processed, digitized, and coded in some relatively "black-box" fashion with respect to most clinicians, including radiologists. Thus, before one makes interpretations of the imaging data, one must understand the semantics and properties of images. It should be noted that the properties of an "image" can change in different settings. For example, in silico properties (numerical representation) differ from the properties of images when they are displayed on a computer monitor (e.g., physical brightness and contrast properties). Thus, before describing how imaging data can be linked to other multiscale biological parameters of a precision medicine model, we first present some important general details of what a medical image represents.

Any given image represents some warped view of reality. It includes noise, artifacts, and spatial distortions and probes the human body for signals in a very restricted manner. The basic unit of a digital image is a "pixel" (picture element) for 2D spatial images and a voxel (volume element) for 3D images. As the name suggests, a pixel exists only within the specialized (artificial) world of an image and can be viewed as the lowest level of abstraction of information in a digital image. In image grammars, a pixel is the minimum unit of image syntax (i.e., the basic building block for how an image is stitched together from a hierarchical organization of scene parts).

Space-Time-Energy The value of a pixel is some function of 2D space: $\text{pixel value} = f(x, y)$. A voxel is a function of 3D space $f(x, y, z)$. The quantity f may be regarded as some physical measureable data containing material/dynamic information about the object under study. It contains information for targeted signals as well as an undesirable noise component. A pixel/voxel value can also be further described as a function of space and time: $\text{voxel value} = f(x, y, z, t)$. Examples of how a voxel value can change with time include things flowing into and out of the voxel (e.g., flow studies), change in from physiological state (e.g., functional magnetic resonance imaging, fMRI) and imaging systems that accumulate counts over time (e.g., nuclear medicine studies, intravenous contrast). A pixel/voxel value can be characterized as some function of space-time-energy as well: $\text{voxel value} = f(x, y, z, t, E)$. We can represent "spectral" dependencies for the value of a pixel at a specified location and time. For instance, we often talk about multispectral acquisitions in magnetic resonance (MR), obtaining T1-weighted pre- and post-contrast, T2-weighted, and FLAIR (fluid-attenuated inverse recovery) sequences for a given patient at a given time. In MR spectroscopy, a voxel can represent a spectrum of targeted metabolite concentrations.

Mathematical Representation Because images are composed of values that vary in space, they are often thought of as fields – a function that varies with position. This abstraction of viewing images as fields brings to light various image processing methods based on the mathematics associated with fields. A pixel value $f(x, y, z)$ can be a *scalar* value, in which case, the image is seen as a scalar field. For example, in x-ray imaging, the primary signal is related to atomic attenuation characteristics of the patient so that the pixel scalar value is related to properties such as electron density, atomic number, and mass density. In routine magnetic resonance imaging, the scalar values are related to the nuclear spin-lattice and spin-spin interaction time constants associated with such physical properties as proton density, nuclear mobility, state of matter (e.g., liquid, solid, gas), molecular size, and local lattice vibration patterns. The intensity value of a digital image $f(x, y, z)$ may represent k-tuples of scalar intensity values across several spectral bands (i.e., it can span over an "energy" space). A pixel value may also be a *vector* quantity that includes a magnitude and direction. For example, a vector field may represent a dynamic contrast flow through a spatially positioned tissue sample as depicted on a coronary magnetic resonance angiography study [123]; it may represent a deformation field used to quantify how a patient's scan differs from a standardized anatomic atlas [124] or lung motion field during the respiratory cycle [83]. Finally, a pixel may be a tensor (more precisely, a tensor of order 2). Defined in 3D space, a tensor is characterized by a 3x3 matrix. Tensor fields are associated with images derived from some property of the imaged media that are anisotropic (i.e., associated with inhomogeneous spatially dependent properties). An example tensor field is the information obtained via diffusion tensor MRI (DTI) which measures Brownian motion of water molecules and is used in applications such as white matter tractography [11].

Spatial Discretization A pixel (and voxel) is a digital representation, being discretized in both its intensity value, f, and its location (x, y, z) within a mathematical lattice. Realistically, a pixel's value is an average over a small neighborhood of points centered about the point (x, y, z). Thus, pixels are not infinitesimal elements and can contain a large number of possible image signal carriers (e.g., in a one millimeter cubic volume, there are on the order of 10^{19} hydrogen nuclei within a typical tissue sample). A pixel ideally is a representation of a relatively homogeneous environment, and we assume that the property we wish to extract from a pixel is also homogenous in the region – but this assumption is often incorrect. There can be both physical inhomogeneities (e.g., chemical) and biological inhomogeneities (e.g., different cells, physiological dynamics), so a pixel value is often generated via a mixture of structures and processes. The degree of heterogeneity is related to the dimensions of the pixel (voxel) under scrutiny. When the spatial resolving power of an imaging system is poor compared to the dimensions of the effect being studied, partial volume effects arise. This is seen, for instance, in studies that attempt to characterize the fiber tracts and neuro micro-architecture of the brain using relatively large voxels.

8.5 Extending the Precision Medicine Knowledge Network for Imaging

Now that some of the very basics of imaging data have been explained, we turn our attention to how inferencing about biological states is performed. Some answers to basic questions need to be formulated including:

- What type of biological inferences can be made from which particular types of imaging studies?
- How strong is this correlation with respect to a particular biological (e.g., genomic) and/or clinical context?
- How can we operationalize this knowledge using mathematical models relating the imaging and clinical data for a given patient?

Thus, the path to building precision medicine decision support applications for imaging, as in other fields of informatics, follows more or less three perspectives of knowledge modeling:

Firstly, at a basic phenomenological level, the biophysicist/medical physicist investigates methods of improving and characterizing the means by which one can infer the chemical/biological state of a tissue sample from a measureable imaging signal (e.g., T1/T2 relaxation times). The focus here is on constructing the signal/image processing chain such that acquisition protocols can be tuned for target physical/biological environments, can be acquired with minimum harm to the patient, can provide required spatial coverage and resolution, and can be completed within the time constraints of a tolerable clinical procedure. Table 8.1 shows examples of various biological parameters associated with the disease GBM. Table 8.2 shows examples of imaging features and their correlation wtih varioius biological states. Section 8.5.1 provides further details.

Secondly, clinical researchers may then conduct a formal trial or observation study to examine the degree to which one class of patients with a known entity (diseased state) may be differentiated from a control group based on imaging findings. These studies help to determine the value of an imaging procedure for a particular patient group with respect to detection, diagnosis, safety, treatment planning, and/or treatment monitoring. The clinical hypothesis for the trial study can be motivated by causality connections (i.e., connection between a biological event/state and a physical imaging signal) or from clinical intuition of imaging patterns seen regularly for a patient class by a radiologist. Clinical trial studies are hypothesis driven, formalized by a deductive inference approach to testing. Section 8.5.2 provides additional details.

Table 8.1 A brief summary of some molecular relationships studied in glioblastoma multiforme

Source	Relationship	Target	References	Comment
+*AKT3*	Gene product is	↑AKT	Cancer Genome Atlas Research Network [35] and Koul [127]	*AKT3* is amplified 2% in GBM.
↑AKT	Effect is	↑Anti-apoptosis, ↓cell cycle regulation	Furnari et al. [68]	ATK decreases FOXO to bypass cell cycle check.
↑AKT	Modulates protein regulator of	↓p53	Furnari et al. [68]	AKT phosphorylates MDM2.
↑AKT	Modulates	↑HIF	Koul [127]	
↑ASCL1	Effect is	↑Stem cell abilities	Verhaak et al. [211], Rheinbay et al. [174], and Ohgaki and Kleihues [155]	ASCL1 inhibits DKK1.
↑CD44	Effect is	↑EMT	Verhaak et al. [211] and Ortensi et al. [156]	
+*CDK4 and 6*	Gene product affects	↓RB	Verhaak et al. [211] and Furnari et al. [68]	This causes ~15% RB inactivity in GBM.
-*CDKN2A*	Gene product affects	↓RB	Verhaak et al. [211], Furnari et al. [68], and Ohgaki and Kleihues [155]	*CDKN2A* mutants are in 50–70% high-grade gliomas.
-*CHD5*	Gene product affects	↓p53	Furnari et al. [68] and Ku et al. [129]	*CHD5* is a tumor suppressor.
↑CHI3L1	Modulates protein regulator of	↑MMP 2 and 9	Ku et al. [129]	
↑CHI3L1	Effect is	↑Adhesion	Ku et al. [129]	
+*EGFR*	Gene product is	↑EGFR	Verhaak et al. [211], Cancer Genome Atlas Research Network [35], Furnari et al. [68], and Ohgaki and Kleihues [155]	*EGFR* is amplified in ~40% GBM.
↑EGFR	Effect is	↑Growth stimulation	Verhaak et al. [211], Cancer Genome Atlas Research Network [35], Furnari et al. [68], and Ohgaki and Kleihues [155]	Mutant EGFR marks tumorigenic behavior, radiation resistance, and drug resistance.

(continued)

Table 8.1 (continued)

Source	Relationship	Target	References	Comment
↑EGFR	Modulates	↑HIF	Furnari et al. [68] and Ohgaki and Kleihues [155]	EGFR correlates with edema and angiogenesis.
↑EGFR	Modulates protein regulator of	↑AKT	Furnari et al. [68] and Ohgaki and Kleihues [155]	Activated EGFR leads to AKT.
↑GLI 1 and 2	Effect is	↑Stem cell abilities	Verhaak et al. [211] and Gupta et al. [86]	SMO, overexpressed in GMB, promotes GLI.
↑HIF	Modulates transcription of	↑VEGF	Furnari et al. [68]	
–*IDH 1* and *2*	Gene product is	↓IDH 1 and 2	Verhaak et al. [211] and Cohen et al. [41]	IDH mutation is an early tumorigenesis event.
↓IDH 1 and 2	Modulates protein regulator of	↑HIF	Cohen et al. [41]	IDH promotes HIF accumulation.
↓IDH 1 and 2	Cell affect is	↓Oxidative stress defense	Cohen et al. [41]	IDH regulates cell redox state.
+*MDM2*	Gene product affects	↓p53	Cancer Genome Atlas Research Network [35] and Furnari et al. [68])	*MDM2* is amplified in 14% of GBM.
+*MDM4*	Modulates transcription of	↓p53	Cancer Genome Atlas Research Network [35] and Furnari et al. [68]	MDM4 also enhances MDM2.
+*MET*	Gene product is	↑MET	Verhaak et al. [211], Cancer Genome Atlas Research Network [35], and Joo et al. [118]	*MET* is amplified in 4% of GBM.
↑MET	Effect is	↑Growth stimulation, ↑invasion, migration, motility	Joo et al. [118]	HGF (hepatocyte growth factor) is a ligand to MET.
-*MGMT*	Gene product is	↓MGMT	Ohgaki and Kleihues [155] and Hegi et al. [92]	*MGMT* promoter methylation occurs often in GBM.
-*MGMT*	Effect is	↓DNA repair	Ohgaki and Kleihues [155] and Hegi et al. [92]	MGMT removes mutagenic alkyl groups in DNA.
-*NF1*	Gene product affects	↑AKT	Verhaak et al. [211], McGillicuddy et al. [147], and Altomare and Testa [5])	NF1 activates Ras leading to ATK activation.

(continued)

Table 8.1 (continued)

Source	Relationship	Target	References	Comment
↑NF-kB	Effect is	↑Anti-apoptosis, ↑necrosis, ↑invasion, migration, motility	Verhaak et al. [211]	TNFR1 and RELB are overexpressed in GBM and promote NF-kB.
↑NOTCH	Effect is	↑Stem cell abilities	Verhaak et al. [211], Fan et al. [66], and Turchi et al. [210]	JAG1 and DDL3 expressions are altered in GBM to promote NOTCH.
↑OLIG2	Effect is	↓Cell cycle regulation	Verhaak et al. [211] and LeCun et al. [137]	
-p53	Gene product is	↓p53	Verhaak et al. [211], Furnari et al. [68], and Ohgaki and Kleihues [155]	*p53* mutation is the highest in GBM.
↓p53	Modulates transcription of	↓PTEN	Furnari et al. [68] and Ohgaki and Kleihues [155]	Normally, p53 enhances PTEN transcription.
↓p53	Effect is	↓Cell cycle regulation, ↑anti-apoptosis	Furnari et al. [68] and Ohgaki and Kleihues [155]	p53 regulates > 2500 genes as key tumor suppressor by balancing cell growth and death.
↓p53	Modulates transcription of	↑p110α	Furnari et al. [68]	
↑p110α	Effect is	↑Growth stimulation	Furnari et al. [68] and Weber et al. [215]	p110α is a part of the PI3K mitogenic pathway and promotes ECM and adhesion protein expression.
+PIK3CA	Gene product is	↑p110α	Verhaak et al. [211] and Furnari et al. [68]	*PIK3CA* is an oncogene.
+PDFGRA	Gene product affects	↑PDGFR	Verhaak et al. [211] and Heldin [93]	
↑PDGFR	Effect is	↑Growth stimulation, ↑EMT, ↑angiogenesis	Heldin [93]	PDGFR balances between cell growth and death.
↑PDGFR	Modulates transcription of	↑SOX2	de la Rocha et al. [50]	PDGF is PDGFR's ligand.
-PTEN	Gene product is	↓PTEN	Verhaak et al. [211], Cancer Genome Atlas Research Network [35], and Furnari et al. [68]	PTEN mutation occurs in 36% of GBM.

(continued)

Table 8.1 (continued)

Source	Relationship	Target	References	Comment
↓PTEN	Effect is	↑Unstable genome, ↑angiogenesis, ↑growth stimulation	Furnari et al. [68] and Ohgaki and Kleihues [155]	PTEN suppresses anchorage-independent growth.
↓PTEN	Modulates protein regulator of	↑AKT	Furnari et al. [68] and Ohgaki and Kleihues [155]	PTEN inhibits the PI3K pathway, upstream of AKT.
-*RB1*	Gene product is	↓RB	Furnani et al. [68] and Verhaak et al. [211]	*RB1* loss transitions the cancer to intermediate-grade glioma.
↓RB	Effect is	↓Cell cycle regulation	Furnari et al. [68]	RB is a tumor suppressor.
↑VEGF	Effect is	↑Angiogenesis	Jackson et al. [113]	

Italicized symbols are genes and un-italicized symbols are proteins. Tables 8.1 and 8.2 complement Fig. 8.3. See Table 8.2 for abbreviations.

Thirdly, the clinical practitioner team requires an integrated model that can perform inductive inferencing on observational data. A data modeling team must synthesize a model based on fragments of knowledge from observed statistics, causality relations, decision options, and costs associated with false-positive or false-negative interpretations. The integrated model should include a comprehensive modeling of variables, relationships, and their strengths and should be able to guide decisions based on individual patient context. The model must be able to perform well for patient management decisions that involve complex patient cases since traditional rule-based guidelines are often overly simplistic [69]. Section 8.5.3 provides further discussions of this level of modeling.

8.5.1 Physical Characterization of Imaging Signals

Figure 8.2 shows the general idea regarding how imaging scientists make biological inferences from physical imaging signals (we will limit the discussion to x-ray and magnetic resonance imaging). To begin with, imaging signals from magnetic resonance and x-ray imaging arise inherently from low-level nuclear and atomic phenomena.[2] Properties such as electron density, atomic number, nuclear proton density, spin-spin relaxation time, spin-lattice relaxation time, and nuclear magnetic resonance (NMR) chemical shift are direct properties related to the physical process generating the imaging signals. Medical imaging physicists study the

[2]An exhaustive review of the physics of image signal acquisition is beyond the scope of this chapter. Suggested references include Bushberg [34], Curry et al. [45], and Smith and Webb [190].

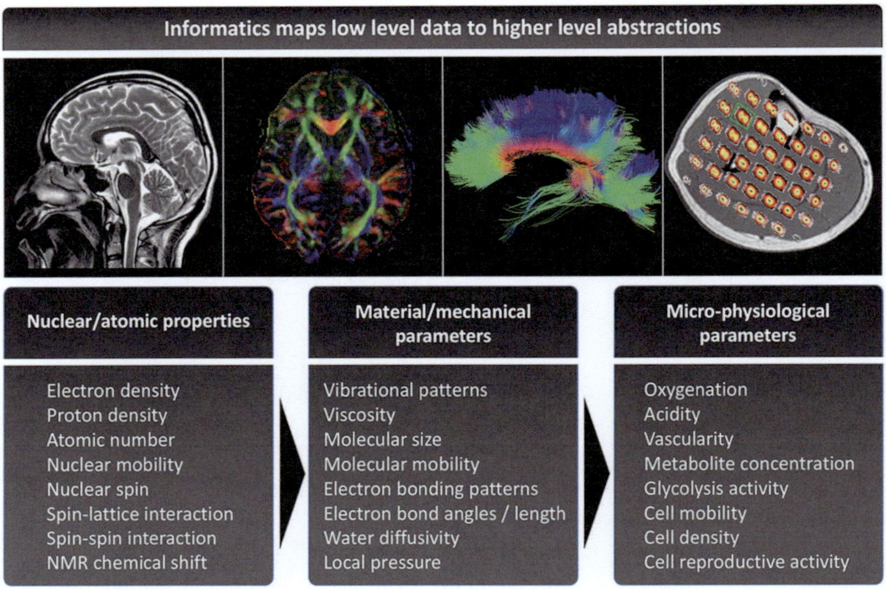

Fig. 8.2 Imaging scientists investigate the correlation between low-level properties and higher-level, more informative physical/biological states

relationship between these low-level properties to slightly higher-dimensional properties such as lattice vibration patterns, molecular mobility, molecular size, electron bonding patterns, and proton diffusivity. In MR imaging, magnetic spin nuclei in different physicochemical environments exhibit different signal patterns [114, 145]. For example, protons in a hydrocarbon-rich setting will have a significantly different MR signal as compared to an aqueous environment. The biophysicist may explore relationships to a still higher-scale state space, making connections with microphysiologic/structural patterns such as cell mobility, cell density, metabolite concentration, and oxygenation levels. The design of magnetic resonance contrast agents that (1) provide strong MR signals (T1/T2) and (2) high sensitivity and specificity to bind with molecular targets is an intense area of research [73, 220]. Paramagnetic metal cations such as chelated gadolinium or superparamagnetic nanoparticles have been used as contrast agents and have shown great potential for detecting physiological/molecular states over broad patient volumes. A variation of conventional MRI methods is diffusion-weighted MRI (DW-MRI) which exploits the translational mobility of water molecules to infer diffusivity properties associated with the tissue, microstructure, and/or pathological environment of the water molecules [15, 57]. Biological factors that can affect DW-MRI signals include diffusion within intracellular fluid (cytoplasm or organelles), extracellular fluid (e.g., interstitial, intravascular, lymphatic, and ventricular), and/or between intra- and extracellular compartments. Characterization of these biological properties can help characterize/inform various pathological conditions such as tumor cellularity, tumor heterogeneity, and neurofiber connectivity [13, 74, 126]. Perfusion-weighted MRI is another variation of MRI that relates to either blood delivery to a tissue element or capillary blood flow. These methods allow characterization of blood volume, blood velocity, and blood oxygenation [64] and have been applied to characterize early stroke and penumbra as well as vascularity of tumor. Yet another powerful variation of MRI is MR spectroscopy imaging which can depict metabolite concentrations (e.g., choline, NAA, and glutamine) that can be correlated with various types of genetic expressions known for given pathologies (specific examples are given below).

To optimize the information that can be derived from a clinical imaging study, the first step is to normalize the imaging data as best as possible. This can involve standardization of acquisition protocols, calibration of imaging signals, and/or calibration of spatial information. The motivation for this step is to reduce variability of image data with respect to a given clinical query, with respect to studies performed on a single patient, and/or with respect to a population of patients. Details of the various aspects of imaging standardization follow.

Protocol Standardization For a given clinical query, the standardization of imaging protocols is an important part of image data collection for clinical trial research and routine longitudinal patient assessment (e.g., for chronic disease, such as cancer) in order to reduce technical dependence on pixel signal values. The protocol may include specifications for acquisition parameters (e.g., kVp, scan time, pulse sequences, contrast agents, sampling parameters, patient positioning, and reconstruction algorithms). Specialized nonuniformity correction methods are often employed in nuclear medicine. In MR examinations that suffer from poor signal fluctuations and/or spatial distortions due to nonuniform magnetic fields, application of shimming operations can be used to correct field inhomogeneities on a patient-to-patient exam level [138, 213]. The use of standardized imaging protocols is important to the performance of quantitative image analysis algorithms, such as for object segmentation or tissue characterization, which may have appearance models based on pixel values. The disadvantage of standardizing imaging protocols lies in the difficulty of enforcing the acquisition method across time, scanners, and institutions.

Signal Calibration In this context, pixel values are mapped from a value that is technique dependent to one that is (relatively) technique independent. For example, MR images can have a wide range of intensities depending on a number of factors, such as magnetic field strength/quality, pulse sequence, and coil design. Even comparison of images acquired with the same weighted contrast (e.g., T1) is limited, as pixel intensities will depend on the specifics of the acquisition method. There are a number of methods for calibrating imaging signals including the use of phantoms [184], the physics-based models, and the re-representation of signals in terms of a signal responses with respect to a set of basis materials (e.g., dual-energy x-ray imaging). For some imaging modalities such as MR [55, 154, 183, 189], physics-based signal models are theoretically known so that actual physical parameters (e.g., T1 and T2) can be estimated from their measured raw signal data and their corresponding technique parameters. Compared to non-calibrated pixel intensities, calibrated MR signals of true T1 and T2 values reduce the variability of MR signal correlation to different types of tissue structures, making possible more accurate statistical characterization. The disadvantage of physics-based models lies in the difficulty of estimating analytic solutions from routine protocols that operate under real-world clinical constraints (e.g., short scan times, reasonable signal-to-noise ratios (SNR), sufficient spatial resolution and anatomic coverage). Additionally, the models are often simplified and ignore secondary signal generating sub-phenomena such as eddy currents and field nonuniformities.

Spatial Calibration Imaging data can also be spatially standardized in certain cases meaning the spatial coordinate (x, y, z) is mapped to a standardized anatomic coordinate system. This is routine in neuroimaging research [201]. Spatial calibration involves developing a prototype anatomic model (i.e., digital atlas) for a given patient cohort (e.g., sex-age group). A reference imaging exam (e.g., MR) for each group is obtained, and a number of in-class patient scans are then spatially registered to the reference scan using some warping algorithm. Registration methods can be roughly classified along the dimensions of being parametric *vs.* nonparametric, global versus local, and/or linear versus nonlinear approaches. The scoring function describing the quality of a transform includes mutual information, correlation ratio, cross-correlations, and sum of squares. Details of registration methods can be found in [148, 201]. Registration of imaging datasets is useful in analyzing various combinations of studies including:

- Same Patient, Different Study Type (Same Time): Registration of the imaging datasets can help to provide multiple independent pieces of evidence for a single spatial location. For example, PET (positron emission tomography) and MRI information can provide a view of the patient that combines functional and anatomical information.
- Same Patient, Same Study Type (Different Time): A fundamental task of radiologists is to monitor temporal changes in image appearance and to compare these findings to some baseline (e.g., a prior study).
- Different Patients, Same Disease: Probabilistic atlases compiled from patients belonging to healthy and/or diseased states are being developed to investigate statistical trends related to difference between anatomic morphology and imaging signal levels between comparison groups. Finally, spatial calibration of scans to a reference atlas can greatly assist in anatomic segmentation and feature extraction tasks [42].

8.5.2 Imaging Signals to Biological Parameters in a Disease Context

There have been a large number of clinical scientific studies related to whether there is a statistically significant detectable difference in imaging patterns of patients in a diseased state as compared to a control state. A wide range of hypotheses have been tested correlating imaging data to biological parameters including those that are causally proximate (e.g., imaging features to vascularity) as well as those that are causally distant (imaging features to survival). Preclinical testing of new imaging approaches need to be optimized and validated within a controlled clinical setting to objectively determine their impact on patient care. Clinical trial studies in imaging involve assessing various imaging protocols, contrast agents, kinetic modeling approaches, and other related imaging aspects in order to better characterize their performance with respect to understanding the biology and pathophysiology of disease. Molecular and functional imaging methods, for instance, are being investigated to study the effects of a given treatment and to study appropriate sampling frequencies and/or trigger events for anticipatory monitoring and thus be proactive in management strategies [70, 72].

An important aspect of imaging informatics then is to create knowledge bases documenting the results of such studies such that a global model of the relation between imaging and important biological parameters can be appreciated and used for scientific experiment planning, development of clinical guidelines, and assessment of quality of conclusions and serve as an evidence-based knowledge source for patient management. As in other areas of medical informatics, creating an ontologic representation of study results is crucial for the following reasons:

1. *Improved Assessment of Quality/Contribution of a Single Research Study*: A reader can be naively led to assume that a correlation is real and can be used in practice since it was published in a reputable journal. Ioannidis, however, comments that most research findings reported in the literature are not totally accurate in the conclusions they draw due to methodological faults related to experimental framework (e.g., follow-up confirmation studies), bias (e.g., selective reporting, conflicts of interest, faulty randomization), lack of independent teams, and lack of statistical power (low number of samples and large state spaces) [112]. Additionally, study propositions are reported at different levels of detail and different pathways to effects. The more general the claim, the more straightforward its application appears although possible patient-specific contextual features may be marginalized out. A proper modeling of the details of a particular investigative study can improve our ability to compare competitive, corroborative, and/or contradictory studies.
2. *Integrated View of Causal and Associative Relations Across Clinical and Biological Variables*: The global integrated model includes an inventory of variables, relationships, and their strengths that guide decisions as to the extent to which clinicians may generalize research findings for their own patient context. Physicians will often read a journal article and then base decisions on some evidence that is in fact only a partial piece of the picture [90]. In fact, a complete understanding of how to diagnose, treat, and/or manage a disease may require a more comprehensive understanding of the complex causal chain and interaction dynamics of a disease process. It is important as knowledge is accumulated regarding a particular disease to be able to synthesize the information and to see higher-order patterns of informatics that may reveal hidden complexities of the disease.

Thus, an important focus of imaging informatics is to provide a realistic objective description of the capabilities, contextual factors, limitations, and uncertainties associated with various imaging modalities, protocols, and processing methods. For example, although conventional MRI methods may be the most sensitive modality available for the detection of brain tumors, its specificity is low, meaning that other disease processes (e.g., primary versus secondary tumor versus abscess) may share a similar MRI appearance. And while sensitivity is considered high, studies such as those by Suzuki et al. showed that the limitations of current MR sensitivity to identify all brain areas that have been infiltrated with tumor cells are significant or vast [196]. Statistical significance measures based on, for example, *p*-values are controversial, given the many factors that are often ignored may have a great impact on its calculated value [80, 111]. All these limitations must be considered when applying such knowledge to the final interpretation of a particular patient imaging study. For more details related to general approaches to standardizing the representation of research studies, see Sim et al. [187] and Tong et al. [208]. One potential formalism is the adaptation of methods used by ResearchMaps, a graph-based representation of causal experimental results in neurobiology. Details of this paradigm can be found in [146, 186].

While it is not possible to provide a comprehensive description of all results of observational and controlled studies within medical imaging, several examples of the use of MRI for precision medicine applications follow with focus on glioblastoma multiforme (GBM).

GBM was the first cancer to be sequenced by The Cancer Genome Atlas (TCGA) research network and has been a focus area of radiogenomics. Verhaak, in 2010, described a molecular classification system elucidating four subtypes that were based on genetic expression. Using the prior naming classification scheme of *proneural*, *neural*, *classical*, and *mesenchymal* subtypes, four aberrations of genetic expression were correlated: the PDGFRA, IDH1, EGFR, and NF1 groups. Several aspects of the disease however make it particularly difficult to manage. Firstly, prognosis of the patient is highly dependent on both the proliferation rate and invasiveness of the tumor cells. GBMs are aggressive and may widely infiltrate the nervous system, while areas in the brain that have malignant involvement may actually still function normally. Diffuse invasion of the lesion can be significant by the time the lesion is detected and characterized. Extensive infiltration into surrounding brain tissue presents substantial challenges to achieving a desirable therapeutic index (comparison of dose of therapeutic agent required for desired effect versus dose that causes toxicity) and increases the probability of nontarget toxicity to vital brain tissues, thereby affecting the patient's quality of life. Secondly, monitoring subclinically diffuse tumor in the surrounding normal tissue is difficult with current MR imaging methods. Conventional MR technology cannot accurately demarcate the extent of invasion by tumor cells peripheral to the bulk tumor mass. Thirdly, clonal heterogeneity due to mutations within a tumor has been shown to have very pronounced effects on treatment efficacy [122]. The inclusion of even one secondary strain is sufficient to significantly alter the overall growth dynamics of a brain tumor. This results in varying degrees of sensitivity among a tumor cell population with respect to chemotherapy. Fourthly, brain tumors are unique among human cancers because of their complex interaction with the brain itself, which greatly complicates the use of existing therapies. Dose delivery is complicated by the blood–brain barrier. Accurate estimations of how much drug intervention to administer for an expected tumor site uptake are difficult and can be variable on a case-by-case basis. Table 8.1 summarizes a number of important precision medicine relationships for GBM.

Although current imaging methods are largely still in various stages of development, testing, and deployment, imaging protocols are rapidly improving in characterizing various biological states that can be highly informative in a precision medicine practice. Below are a few such associations.

Genotype Several studies have attempted to identify genotypic imaging characteristics, which may be useful in targeting therapeutic regimens. For instance, it has been shown that the allelic loss of 1p and 19q is associated with an indistinct border on T1 MRI images and mixed signal intensity on T1 and T2 [2, 117, 211]. A correlation of genotype in oligodendroglial tumors with MR imaging characteristics has also been identified, connecting longer survival and increased therapeutic responsiveness of certain tumor genotypes (e.g., −1p/−19q) [117]. The first large-scale MR imaging microRNA-mRNA correlative study in GBM was published by Zinn et al. in 2011 [230]. The use of novel contrast agents in the form of MRI reporter genes is an area of promising clinical potential and may soon become a routine method to monitor gene expression levels and/or monitor dynamic molecular interactions between proteins [75, 87].

Histopathology Dean et al. [51] investigated MR characteristics that could be used to classify supratentorial gliomas into low-grade astrocytoma, anaplastic astrocytoma, and GBM, finding that mass effect, cyst formation, and necrosis as seen on MR were statistically significant in distinguishing low- and high-grade tumors. Later studies also found heterogeneity of contrast enhancement, edema or mass effect, cyst formation/necrosis, and flow void to be significantly higher in GBM [10, 165]. Other suggested MR findings include: correlation between contrast enhancement and the volume of peritumoral edema in GBMs (*vs.* meningiomas which are benign intracranial tumors and typically enhance homogeneously and have no edema) [102, 151].

Metabolism Cellular functions are mediated through complex systems of macromolecules and metabolites linked through biochemical and physical interactions [226]. MR spectroscopy relies on an atomic/nuclear phenomenon known as chemical shift (resonance frequency shift) dependent on electron distribution properties (e.g., binding partners, bond lengths, bond angles) local to a nonzero spin nuclear isotopes (e.g., ^{1}H, ^{13}C, ^{15}N, ^{15}N, ^{23}Na, ^{31}P, ^{39}K) in order to infer certain metabolite (e.g., N-acetylaspartate, lactate, creatine/phosphocreatine, choline, glutamate, glutamine, glucose) concentrations in a localized voxel region. Thus, it can provide a coarse map of the spatial distribution of metabolites in an imaging volume. Various studies have shown that metabolite spectrums within a region can effectively distinguish pathological and normal tissues in vivo [85, 104, 150]. Cordova and colleagues reported a method for identifying infiltrating margins in GBM patients using whole-brain proton spectroscopic MRI [43]. They obtained surgical core samples from the periphery of the visible enhancing tumor prior to initial debulking surgery that demonstrated abnormal choline to N-acetylaspartate ratios, and these areas predated contrast-enhancing recurrent tumor. Such metabolic markers that identify infiltration and areas at risk for recurrence can guide initial surgery or postoperative adjuvant therapy if added to the preoperative MRI assessment of GBM patients.

Vascular Proliferation/Angiogenesis Tumor angiogenesis is a process whereby blood vessels are formed to supply nutrients and oxygen for tumor growth. As gliomas dedifferentiate into more malignant tumors, they express increasing levels of proangiogenic cytokines like vascular endothelial growth factor (VEGF) which promotes angiogenesis [71]. This angiogenic cytokine is responsible for proliferation of highly permeable tumor vessels that are missing the tight gap junctions found in normal cerebral blood vessels. These leaky tumor vessels have larger than normal diameters and are heterogeneously distributed within the tumor as it grows, providing ample flow to some portions of the tumor while providing inadequate blood flow to other areas of the tumor. The leakiness of these vessels allows intravenously delivered contrast agents to permeate into the interstitial spaces resulting in the contrast enhancement seen on T1W MRI images. It also allows for the weeping of serum plasma from the intravascular space into the interstitial spaces causing edema, detectable on MRI images as surrounding T2 hyperintensity. Thus, these two findings as seen on MRI images serve as surrogate measures for malignancy [119]. Somewhat ironically, the edema caused by these leaky vessels raises the interstitial pressure compromising the blood flow through the vessels and consequently compromising the blood supply to the tumor itself resulting in areas of necrosis. The anti-VEGF compounds are effective in part due to the inhibition of formation of the immature vessels and the normalization of tumor neovascularity such that they have more normal gap junctions and are not as weak or leaky. This improved vascular integrity reduces the incidence of regional necrosis and provides for better delivery of chemotherapeutic agents to all parts of the tumor. As might be expected, the therapy also results in less contrast enhancement and less edema, a condition known as pseudo-normalization, rendering the imaging biomarkers a bit less reliable in the absence of a good history of treatment [218].

Necrosis The presence of necrosis is one of the diagnostic hallmarks of GBM (see Fig. 8.4). Necrosis is likely related to the heterogeneous distribution of tumor vascularity with tumor dying in areas of inadequate blood supply. In 2004, Raza et al. looked at gene expression in 15 GBM patients and found 9 genes that positively correlated with necrosis and 17 that negatively correlated with necrosis [173]. In 2005, Pope and colleagues demonstrate a prognostic significance to the presence of necrosis on MRI, showing that patients with histologically proven grade 3 glioma (as opposed to GBM) containing visible necrosis on an imaging study had survival times similar to GBM patients [169]. Although necrosis is an inherent feature of GBM, it can also be manifested as a response to therapy. MR spectroscopy studies have also shown that a decrease in choline-containing compounds (Cho) and an increase in lactate (Lac) and/or lipids are indicative of response to therapy and reflect tumor necrosis [166]. Moreover, a total absence of metabolites in the former tumor region is indicative of necrotic tissue.

Cellularity There is increasing evidence to suggest that DWI and the calculated ADC values correlate with tissue cellularity [125]. Many cancers demonstrate restricted diffusion (or high signal intensity on DWI), which has classically been attributed to the increased cellular density seen within tumors, but other factors are likely to play a role, such as the tortuosity of the extracellular space, extracellular fibrosis, and the shape and size of the intercellular spaces [159]. In GBM patients, DWI/ADC values have been shown to differ within various regions of the tumor. There is restricted diffusion and lower ADC values in those portions most densely packed with cells and where there is high nuclear to cytoplasmic ratio. DWI/ADC values can also help to distinguish non-enhancing tumor (high cellular density) from peritumoral edema, both of which may appear bright on T2W MR images [206]. Higano et al. [98] showed a significant negative correlation between minimum MR apparent diffusion coefficient (ADC) values and the immunohistology Ki-67 labeling index. This trend translated into a lower mean minimum ADC value in tumors found in the patient group with progressive disease compared to a stable disease group.

Proteomics One of the earliest reports of correlation of imaging features with proteomics was written by Hobbs et al. in 2003 [101]. They looked at the gene expression of approximately 100 proteins in tumor samples of four patients with glioblastoma multiforme and compared the protein levels within contrast and non-contrast enhanced portions of tumor on MRI scans. Not only did they discover that protein levels were different between the two regions within any given patient, they also noted that protein concentrations were highly variable across the enhancing regions that had similar histology at light microscopy level. The non-enhancing portions of tumor were even more similar between patients. This paper was one of the first to associate genetic profiling with imaging features and to confirm the highly heterogeneous nature of GBM. In 2010, Barajas and colleagues compared gene expression in the enhancing central portions of GBMs to the non-enhancing periphery and found that the enhancing portions demonstrated higher concentration of vascular hyperplasia and cellular density and increased hypoxia which is thought to be the underlying mechanism behind necrosis and upregulation of VEGF and ultimately angiogenesis [16]. There was twice the expression of over 800 proteins within the enhancing core compared with the non-enhancing periphery.

Oxygenation Oxygen levels have an established role in regulating cell proliferation and motility [163]. Hypoxia, through the activation of transcription factors, induces cellular proliferation, tumor neovascularity, and necrosis. Hypoxic tumors become increasingly aggressive and are more resistant to chemotherapy and radiation therapy [7]. An experiment done in 2008 by Spence and colleagues on

22 GBM patients prior to undergoing radiotherapy showed that the volume and intensity of hypoxia within the tumor as determined on 18F fluoromisonidazole PET and blood venous sampling strongly correlated with poorer time to progression (TTP) and shorter overall survival [193].

Table 8.2 summarizes some of the important relationships investigated, connecting imaging findings to biological states. Figure 8.3 summarizes the relations using an association map to allow a more global visualization. Note that although Fig. 8.3 provides a succinct first-order summarization of the global picture of the disease, in practice more highly contextual factors should be considered in inferring multi-link pathways.

8.5.3 Imaging Data Analysis: Radiomics

With modern imaging methods, we are rapidly advancing the means of routinely probing into the structure, function, and pathology of the human body. Traditionally, imaging protocols were tailored for optimizing and/or highlighting certain types of signals present in collected data (e.g., T1/T2 recovery times, bone, soft tissue, etc.). Special weighting algorithms and/or filters were applied within the image acquisition computer, and images were then presented to the radiologist either on a static film and/or displayed on a digital monitor with limited processing capabilities. The value of the imaging data was then based on how well a radiologist could visualize and detect spatial-temporal-spectral light patterns depicted on a visualized image dataset and correlate these imaging patterns to various disease and normal states.

However, the rich field of medical image analysis has shown many examples in which a computer's ability to detect subtle mathematical patterns can exceed a human's ability to detect visual patterns. That is, information from modern scanners is more than just simply pictures for viewing by radiologists; rather they represent data that may be analyzed mathematically in various ways to infer greater levels of information related to the image generation process [76]. For example, a computer cannot only perform analysis at the signal voxel level, but the analysis of imaging data can also continue to higher-order image element structures ranging from small patches to ever-increasing coherent regions corresponding to semantically distinct areas of interest. Group properties associated with regional pixel characteristics (e.g., size, shape, signal intensity distribution) can be computed to infer still higher-scale systems biology properties (tumor size, growth rate, pressure, vascularity, genetic heterogeneity, *etc.*). It is thus becoming clear that there is a need to complement a radiologist's arsenal for image interpretation beyond visual detection of patterns on spatial brightness maps, expanding the image feature space to include those that can be identified by a computer algorithm.

These and other image interpretation motives fall into what has been termed *radiomics*, which is the study of quantifiable radiological image phenotypes [1, 76,

Table 8.2 A brief summary of associations found between imaging signals and cell or tissue states

Upstream event(s)	Downstream event	Imaging detection	References	Comment
↑Proliferation	↑Cellular density	DWI, T2	Huang et al. [110] and Barajas et al. [17]	T2 signal decreases, diffusion is restricted (also on ADC maps and DWI), and has a high nuclear to cytoplasmic ratio. H&E can indicate cellularity.
1. ↑Edema 2. ↑Poor architecture	↓Blood flow	ASL, DSC, DCE, BOLD, VASO	Huang et al. [110], Furnari et al. [68], Lacerda and Law [132], Heldin et al. [94], and Thompson et al. [202]	Poorly formed blood vessels are convoluted, slowing down blood flow. Increased hydrostatic gradient causes edema. Cell growth can compress vessels to reduce flow. Inflowing blood is seen on arterial spin labeling (ASL) perfusion imaging. Some imaging parameters that change include CBF, MTT, CBV, PH, RF, and PSR.
↑Vascular permeability	↓Blood–brain barrier	DCE, DSC, T1 + contrast	Huang et al. [110], Furnari et al. [68], Lacerda and Law [132], and Heye et al. [97]	Improper cell-cell adhesions lead to an improper blood–brain barrier detectable after contrast administration.
↑Abnormal metabolism	↓Creatine	MRS	Bulik et al. [33]	Tumor tissues have high metabolic demand, which is associated with a low creatine level.
↑Hypoxia → ↑necrosis → ↑death	↓Choline and ↓NAA	MRS	Bulik et al. [33]	Choline is a biomarker for cell membrane integrity and is lowered in cell death. NA is a marker of neuronal viability.

(continued)

Table 8.2 (continued)

Upstream event(s)	Downstream event	Imaging detection	References	Comment
↓Blood–brain barrier	↑Edema	FLAIR, T2	Huang et al. [110], Furnari et al. [68], Zinn et al. [230], Barajas et al. [17], and Hatanpaa et al. [89]	Interstitial edema increases when the blood–brain barrier leaks. FLAIR and T2 brightness show tumor edema.
↓Blood flow	↑Hypoxia	MRS	Furnari et al. [68] and Barajas et al. [17]	Poor blood flow decreases oxygen diffusion into tissues, promoting the expression of CA-9.
1. ↑ECM remodeling 2. ↑Adhesion 3. ↑EMT	↑Invasion, migration, motility	DCE, DSC, DWI, FLAIR, MRS, T1 + contrast, T2	Huang et al. [110], Furnari et al. [68], Zinn et al. [230], Barajas et al. [17], Thompson et al. [202], and Bulik et al. [33]	ECM remodeling, adhesion, and EMT promote invasion. CBV is higher in high-grade gliomas and metastases than abscess. ADC value is lower in abscess. Low NAA differentiates primary tumor from metastases. T2 and FLAIR can show infiltrative non-enhancing tumor. T1 + contrast non-enhancement contains infiltrative cells.
1. ↑Anti-apoptosis 2. ↑Growth stimulation 3. ↓Cell cycle regulation 4. ↑Choline	↑Proliferation	DWI, MRS	Higano et al. [98], Barajas et al. [17] and Bulik et al. [33]	IHC stains for Ki-67, which is present in proliferating cells. ADC values are inversely correlated to Ki-67 expression. Choline is a biomarker for cell membrane density, which is increased in proliferation.
↑Hypoxia → ↑necrosis → ↑anaerobic glycolysis	↑Lactate	MRS	Bulik et al. [33]	Higher-grade gliomas perform anaerobic glycolysis. Lactate is a biomarker for anaerobic glycolysis.

↑Poor architecture	↑Micro-hemorrhage	T2	Furnari et al. [68]	Tumor blood vessels with tortuous turns can lead to thrombosis and hemorrhage.
↑Hypoxia	↑Necrosis	DWI, T1 + contrast	Garcia-Figueiras et al. [72] and Barajas et al. [17]	Lack of oxygen diffusion causes cell lysis. Cell membrane integrity and necrosis is extracted from DWI. T1 + contrast non-enhancement within tumor region is necrosis.
1. ↑Cellular density 2. ↑Vascular density	↑Poor architecture	DWI, T1 + contrast	Garcia-Figueiras et al. [72] and Barajas et al. [17]	Architecture factors on DWI (e.g., ADC, FA) include cell size, cell density, extracellular space, and tortuosity. T1 + contrast enhancement represents viable tumor cells. IHC can visualize brain cell architecture via SMI-31 antibody, which binds to axons.
↑Necrosis	↑Pseudo-palisades			Hypercellular tissue surrounding the necrotic foci called pseudopalisades, identifiable in H&E.
	↑Pseudo-progression	DCE, DSC, MRS	Huang et al. [110], Barajas et al. [17], and Thompson et al. [202]	DCE- and DSC-derived parameters may delineate radiation necrosis from recurrence (e.g., PSR, relative PH, and relative CBV). Parametric response maps have been linked to making this distinction. Ratios between choline, NAA, and creatine show tumor progression.

(continued)

Table 8.2 (continued)

Upstream event(s)	Downstream event	Imaging detection	References	Comment
	↑Stem cell abilities	DCE, DSC, T1 + contrast	Barajas et al. [17] and Thompson et al. [202]	Raised CBV can predict malignant dedifferentiation sooner than enhancement on T1W MR images.
1. ↑Hypoxia → ↑HIF → ↑VEGF → ↑angiogenesis 2. ↑Factor VIII → ↑angiogenesis	↑Vascular density	DCE, DSC, VASO	Garcia-Figueiras et al. [72], Furnari et al. [68], Jackson et al. [113], Barajas et al. [17], Lacerda and Law [132], and Thompson et al. [202]	Hypoxic conditions stabilize HIF proteins to promote vascular growth. Signs of angiogenesis are increased vascularity and increased CBV, recirculation abnormality in DSC, and factor VIII from IHC.
1 ↑Angiogenesis → ↑leaky tight junctions 2. ↑VEGF → ↑leaky tight junctions	↑Vascular permeability	DCE, DSC, T1 + contrast, VASO	Garcia-Figueiras et al. [72], Furnari et al. [68], Heldin et al. [94], and Thompson et al. [202]	Endothelial cells of tumor blood vessels do not form proper tight junctions. VEGF can increase vessel leakiness. Many parameters such as permeability surface area, K^{trans}, can be calculated from DCE.

Abbreviations: *ADC* apparent diffusion coefficient, *ASL* arterial spin labeling, *BOLD* blood-oxygen-level dependent, *CA-9* carbonic hydrase-9, *CBF* cerebral blood flow, *CBV* cerebral blood volume, *DCE* dynamic contrast enhanced, *DSC* dynamic susceptibility contrast enhanced, *DWI* diffusion-weighted imaging, *ECM* extracellular matrix, *EMT* epithelial-mesenchymal transition, *FA* fractional anisotropy, *FISH* fluorescence *in situ* hybridization, *FLAIR* fluid-attenuated inversion recovery, *H&E* hematoxylin and eosin, *IHC* immunohistochemistry, *MRS* magnetic resonance spectroscopy, *MS-MLPA* methylation-specific multiplex ligation-dependent probe amplification, *MTT* mean transit time, *PCR* polymerase chain reaction, *PH* peak height, *PSR* percentage of signal intensity recovery, *RF* recovery factor, *VASO* vascular space occupancy.

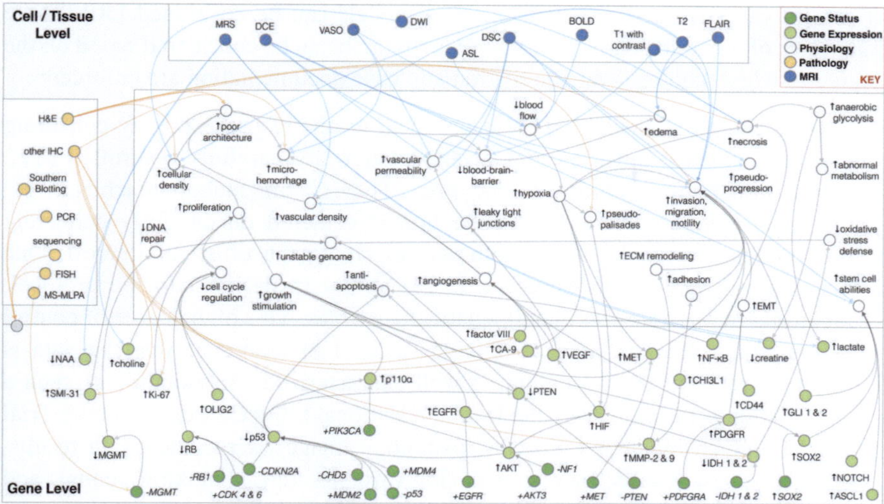

Fig. 8.3 A first-order association map summarizing magnetic resonance signals and their suggested relationships to genetic, molecular, and cellular properties in glioblastoma multiforme. Nodes are physical entities or processes. *Curved arrows* indicate direction of affect from one node to another. *Rectangular boxes* separate various physiological scales on which nodes are observed in. See Tables 8.1 and 8.2 for more detailed descriptions, abbreviations, and references. Notation: *plus symbol* (gain of function in a gene), *minus symbol* (loss of function in a gene), *up arrow* (increased expression or presence), *down arrow* (lowered expression or presence), *italicized gene symbol* (gene), *regular gene symbol* (protein). (Figure and related tables are courtesy of Nova Smedley)

95, 130, 133]. The goal then is to relate radiomics features (in conjunction with other evidence derived from clinical observables) to biological states in order to support precision medicine type queries. A particular subclass of radiomics is *radiogenomics* which is the study of the association of imaging patterns to genetic states.

Radiomics metrics must be learned from observational data. Researchers rely heavily on methods from computer vision and machine learning, fields based on technologies borrowed from computer science, cognitive science, physics, applied mathematics, electrical engineering, and statistics. The typical steps for radiomics investigations are summarized in Fig. 8.1. The main issues include: (1) curating a large pool of imaging cases relevant for a particular biologic investigation, (2) normalization of the pool of images (preprocessing), (3) identification of target regions of analysis, (4) feature engineering, (5) integration of features for inferencing target states, and (6) model validation.

8.5.3.1 Case Pool for Training and Testing

As in other areas of informatics, Big Data analytic methods are critical for learning probabilistic trends under varying contextual patient situations. Thus, there is a growing need for curating large sample databases that store comprehensive records of patient cases from which mathematical models can be developed [30]. Three categories of compiled cases for radiomics research can be considered based on the degree to which control over acquisition and case documentation are enforced:

1. Controlled data collection. This is the ideal means of prospectively collecting data, for which acquisition and documentation are carefully controlled (e.g., within a randomized clinical trial setting). The experimental approach is investigational from the start. Data collection is centered on carefully collecting observations driven by the hypothesis describing the effect of interest. One major advantage of this approach is the ability to design into the protocol, recording of possible confounding patient variables. This high degree of control in the data collection process provides a level of assurance that the data is reproducible, and the population is well characterized. An example of such a compilation is the data available from the National Lung Cancer Screening Trial (NLST) which includes CT scans, pathology images, screening exam results, diagnostic test results, patient behavior variables (e.g., smoking status), and mortality information from over 50,000 cases [198].
2. Clinical multi-institutional standardized data collection. Multi-institutional efforts (e.g., ENIGMA [203]) provide the capability of compiling large databases sampled from a wide range of machine models, over large geographic populations, and diverse clinical settings. These are typically clinically oriented scans from institutions forming a consortium that agrees to scan patient populations in a consistent way. The consortia provide resources to securely store the data in a manner that respects patient privacy (e.g., HIPAA compliance, institutional review board approval, and informed patient consent) but whose main vision is broad dissemination with as few restrictions as necessary to allow qualified researchers to objectively analyze the data using a variety of methods. The acquisition protocol may include information related to hardware considerations (e.g., magnetic field strength and coil selection), interrogation parameters (e.g., pulse sequences), patient positioning details, sampling parameters (space and time), contrast agent information (agent name, dose, injection rate), reconstruction details (e.g., kernels), parameter estimation methods (e.g., DCE signal intensity to contrast concentration), and performance timing information (temporal relation of the exam relative to a reference event). Example efforts include: ENIGMA (Enhancing Neuroimaging Genetics Through Meta-Analysis) [203], QuIC-ConCePT (Quantitative Imaging in Cancer: Connecting Cellular Processes with Therapy) European Consortium, TCIA (The Cancer Imaging Archive), OAI (Osteoarthritis Initiative), QIN (Quantitative Imaging Network, TCGA (The Cancer Genome Atlas), and QIBA (Quantitative Imaging Biomarker Alliance).
3. Natural data collection. Natural data collection refers to the gathering of imaging studies from a routine (i.e., natural) setting, as seen in daily clinical practice. This observational approach does not involve intervening or controlling how the data is generated. The advantage of natural data collection is the potentially large number and diversity of cases that can be collected. The downside is that such data can be messy: in routine practice, images along with the associated

metadata and documentation are often highly variable, incomplete, imprecise, and inaccurate. Clinical data is subject to various sampling concerns (precision, randomness, missing data, selection biases) and representational problems (heterogeneous representations, missing context, accuracy of summarizations, source documentation errors, *etc.*).

Regardless of the source of images, all efforts should be made to take a disciplined, meticulous approach to data collection to ensure its utility for model building. In creating such databases, it is important to understand the quality of the data in order to best decide how to model the data. The paper by Ellingson et al. [62] showed that data collection, even with precise protocols in place, can result in highly variable quality of data. In that study involving a 24-site investigation of biomarkers associated with diffusion-weighted MR images (DWI) for patients with recurrent GBM patients, only 68% of the patients had usable data, and only 47 % had high quality data. Thus, the quality of data used for radiomics research can have a significant impact related to feature selection and model building. Typically there is a trade-off between the quality of data and number of samples within a corpus. In some applications, the label noise (i.e., the quality of ground truth) can be somewhat compromised if a sufficiently large corpus of training data can be compiled [131]. This may open up the opportunity to crowdsource some tasks in limited applications.

8.5.3.2 Preprocessing of Data

Radiomics involves the statistical analysis of a large number of imaging cases. It strives to identify quantifiable detectable patterns that can be correlated with a precision medicine parameter. It is thus critical to first harmonize the imaging data in such a way as to reduce statistical variations due to differences in acquisition rather than patient-specific variations. Traditionally, standardization of acquisition parameters allows physicians to develop a subjective calibration of brightness and contrast level patterns allowing them to establish mental models of the limitations in various image properties (e.g., resolution, dynamic range, inherent noise, *etc.*). Similarly, computer models which learn to recognize some semantic aspect of a class of images will benefit (e.g., higher precision and robustness) from reducing such variability and provide definitive constraints on the class of imaging data for which a developed technique can be applied. Common types of preprocessing operations include: (1) signal dynamic range normalization [21, 115, 229], (2) noise suppression [91], (3) view standardization, (4) voxel size standardization, (5) motion correction, (6) bias field correction [143], and (7) spatial registration of volumetric datasets. Calibration methods (see Sect. 8.5.1) should be applied whenever possible in order to regularize the interpretation of the data across imaging centers and time (i.e., day-to-day variations).

There are two important notes related to preprocessing. Firstly, one must always keep in mind the physical correspondence between a pixel's value and what the value physically represents. Preprocessing operations which are guided by knowledge of the generating image signal/noise (e.g., noise model, signal generation model, normative distributions, *etc.*) are likely to better preserve the physical characterization of the imaged subject. Secondly, each preprocessing operation may have a varied effect on the performance value of a selected radiomics feature set. An investigation of the degree to which candidate preprocessing operations perturb a particular model should be performed. For example, Tourassi et al. demonstrated the effect of various preprocessing noise algorithms on the performance ability of a computer-aided diagnosis system for mammography [209]. Texture analysis methods exploring some aspect of microstructure can be poorly trained without the distinction of imaging data with an inherently specified spatial resolution as compared to those that are artificially rescaled. Similarly, inherent signal strength and noise properties are strongly tied to acquisition voxel dimensions. Thus, expectations of noise and signal properties can be greatly distorted if artificial rescaling operations are not evaluated per algorithm.

8.5.3.3 Identifying Regions of Interest

It is often the case that relevant quantifiable measures are computed over an imaging region of interest (ROI). For example, these ROIs may outline an anatomic structure, a tumoral mass, or relevant subregions within a mass (e.g., necrotic core and peripheral area of proliferation; see Fig. 8.4). The problem of segmenting target ROIs in medical images has been a long-standing challenge in the image understanding community. Approaches have been seen along many dimensions.

Degree of Automation This aspect can range from fully manual to fully automated. Manual segmentation may further require various degrees of domain expertise. Depending upon the difficulty of the task and skill and discipline of annotators, inter-operator consistency of manual annotations can be quite variable. Manual efforts, while typically serving as the gold standard of quality, are time-consuming. Automated methods, while typically less accurate, result in generally more consistent repeatable results. Semiautomated methods (e.g., manually specified seed point approaches with iterative refinement [82, 105, 168]) fall somewhere in between the two extremes.

Number of Input Channels Multispectral/multi-parametric data in which multiple acquisition sequences obtained within a relatively short time period for a given patient which are spatially registered can be used in order to provide improved characterization of pixel data. For example, T1, T2, and DWI MR datasets can be co-registered for use for a segmentation algorithm. Cross-modality registration is also common (e.g., CT-MR, CT-PET, MR-ultrasound, etc.).

Degree of Compositionality Image pattern recognition algorithms often rely on the concept of compositionality in order to reduce the dimensionality of an image interpretation problem. Various types of image grammars have been explored

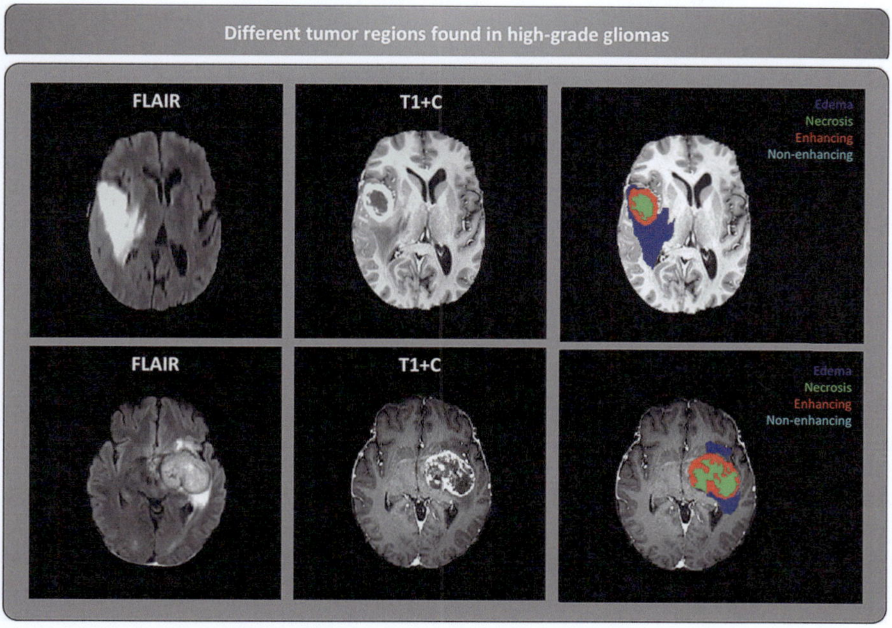

Fig. 8.4 Example regions of interest from two patient cases (*row 1* and *row 2*) with GBM. From these regions of interest, different radiomics features can be quantified. FLAIR images demonstrate the extent and distribution of edema surrounding the bulk portion of the tumor related to leakiness of the abnormal tumor vascularity that results from angiogenesis and VEGF production by the tumor. T1W gadolinium enhanced images (T1 + C) demonstrating better anatomic detail about the inherent variability in tumor composition including the enhancing margin which is a result of breakdown of the blood–brain barrier from the development of abnormally leaky tumor vessels and the central necrosis that results from inhomogeneous distribution of tumor vascularity (Courtesy of Edgar Rios Piedra)

utilizing both syntactic (e.g., pixels, edges, textures, contours, *etc.*) and semantic constituents (e.g., membranes, borders, vessels) [161, 227]. Typically, the interpretation process is hierarchical, with upper levels corresponding to richer semantic abstractions. The basic issues include determining the types of constituents at each level and providing a means for specifying how these constituents are related. Composition rules can be specified by context-free and/or context-sensitive grammars. Bayesian statistical methods are often employed to maximize the interpretation probability for a given image instance as specified by the grammar. A special processing class of algorithms for segmentation is deep learning convolution neural networks (CNN) [128, 137, 188]. CNNs utilize a distributional representation and employ methods for learning data-driven features via a multilayer architecture. The learning of these features is largely unsupervised and is achieved by hierarchically compressing information from lower levels of the network. Successive layers encode a richer, less redundant set of features that roughly correspond to different types of semantic object primitives (e.g., edges, textures, borders, vessels, membranes, *etc.*). Neuron weights are learned in training via batch gradient descent and backpropagation [136]. The output layer of the CNN is a set of complex features that best represent the characteristics of the input image and which can be used as features into a standard machine learning classifier (e.g., maximum entropy and maximum margin methods).

Degree of Domain Knowledge/Prior Expectations As part of the segmentation algorithm, various application-specific knowledge can be used for probabilistically constraining various attributes of image constituents. This may include, for example, constraints on intensity levels, parametric values, shape bounds, positional semantics, homogeneity measures, volume range of target structures, curvature bounds for boundaries, and smoothness.

Degree of Training Set Quality Segmentation algorithms can be highly sensitive to the quality of training data. Considerations include number of training samples, diversity of samples, sampling pool population, sampling strategy, definition of output label set (e.g., inside, edge, outside tumor object), and quality of gold standard truth labels. Various algorithms often have strengths and weaknesses with respect to the quality of these attributes. The combination of the quality of these attributions can significantly modify the performance of a given instantiated segmentation program.

Processing Resources/Time Many image segmentation algorithms require a large amount of processing resources due to the vast input and solution space which must be globally searched. Some algorithms can be formulated to take advantage of parallel processing architectures and/or specialized hardware (e.g., GPUs, computing clouds). Different applications may have different time constraint requirements based on clinical workflow issues. For example, semiautomated methods may need to provide user feedback in near real time.

8.5.3.4 Computed Features

At its most fundamental level, an image finding can be seen as a cluster of pixel data that represents some spatial, temporal, and/or spectral pattern that reveals some signature of the nature, the extent, and/or the dynamics of a disease. Traditionally, such features are reported either subjectively (with or without standard terminologies) and/or semiquantitatively, leading to difficulties in objectively assessing change, similarity, and/or severity. Below, we give some examples of quantifiable features that have been used in various radiomics assessments tasks.

Voxel-Level Signal Intensity-Related Features Given a biological phenomenon of interest (e.g., vascularity), the best discriminating imaging signal type(s) needs to be determined (see Sect. 8.5.2). Depending on the modality used (e.g., CT, T1-MR, DCE-MR, PET, etc.), the resulting conclusions about a given radiomics state may differ. Signal intensities are most useful when they can be placed in the context of a target region of interest or tissue type (e.g., anatomic structure, diseased area or subregion). Prior distributions for different states can then be established from population studies and used for comparison testing and classification. Voxel values in the assessment of stroke are often compared to their corresponding contralateral values in the brain [204]. In complex voxel-level signals, such as those in dynamic contrast studies and/or diffusion tensor studies, voxel value representations are often summarized by various parameterizations. In DCE-MR, for example, the data is fit to a pharmacokinetic model to characterize flow which includes the parameters K^{trans} (transfer constant of a paramagnetic contrast agent from blood to issue), v_p (fractional plasma volume), and k_{ep} (reflux rate constant) [36, 207]. Normal and disease variant distributions for these parameter values again can be compiled and used in classification models (e.g., tumor vascularization) and/or for the identification of outlier (out-of-possible range) pixels. In DTI-MR, voxel data are in the form of a tensor, which can be parameterized using eigenvectors. The eigenvalue magnitudes often reflect changes in local tissue microstructure associated with tissue injury [177], disease (e.g., Alzheimer's disease) [195], or normal physiological changes (i.e., aging) [214]. Finally, a voxel value can be represented in what is known as a distributed representation as in a neural network (e.g., sparse auto-coding [221]) which provides computational advantages in algorithms that compare and combine imaging data in a hierarchical compositional manner [100].

Regional Intensity-Based Features From voxel-level features, various summarizations of the data within an ROI can be used for discrimination. A common class of features are those derived from a histogram of the voxel-level features. From the histogram distribution, one can compute features such as mean, median, standard deviation, maximum, minimum, moments, *etc*. For example, skewness of T2W MR signal histograms has been used for classifying cancerous versus normal tissue for prostate cancer [162].

Size Tumor size is an important indicator of therapeutic response and overall survival time [40]. The most accurate estimation follows from a 3D volumetric analysis [54]. However, given that any segmentation procedure is not 100 % accurate, some estimation of error with respect to the true volume is needed in order to objectively assess change. This can be quite a concerning task given tumor margins may be rather radiographically indistinct, borders may have complex shapes, and/or surrounding processes (e.g., edema) may obscure radiographic distinction of the tumor bounds. Furthermore, the measurement accuracy can be quite different for the same patient study given different sequences (e.g., T1W, T2W, DCE) [200]. One should also note that all imaging scanners produce some level of spatial distortion. Considering all these aspects, for some scenarios, medical societies have specified certain criteria for thresholds for clinically significant size change (e.g., 25% increase in a linear reference dimension). These general guidelines however may result in a delay in the declaration of a positive size change that could affect the timely reassessment of therapeutic interventions. Athanasiou et al. demonstrated error propagation for the characterization of plaque area in MR imaging data [12]. Methods for bounding size measurement errors due to variations in image segmentation software are also being explored [176]. This estimation of measurement error can be used for applying hypothesis testing for change (e.g., use of a test statistic and computation of a p-value).

Shape Shape descriptors have been used as important features for assessing tumor heterogeneity, tumor proliferation, and overall expected survival. For example, irregularly shaped tumor margins (e.g., degree of spiculation) that show lower convexity shape scores have been correlated with poor patient outcomes [81]. Shape irregularities along tumor perimeters are seen in heterogeneous tumors which contain cell lines with differences in growth patterns. Proliferative tumors have strong interactions with their environment often manifested by fingerlike projections into the parenchyma indicative of poor prognosis [103]. Several types of shape descriptors have been proposed in the literature; a review of common general shape-based metrics can be found in [224]. Common simple shape descriptors include area, perimeter, eccentricity, elongation, and orientation. Geometry-based features can be divided twofold: (1) contour-based, in which the calculation only depends on knowing the shape boundary, and (2) region-based, wherein the boundary and the internal pixels must be known. Examples of the boundary-based metrics include perimeter, compactness, eccentricity, and the minimum bounding box. Examples of the latter include area, shape moments (e.g., centroid), and convex hull. Tumor shapes have been characterized, for example, by the ratio between the area of the tumor mask and its convex hull which quantitate the amount of protuberances and depressions along the tumor border. The boundary shape of an object can be represented in various ways including chain codes [46, 153] and Fourier descriptors [63] which have the advantage of well-known, simple equivalents for affine transforms (rotation, scaling, translation, shearing) in the frequency domain. Shen et al., for example, used a Fourier representation of mammographic calcification shape to infer malignancy [185]. Topological features refer to properties of shape that don't change, so long as perturbations do not tear or join parts. A useful topological descriptor is the Euler number which is the number of connected components minus the number of holes in the object and has been used for evaluating therapy outcome [61]. Shape comparison is often performed based on their statistical properties. One approach is simply to evaluate distances between their feature vectors [60]. More complex methods represent shapes as a probability distribution sampled from a shape function. The dissimilarity between comparison objects can then be evaluated using common metrics between distributions (e.g., L_N norm) [157].

Global Morphology Tensor-based morphometry has been used to measure size and shape changes ongoing in the brain in the context of normal and diseased states. Spatial registration algorithms between the two imaging dataset are commonly used via various nonlinear transformations [192]. The determinant of the Jacobian tensor of the transformation is then a measurement of the local tissue contraction or dilation [139]. Tensor-based morphometry has been used to study the dynamics of Alzheimer's disease (atrophy) and other conditions in the brain [108, 109]. Morphologic instability of tumor masses may provide an indication for tumor tissue invasion [44]. The use of geodesic regression in diffeomorphisms for tensor-based morphometry is a promising emerging method which can extend nonlinear transformation methods to analysis of a number of time-sampled imaging studies for a patient (>2) and allows interpolation and extrapolation in time in a more theoretically principled manner [19, 67].

Texture Texture is a property associated with a statistically repeated spatial pattern in an image. It is assumed that these spatial regularities come from the underlying structural properties of the tissue being imaged. Texture features have been correlated with EGFR mutation status, intra-tumoral heterogeneity, angiogenesis, predication of treatment response, and prognosis in a number of different cancers and with various imaging modalities (e.g., CT, DCE-MRI, PET, US) [4, 29, 38, 58, 158, 199]. Haralick's method of using a gray-level co-occurrence matrix (GLCM) to analyze image texture is a widely used analysis method [88]. Other local texture representations include run-length matrix (RLM) and neighborhood gray-tone difference matrix (NGTDM) [6]. Measures that can be computed from such a matrix representation include energy, entropy, contrast, homogeneity, and correlation measures. Geometric texture analysis aims for decomposing the texture of an image area into a set of simple texture elements that can be extracted using various filters (e.g., Gabor and/or Laws filters) [172]. Transform-based methods (e.g., Fourier, wavelet, S, and discrete cosine) are also widely used for characterization of texture [49]. Fractal analysis measures can be used to characterize structural pattern changes as a function of scale [144]. Nagao et al. described a method for quantifying heterogeneous distribution of a radiotracer in SPECT images by 3D fractal analysis in order to quantify severity of pulmonary emphysema [149]. Szigeti et al. noted a correlation between fractal analysis measures and the ability to perform early diagnosis and risk assessment of air pollutants to mice on CT images [197].

Connectivity The communication of cells in the brain can be investigated by high-resolution diffusion tensor imaging with the application of tractography algorithms to infer neural pathways [18, 79]. Given N regions of interest (e.g., cortical), a $N \times N$ connectivity matrix can be constructed by quantifying the proportion of fibers interconnecting a pair of target regions. Given this connectivity network, various local and global metrics based on graph theory can be computed from such a representation [23]. Network connectivity in the brain can be considered at various scales: from the lowest level, in terms of its synaptic connections, to connections between cortico-cortico and cortico-deep gray neurons, or at a more macroscopic level, looking at connections between cortical areas in the form of bundles of white matter tracts [164, 179]. Various neuropsychological deficits have been associated with abnormal disconnection patterns between brain regions. Common themes of investigation include functional integration and segregation of brain regions, characterization of local anatomic circuitry, and characterization of resilience of networks to insult [179]. Different measures of connectivity are used in different studies to describe the integrity of the healthy or diseased human brain network and include the nodal degree, characteristic path length, efficiency, clustering coefficient and "small-worldness" [194]. Other metrics include: "rich club" network property corresponding to high-degree network nodes that are more interconnected than expected by chance and have been found to correlate with cognitive impairment as network complexity degenerates [47].

8.5.3.5 Model Development

Inference models attempt to synthesize imaging features and other clinical context variables in order to predict the states of biological parameters. The task can be difficult, given that in general, there are a large number of dimensions required to characterize the range of image patterns associated with all possible biological interpretations. For example, a lesion may be quite variable and heterogeneous across patients and tumor types and furthermore depend strongly on acquisition and preprocessing operations. Thus, models can range from low-dimensional mappings (e.g., a single image feature to a single target state) to high-dimensional complex input/complex output state associations. Typically, a large number of models can be fitted to a set of training examples, implying that optimization criteria need to be clearly justified. Different modeling approaches have different strengths and weaknesses relative to the nature and availability of the training data and complexity of the inherent modeling problem. Some key considerations in model building include:

Physical/Biological Insights A modeling task can be greatly facilitated if there are known causal/associative connections that can help to impose dependence and independence constraints on the topological structure of the model [160]. Independence constraints are a means of reducing the dimensionality of a modeling problem, allowing a model to be factored into a number of smaller simpler problems.

Correlation of Variables Radiomics models often contain a large number of plausible image features and/or clinical context variables. The problem is that the dimensionality of the data grows very quickly as the variable count increases, requiring impractically large amounts of training data for parameter estimation. Often there are informational overlaps associated with the discriminatory ability of feature combinations. Various feature selection algorithms can be used to reduce the number of features without significantly compromising model accuracy. A review of feature selection techniques (filter, wrapper, eigenspace) can be found in [14, 180, 228]. Eigenspace methods include principal component analysis (PCA), Fisher's linear discriminant analysis (LDA), and independent component analysis which project or transform the original feature space into uncorrelated or independent directions. Assumptions regarding the shape of data distributions (e.g., Gaussian) differ among these techniques [170]. Regularization methods like LASSO and ridge regression automatically perform feature selection as part of the model construction process [140]. The minimax principle approach to feature selection involves the use of entropy measures to search for an optimal combination of features [228]. In this approach, an iterative algorithm is used to progressively add forward the feature that minimizes the entropy of candidate maximum entropy models. Some modeling approaches such as maximum entropy models can principally deal with overlapping variables in a principled way (computation of Lagrangian per feature) [116].

Interaction of Variables Feature selection algorithms often view single variables at a time and do not take into account their possible complex interaction. The topology of a belief network is an intuitive means of visualizing such interactions if they are known. Random forest methods can be used to automatically identify complex feature interactions by randomly testing different configurations of a subset of variables [27]. Modern deep learning methods are achieving some success in this area and have been shown to automatically learn abstract representations of features (a compressed form) that in turn have supported state-of-the-art results in domains such as speech recognition, image classification, and disease characterization [99]. Figure 8.5 shows an example of 64 complex feature maps derived from the third convolution layer of a deep convolution neural network (CNN) trained to perform voxel-wise classification of tumor versus non-tumor differentiation. The

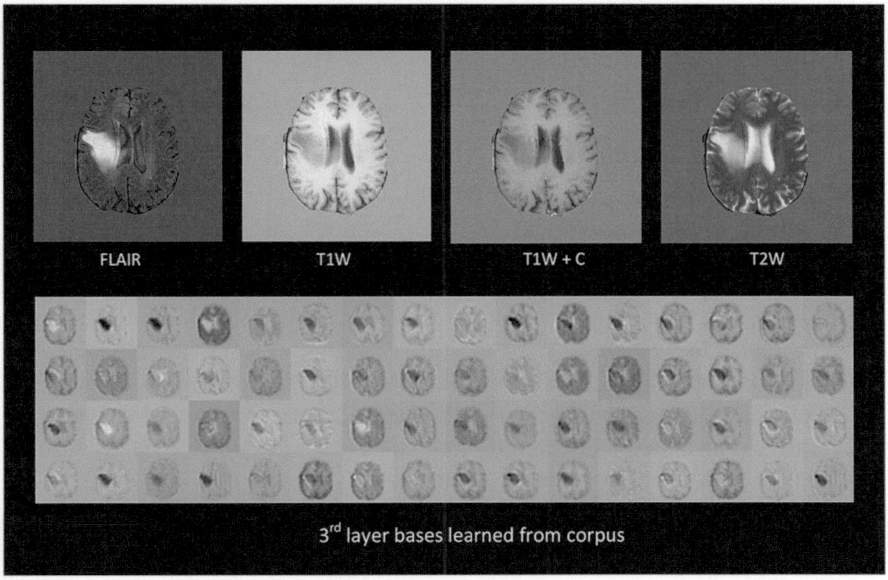

Fig. 8.5 Example of 64 complex feature bases derived from the third layer of a deep convolutional neural network. The feature maps are learned automatically from a set of learning cases consisting of registered FLAIR, T1W, T1W with contrast, and T2W magnetic resonance images labeled at the voxel level as tumor or non-tumor. The deep CNN consists of three sequences of convolutional-nonlinear-pooling layers, followed by two fully connected layers, and a softmax classifier. This figure illustrates the feature maps (outputs) of the third convolutional layer. These rich representations can be combined and used by a softmax classifier to better segment tumor regions (Courtesy of King Chung Johnny Ho)

deep CNN is trained on a concatenation of FLAIR, MR T1-weighted, MR T1-weighted with contrast, and MR T2-weighted image sequences.

Representation of States The state definition of variables can have a significant impact on classifier performance. If continuous values are to be discretized, then some rationale means for identifying intervals should be used [141, 142]. Agreed-upon ontologic definition of states can help to improve comparison testing and application of models.

Quality and Distribution of Training Sets Models are dependent upon the sample cases from which they are trained. Attributes that can affect model performance in this regard include: (1) sampling method (e.g., random, balanced, diversity, highest uncertainty), (2) number of samples, and (3) amount of noise in training labels – truth determination in imaging can be especially difficult in the dynamic ecosystem of a cancerous tumor. For example, in the TCGA dataset, what is the probability that the imaging has taken place after a cancer has undergone (multiple) mutations? What is the probability that the biopsy sample does not reflect the majority of tumor cells? This would mean that due to the time difference, a predicted outcome for a given training case is erroneously labeled for the observed phenotype.

Model Assumptions In general, the performance of different modeling approaches depend upon the degree of concordance between the data and the assumptions inherent to the modeling technique. This could be, for example, assumptions regarding the shape of distributions, complexity of data, reliability of frequency estimations of probabilities, and independence of variables.

8.5.3.6 Evaluation and Acceptance

In evaluating a radiomics metric or model, rigorous unbiased testing is necessary in order to gain scientific trust. Without such evidence, efficacy acceptance by an oversite organization (e.g., FDA or medical society) is unlikely, and hence widespread clinical adoption cannot be realized. Hence, there may be a greater need for comparative effectiveness research for radiomics technology [28, 167]. Evaluation may be with respect to a number of dimensions:

Operational Constraints The process by which a radiomics measure is obtained may be constrained by clinical operational requirements (e.g., effort, cost, and timing). Cost may be related to whether a specially trained person is needed. It is likely that new types of clinical informatics personnel may be required to assist processing and preparing results for clinical interpretation by a physician team.

Accuracy For the given clinical scenario and for the given class of inputs, the accuracy of the algorithm should be estimated. For nonbinary-type decisions, a confusion matrix and/or multiclass receiver operator characteristic (ROC) analysis [135] should be provided. Estimation of error bounds for the quantitative measure should be provided [171].

Model Validation and Reuse An ongoing challenge in predictive modeling is the inability to reuse models in different contexts. For example, models developed at one institution often do not achieve the same level of sensitivity and specificity at another, limiting the clinical utility of such models.

Reproducibility The combination of the type of algorithm and the nature of the data being processed can greatly influence the reproducibility of a radiomics quantitative measure. Influential considerations include the choice of initialization methods, cost functions, convergence rates, operator input, study quality, anatomic location, etc. Part of quality assurance is to be able to determine the degree to which various processing algorithms produce consistent results. Testing this property can be facilitated using image compilations in which clinical scans for the same patient are acquired with the same technique within a very short time period (e.g., 15 min) to reflect only acquisition signal detection fluctuations (so-called "coffee break" study) [225]. Various test statistics can be computed (e.g., concordance correlation coefficient, Pearson correlation coefficient) to quantify this evaluative aspect.

Robustness Robustness in radiomics applications is a quality associated with how well the system can handle a diversity of clinical context, including patient situations that perhaps have not been previously well examined. For example, user input of out-of-bounds/illegal parameters must be well handled by a clinical software application.

Effectiveness Given that the radiomics feature is able to perform at its documented level, effectiveness questions whether there is a clear utility of the imaging biomarker for improving treatment decisions at the current level of performance accuracy. The utility of a biomarker must be evaluated in terms of what it adds in addition to established clinical markers.

Documentation Every representation and model have limitations and can produce artifactual results. It is important to be able to document known assumptions and limitations related to types of input data, error rates (e.g., confusion matrix), and descriptions of known bugs. Interpretations to be aware of should be documented as part of a radiomics study. Buckler et al. describe an early attempt to define an ontologic representation for imaging biomarkers [31]. The constraints on the types of data and the expected error bounds need to be known to the end user involved with patient management.

Software Testing The greater the role a decision support system plays in clinical practice, the greater the need to insure that complex, multisystem software applications are thoroughly tested [120]. Due to the complexity of scientific software and the required specialized domain knowledge needed for its development, scientists/informaticists often develop these programs without being familiar with accepted software engineering practices. Due to the lack of systematic testing of scientific software, subtle faults can remain undetected and affect patient care. Complex interfaces that are poorly designed can also lead to significant user execution errors.

8.6 Clinical Considerations

The use of radiomics analysis as part of routine clinical decision making is currently relatively scarce. A question remains as to whether precision medicine imaging techniques can be practically translated and integrated into clinical radiology. Its widespread application will require the coordination of a variety of individuals, the integration and standardization of a number of software services and representations, and the enforcement of an overarching clinical workflow protocol [30]. In this section, we describe some relevant operational issues associated with the deployment of a precision medicine radiology consultation service.

8.6.1 Request for Examination

The radiology process starts with the request for examination. Typically, the decision regarding which imaging procedure to order for a patient falls on the burden of the treating physician. It has become increasingly difficult however to identify the most appropriate imaging studies given the ever-increasing number of alternatives, methods, and details (views, protocols, modalities, resolution, and contrast agents), which must be specified on a clinical order. Currently, there is no clear guideline on how to define the "reason" for exam, which is the driving basis for the type of exam, analysis methods, and reporting details. With the advent of advanced imaging and analysis methods comes the grave possibility of inappropriate expectations of the capabilities of various imaging protocols. A lack of sufficient knowledge by a requesting physician and/or a lack of knowledge about a patient's case can lead to ordering of examinations that might not be suitable for the underlying clinical questions of concern. It is not uncommon, for example, for the request to originate from physicians who have not even seen the patient and presumably have little insight into the working hypotheses and evidence related to the patient. Radiologists, who likely understand imaging procedures best, typically do not consult with referring physicians on the type of exam to perform for a given patient situation. (A radiologist performing a diagnostic interpretation is not considered a treating physician by most insurance companies.) Reimbursement issues and physician order entry remain a point of discussion between the American College of Radiology and Centers for Medicare and Medicaid Services (CMS).

In general, guidelines tend to be too simplistic and do not capture all possible contexts for selection of an imaging exam. Simplicity is emphasized over comprehensiveness because a team of physicians with various backgrounds and involvement with the patient need to be easily able to interpret these guidelines. The explosion of new methods and protocols being developed in the biophysics imaging community makes the task more daunting as technology rapidly advances. Guidelines, furthermore, are often based on statistical trends in which many important distinguishing contextual features may have been marginalized out. The scope of application for specific guidelines is not always obvious; complex patient cases often do not fit into guidelines created by optimizing some metric across a cohort population. Guidelines typically are represented by a decision tree, which has flaws associated with incorporating uncertainty and decision nodes which are not determinant for a given patient case. Guidelines do not provide threshold probabilities and/or sufficient context for when one should consider one alternative over another. This often leads to conservative medicine and, in some cases, a greater increase in low-value exam requests for a given patient. (For example, the US medicolegal system is such that false-negative results are intolerable, far more so than false-positive findings.) Furthermore, guidelines often follow a one-size-fits-all approach that does not leverage the strengths of a particular institution (e.g., advanced acquisition protocols and/or hardware).

Decision support for physician order entry can be implemented in various forms: (1) built into a computerized medical information system, (2) a routine person-to-person service to be conducted before order completion when test utility is deemed uncertain (i.e., consultative feedback before completion of order), or (3) radiology gatekeeper model where the radiology department is responsible for choosing the right test for the patient for their given current situation. Currently, radiology is seen as a service provider, not a gatekeeper. Thus, service metrics are more aligned with internal concerns such as volume and turnaround time rather than patient outcomes. Regardless of the form of a decision support system, the information required should include [205]: the patient's differential diagnoses, the expected information gain (e.g., is the candidate test redundant with respect to a previous recent exam?), the sensitivity and specificity of each alternative test and corresponding consequences of a false-positive/false-negative/inconclusive diagnosis (diagnosis drives therapy), the procedure risks (contrast allergies, radiation exposure, invasiveness), the patient concerns (e.g., phobias, religious, physical status), the availability of test (scheduling window, location), and the cost/insurance issues.

8.6.2 Imaging and Clinical Context

Precision medicine will require a closer collaboration between diagnostic services (e.g., radiology and pathology) and the referring clinical team. Reporting details and diagnostic conclusions (and hence treatment decisions) can be significantly affected by a lack of clinical information [22, 26, 48]. Currently, radiologists often receive limited clinical information, typically in the form of high-level "reason-for-request" statements. Ideally, however, the radiologist and clinicians should have a consistent canonical framing of the patient's case in a precision medicine representation. This would allow obvious entailment questions to be inferred from the clinical question to be investigated through imaging. Such related questions should be anticipated such that the information provided within the corresponding radiological report can be more synchronized with the cognitive reasoning process of the referring precision medicine practitioner. One could envision a more coordinated consultation as follows:

Step 1 – Formulate Clinical Query The clinical team managing a patient case may have several high-level questions that need to be addressed through imaging. For example, these questions may be related to the specific subtype of the disease, patient's predisposition for disease, prognosis, disease progression, treatment planning, and therapy monitoring. Methods to improve specificity of reason-for-request should be developed that encourage meaningful reporting of results.

Step 2 – Infer Entailment Questions The stated reason for request should be re-factored in terms of logical propositions that can be tested. In this step, consultation with the current working model of the disease is made, and identification of those pieces of information (e.g., biomarkers) that are relevant to answering the clinical question are inferred.

Step 3 – Execute Relevant Imaging Protocols Considering the set of propositions to be tested, decide on an appropriate imaging protocol(s) for the given patient case.

Step 4 – Data Collection and Standardization After data acquisition, regularize the imaging data as necessary for advanced processing and visualization.

Step 6 – Data Analysis Analyze the data to provide probabilities regarding the truth of each of the propositions. Questions related to imaging phenotypes should be recorded and maintained for future evaluations. Understanding the dynamics of a disease (i.e., progression) is an important component of phenotyping.

Step 7 – Study Conclusion Query the theoretical model to infer answers to the high-level clinical question. This can be computer assisted and/or part of the radiologist's expertise. A precise, consistent method of documenting such findings is important to facilitate accurate, unambiguous communication of findings to the referring physician team.

Some examples of the expanded thought processes that are likely in a precision medicine radiology service include:

Diagnosis New classification schemes related to genetic/molecular subtyping of disease are evolving and must be targeted as an endpoint whenever possible during a radiologic diagnostic investigation. Conclusions should be based on as much supporting evidence as possible (e.g., biopsy and imaging confirmation). Limitations of test conclusions related to the reality of the situation should be stated. Furthermore, a more detailed representation for their specification is needed, given, for example, the dynamic complex ecosystem of cancers with respect to space and time. Spatial location should be obtained in biopsy samples (e.g., via image-guided biopsy procedures) in cancers known to be highly heterogeneous.

Surgical Therapy Planning A surgical planning imaging exam can entail many questions related to location, margins of tumors, surgical pathways, and proximity to functional centers of the brain (e.g., language or motor centers). The answers to such questions will lead to issues related to degree of tumor resection, surgical risk, and selection between surgery and other treatments [212, 219]. After debulking, residual tumor margins can be imaged and assessed for heterogeneity which can guide subsequent chemotherapy.

Therapy Monitoring Heterogeneity is critical to understanding a tumor's resilience to therapy and to designing strategies for treatments (combinations of drugs/modalities) that deal specifically with the problem of recurrence. Phenotypic and genotypic trajectories of a cancer can change rapidly in time and these aspects need to be monitored. An awareness of possible resistance mechanisms for a given tumor type and their radiomics signatures need to be anticipated during therapy monitoring assessment. Additionally, monitoring therapeutic side effects and possible causes of death is important (e.g., herniation, systemic illness, brainstem invasion, tumor-induced cytotoxicity, and tumor cell burden).

Information Systems Information systems will have to advance accordingly to meet the new capabilities of a precision medicine radiology practice. Greater integration of clinical and radiology data will be necessary to improve diagnostic and monitoring capabilities. A common information model can serve as a framework for improving context for an exam and reporting requirements. Workflows for integration of processing intense radiomics analysis will need to be enforced. Likely, new types of personnel specializing in various information processing tasks will be needed in order to ensure radiologists are presented with all forms of evidence, computational analysis results, and possible interpretations in order to make their final conclusions for a given study in a timely manner. Various working groups, medical societies, and consortia are actively publishing specifications for reporting to ensure clarity, specificity, and compliance with the information needs of the clinician. These reporting forms serve as the information interface to ensure radiology findings have meaningful clinical utility. For example, in 2010, the Response Assessment in Neuro-Oncology (RANO) Working Group published updated criteria to address this need and to standardize response assessment for high-grade gliomas [39]. Reference ontologies that are situational to a given disease are also assisting in improving clarity and understanding. Finally, interfaces to complex information sources (images, graphs, text, and statistics) required for optimizing imaging study interpretations will have to advance in order to provide visualizations and tools that can greatly enhance the cognitive reasoning skills of the radiologist.

8.7 Challenges and Opportunities

Precision medicine is predicated in part on our ability to characterize patients in new ways, evolving our understanding of a disease across biological/physiological scales, and to learn likely outcomes for an individual. The volume and variety of information captured in the health record and medical images is vast and quickly growing: extracting meaningful insights from clinical big data requires algorithms that not only discover relevant features and their relationships from a large information space but also consider the specific context (i.e., the clinical decision and patient at hand) under which these relationships hold true. The rate limiting step is no longer our ability to generate deep phenotypes [53, 178]: the advent of radiomics and other high-throughput feature extraction techniques have yielded a large number of potential phenotypes that may correlate with biological processes. Rather, the overarching challenge is selecting and interpreting features that are robust and informative to a clinician in deciding on a diagnosis or treatment for a patient and, thereby, fully achieving the promise of precision imaging. Summarizing the developments presented in previous sections, we highlight three emerging areas where developments in biomedical imaging informatics play a significant role in enabling precision medicine.

Infrastructure for High-Throughput Phenotyping While the size of generated data per patient is continually growing, this information is not captured in a manner that permits its reuse for research [96]. Contributions from the informatics community have started addressing information gaps by capturing provenance information and contextual details about each observation [217]. Furthermore, new insights into genotype-phenotype connections are being driven by the establishment of imaging-based observational databases [32]. For example, at our institution, a significant effort is underway to link clinical, radiologic, pathologic, and genomic information about various cancers at the lesion level: segmentations identified on CT/MR imaging are correlated with whole-mount pathology slides, image-guided biopsies, and information about patient symptoms, comorbidities, and their outcomes. This repository supports a wide variety of studies examining how to improve the sensitivity of detecting prostate lesions on MR as well as the ability to accurately classify the aggressiveness of a lesion. Collecting this information longitudinally allows researchers to understand how the biological properties of a lesion change over time and how such changes are reflected in the clinical and imaging observations. Moving forward, consortia such as ENIGMA and TCIA serve as successful models to attain the necessary statistical power to discover new genetic underpinnings of disease processes using large shared datasets.

Deep Hierarchical Modeling Matching the growth of available biomedical data is the need for machine learning techniques that are capable of learning robust patterns from heterogeneous input data. Deep learning is an emerging class of techniques that has shown promise in learning latent multilevel hierarchical representations adaptively from raw data with multiple processing layers [20]. However, unlike natural images, deep learning techniques such as CNNs, DBMs, and autoencoders require a large amount of labeled data, and acquiring such a training set is challenging given that the data are often retrospectively collected and require manual annotation by domain experts. Furthermore, it is impractical to obtain data to train a new model each time the model is applied to a new dataset or to perform a related – but different – classification task (e.g., lung nodule detection versus lung nodule classification/diagnosis). Further exploration of techniques such as data augmentation, regularization, and transfer learning may lead to alternatives for manually labeling additional data. Moreover, building deep networks is an art form: a variety of hyperparameters and architectures need to be selected, and their selection greatly influences model performance. Interpretation of intermediate layers or the effect of changing network parameters is notoriously difficult to understand by a domain expert [175, 223]. As such, more tools that permit the clear understanding of intermediate layers and training on smaller, imperfect datasets are crucial.

Knowledge Repositories of Biological Imaging Associations While our knowledge about genetic and epigenetic factors in cancer continues to increase, a knowledge gap exists in understanding the significance of these findings and their clinical manifestations. For example, while the presence of a BRCA1/BRCA2 gene in a woman imparts an increased risk of breast cancer, it is the environmental factors that determine whether a woman will ultimately develop cancer. We lack a clear understanding between the interplay of genetic, molecular, and cellular properties with information about the microenvironment and environmental exposures that can be observed using biomedical imaging and clinical evaluation. Knowledge repositories constructed from models such as the one for GBM depicted in Fig. 8.3 are necessary to explicitly formalize our growing knowledge of pathways and interactions among risk factors, bridging genotypic and phenotypic information. One important role of such a knowledge repository is to provide a basis for re-characterizing disease subtypes [152]. Current pathologic staging systems are insufficient in capturing the growing number of disease subtypes with some subtypes needing to be reclassified as different diseases. Development of a knowledge repository that characterizes diseases across clinical, radiologic, pathologic, and molecular observations will facilitate the discovery of new signatures across these data sources that uniquely identifies the disease subtype, whether it is indolent or aggressive and whether it will respond to a specific drug therapy. Furthermore, this information can be combined with a growing number of pharmacogenomics databases to reveal potential drug targets and imaging biomarkers that predict treatment response.

Application of Imaging-Derived Insights to Drive Targeted Therapies Ultimately, the development of these high-throughput image analysis methods and knowledge repositories is to generate better insights from images and image-derived information from which clinical decisions can be made. Operationalizing this information is not trivial: for example, how should the increasing amount of evidence extracted from imaging and other biomedical data sources be conveyed both to the image interpreter and referring physician? Reporting will fundamentally change to not only convey an imager's qualitative assessment but also perspectives from multiple disciplines (e.g., radiology, pathology) in an effort to provide more consistent and actionable information to referrers [9]. In addition, clinical departments are realizing the importance of leveraging data beyond imaging to perform quality assurance [107], demonstrate the utility and potential cost savings of performing imaging exams [121], and improve how imaging exams are ordered [25]. Development of novel user interfaces that integrate imaging and other clinical information [106] as well as provide users with some sense of potential sources of errors and variability [176] is needed to better aid in the validation and interpretation of clinically significant findings.

Imaging informatics continues to evolve at a rapid pace to meet the evolving need for improved methods to harness images and image-derived information to inform clinical decision making. As evidenced by the examples and developments highlighted in this chapter, biomedical imaging informatics contributes a valuable and unique perspective about disease state. When interpreted alongside information from other biomedical data sources, imaging provides a critical link between observed clinical phenotypes and their genetic and environmental causes. Recent developments in imaging informatics have provided the basis for constructing a high level framework for applying radiomics in clinical practice, illustrated in Fig. 8.1. Moving forward, further progress in areas such as (1) advancing radiomic techniques to extract meaningful features from imaging data; (2) utilizing a systems modeling approach to relate clinical, imaging, and genomic features; and (3) validating models in their ability to generate relevant diagnoses and treatment recommendations are necessary to move precision medicine forward.

Acknowledgments The authors would like to thank the following medical imaging informatics doctoral students for their valuable intellectual contributions to this chapter: Nova Smedley, Nicholas J. Matiasz, Edgar Rios Piedra, and King Chung (Johnny) Ho. We would also like to thank Lew Andrada for his proofreading and general editing services.

References

1. Aerts HJ, Velazquez ER, Leijenaar RT, Parmar C, Grossmann P, Carvalho S, Bussink J, Monshouwer R, Haibe-Kains B, Rietveld D, et al. Decoding tumour phenotype by noninvasive imaging using a quantitative radiomics approach. Nat Commun. 2014;5:4006.
2. Aghi M, Gaviani P, Henson JW, Batchelor TT, Louis DN, Barker 2nd FG. Magnetic resonance imaging characteristics predict epidermal growth factor receptor amplification status in glioblastoma. Clin Cancer Res. 2005;11(24 Pt 1):8600–5.
3. Aktipis CA, Boddy AM, Gatenby RA, Brown JS, Maley CC. Life history trade-offs in cancer evolution. Nat Rev Cancer. 2013;13(12):883–92.
4. Alic L, Niessen WJ, Veenland JF. Quantification of heterogeneity as a biomarker in tumor imaging: a systematic review. PLoS One. 2014;9(10):e110300.
5. Altomare DA, Testa JR. Perturbations of the AKT signaling pathway in human cancer. Oncogene. 2005;24(50):7455–64.
6. Amadasun M, King R. Textural features corresponding to textural properties. IEEE Trans Syst Man Cybernet. 1989;19:1264–73.
7. Amberger-Murphy V. Hypoxia helps glioma to fight therapy. Curr Cancer Drug Targets. 2009;9(3):381–90.
8. Andriole KP, Morin RL, Arenson RL, Carrino JA, Erickson BJ, Horii SC, Piraino DW, Reiner BI, Seibert JA, Siegel E, et al. Addressing the coming radiology crisis-the Society for

Computer Applications in Radiology transforming the radiological interpretation process (TRIP) initiative. J Digit Imaging. 2004;17(4):235–43.
9. Arnold CW, Wallace WD, Chen S, Oh A, Abtin F, Genshaft S, Binder S, Aberle D, Enzmann D. RadPath: a web-based system for integrating and correlating radiology and pathology findings during cancer diagnosis. Acad Radiol. 2016;23(1):90–100.
10. Asari S, Makabe T, Katayama S, Itoh T, Tsuchida S, Ohmoto T. Assessment of the pathological grade of astrocytic gliomas using an MRI score. Neuroradiology. 1994;36 (4):308–10.
11. Assaf Y, Pasternak O. Diffusion tensor imaging (DTI)-based white matter mapping in brain research: a review. J Mol Neurosci. 2008;34(1):51–61.
12. Athanasiou LS, Rigas G, Sakellarios A, Bourantas CV, Stefanou K, Fotiou E, Exarchos TP, Siogkas P, Naka KK, Parodi O, et al. Error propagation in the characterization of atheromatic plaque types based on imaging. Comput Methods Programs Biomed. 2015;121(3):161–74.
13. Baehring JM, Bi WL, Bannykh S, Piepmeier JM, Fulbright RK. Diffusion MRI in the early diagnosis of malignant glioma. J Neurooncol. 2007;82(2):221–5.
14. Bagherzadeh-Khiabani F, Ramezankhani A, Azizi F, Hadaegh F, Steyerberg EW, Khalili D. A tutorial on variable selection for clinical prediction models: feature selection methods in data mining could improve the results. J Clin Epidemiol. 2016;71:76–85.
15. Bammer R. Basic principles of diffusion-weighted imaging. Eur J Radiol. 2003;45 (3):169–84.
16. Barajas Jr RF, Hodgson JG, Chang JS, Vandenberg SR, Yeh RF, Parsa AT, McDermott MW, Berger MS, Dillon WP, Cha S. Glioblastoma multiforme regional genetic and cellular expression patterns: influence on anatomic and physiologic MR imaging. Radiology. 2010;254(2):564–76.
17. Barajas Jr RF, Phillips JJ, Parvataneni R, Molinaro A, Essock-Burns E, Bourne G, Parsa AT, Aghi MK, McDermott MW, Berger MS, et al. Regional variation in histopathologic features of tumor specimens from treatment-naive glioblastoma correlates with anatomic and physiologic MR Imaging. Neuro Oncol. 2012;14(7):942–54.
18. Basser PJ, Pajevic S, Pierpaoli C, Duda J, Aldroubi A. In vivo fiber tractography using DT-MRI data. Magn Reson Med. 2000;44(4):625–32.
19. Beg MF, Miller MI, Trouve A, Younes L. Computing large deformation metric mappings via geodesic flows of diffeomorphisms. Int J Comput Vision. 2005;61(2):139–57.
20. Bengio Y, Courville A, Vincent P. Representation learning: a review and new perspectives. IEEE Trans Pattern Anal Mach Intell. 2013;35(8):1798–828.
21. Bergeest JJ. A comparison of five methods for signal intensity standardization in MRI. In: Tolxdorff B, Deserno H, Horsch M, editors. Bildverarbeitung für die Medizin (Bildverarbeitung für die Medizin 2008, Algorithmen, Systeme, Anwendungen, Proceedings des Workshops vom 6. bis 8., 2008). Berlin: Springer; 2008.
22. Berlin L. Radiologic errors and malpractice: a blurry distinction. AJR Am J Roentgenol. 2007;189(3):517–22.
23. Biggs N. Algebraic graph theory. 2nd ed. Cambridge mathematical library. Cambridge: Cambridge University Press. 1993. vi, 205 p.
24. Boland MR, Hripcsak G, Shen Y, Chung WK, Weng C. Defining a comprehensive verotype using electronic health records for personalized medicine. J Am Med Inform Assoc. 2013;20 (e2):e232–8.
25. Boland GW, Duszak Jr R, McGinty G, Allen Jr B. Delivery of appropriateness, quality, safety, efficiency and patient satisfaction. J Am Coll Radiol. 2014;11(1):7–11.
26. Brady A, Laoide RO, McCarthy P, McDermott R. Discrepancy and error in radiology: concepts, causes and consequences. Ulster Med J. 2012;81(1):3–9.
27. Breiman L. Random forests. Mach Learn. 2001;45(1):5–32.
28. Brenner DJ, Hall EJ. Computed tomography--an increasing source of radiation exposure. N Engl J Med. 2007;357(22):2277–84.

29. Brynolfsson P, Nilsson D, Henriksson R, Hauksson J, Karlsson M, Garpebring A, Birgander R, Trygg J, Nyholm T, Asklund T. ADC texture – an imaging biomarker for high-grade glioma? Med Phys. 2014;41(10):101903.
30. Buckler AJ, Bresolin L, Dunnick NR, Sullivan DC, Group. A collaborative enterprise for multi-stakeholder participation in the advancement of quantitative imaging. Radiology. 2011;258(3):906–14.
31. Buckler AJ, Paik D, Ouellette M, Danagoulian J, Wernsing G, Suzek BE. A novel knowledge representation framework for the statistical validation of quantitative imaging biomarkers. J Digit Imaging. 2013;26(4):614–29.
32. Bui AA, Hsu W, Arnold C, El-Saden S, Aberle DR, Taira RK. Imaging-based observational databases for clinical problem solving: the role of informatics. J Am Med Inform Assoc. 2013;20(6):1053–8.
33. Bulik M, Jancalek R, Vanicek J, Skoch A, Mechl M. Potential of MR spectroscopy for assessment of glioma grading. Clin Neurol Neurosurg. 2013;115(2):146–53.
34. Bushberg JT. The essential physics of medical imaging. 3rd ed. Philadelphia: Wolters Kluwer Health/Lippincott Williams & Wilkins; 2012. p. xii. 1030 p.
35. Cancer Genome Atlas Research Network. Comprehensive genomic characterization defines human glioblastoma genes and core pathways. Nature. 2008;455(7216):1061–8.
36. Cao Y, Li D, Shen Z, Normolle D. Sensitivity of quantitative metrics derived from DCE MRI and a pharmacokinetic model to image quality and acquisition parameters. Acad Radiol. 2010;17(4):468–78.
37. Chen R, Snyder M. Promise of personalized omics to precision medicine. Wiley Interdiscip Rev Syst Biol Med. 2013;5(1):73–82.
38. Chen W, Giger ML, Li H, Bick U, Newstead GM. Volumetric texture analysis of breast lesions on contrast-enhanced magnetic resonance images. Magn Reson Med. 2007;58 (3):562–71.
39. Chinot OL, Macdonald DR, Abrey LE, Zahlmann G, Kerloeguen Y, Cloughesy TF. Response assessment criteria for glioblastoma: practical adaptation and implementation in clinical trials of antiangiogenic therapy. Curr Neurol Neurosci Rep. 2013;13(5):347.
40. Chow KL, Gobin YP, Cloughesy T, Sayre JW, Villablanca JP, Vinuela F. Prognostic factors in recurrent glioblastoma multiforme and anaplastic astrocytoma treated with selective intra-arterial chemotherapy. AJNR Am J Neuroradiol. 2000;21(3):471–8.
41. Cohen AL, Holmen SL, Colman H. IDH1 and IDH2 mutations in gliomas. Curr Neurol Neurosci Rep. 2013;13(5):345.
42. Cootes TF, Taylor CJ. Anatomical statistical models and their role in feature extraction. Br J Radiol. 2004;77(Spec No 2):S133–9.
43. Cordova JS, Shu HG, Liang Z, Gurbani SS, Cooper LA, Holder CA, Olson JJ, Kairdolf B, Schreibmann E, Neill SG et al. Whole-brain spectroscopic MRI biomarkers identify infiltrating margins in glioblastoma patients. Neuro Oncol. 2016;18(8):1180–9.
44. Cristini V, Frieboes HB, Gatenby R, Caserta S, Ferrari M, Sinek J. Morphologic instability and cancer invasion. Clin Cancer Res. 2005;11(19 Pt 1):6772–9.
45. Curry III TS, Dowdey JE, Murry Jr RC. Christensen's physics of diagnostic radiology. 4th ed. Philadelphia: Lippincott Williams and Wilkins; 1990.
46. Dai XL, Khorram S. A feature-based image registration algorithm using improved chain-code representation combined with invariant moments. IEEE Trans Geosci Remote Sens. 1999;37 (5):2351–62.
47. Daianu M, Jahanshad N, Nir TM, Toga AW, Jack Jr CR, Weiner MW, Thompson PM, Alzheimer's Disease Neuroimaging Initiative. Breakdown of brain connectivity between normal aging and Alzheimer's disease: a structural k-core network analysis. Brain Connect. 2013;3(4):407–22.
48. Dalla Palma L, Stacul F, Meduri S, Geitung JT. Relationships between radiologists and clinicians: results from three surveys. Clin Radiol. 2000;55(8):602–5.

49. Davnall F, Yip CS, Ljungqvist G, Selmi M, Ng F, Sanghera B, Ganeshan B, Miles KA, Cook GJ, Goh V. Assessment of tumor heterogeneity: an emerging imaging tool for clinical practice? Insights Imaging. 2012;3(6):573–89.
50. de la Rocha AM, Sampron N, Alonso MM, Matheu A. Role of SOX family of transcription factors in central nervous system tumors. Am J Cancer Res. 2014;4(4):312–24.
51. Dean BL, Drayer BP, Bird CR, Flom RA, Hodak JA, Coons SW, Carey RG. Gliomas: classification with MR imaging. Radiology. 1990;174(2):411–15.
52. del Sol A, Balling R, Hood L, Galas D. Diseases as network perturbations. Curr Opin Biotechnol. 2010;21:566–71.
53. Delude CM. Deep phenotyping: the details of disease. Nature. 2015;527(7576):S14–15.
54. Dempsey MF, Condon BR, Hadley DM. Measurement of tumor "size" in recurrent malignant glioma: 1D, 2D, or 3D? AJNR Am J Neuroradiol. 2005;26(4):770–6.
55. Deoni SC, Rutt BK, Peters TM. Rapid combined T1 and T2 mapping using gradient recalled acquisition in the steady state. Magn Reson Med. 2003;49(3):515–26.
56. Diehn M, Nardini C, Wang DS, McGovern S, Jayaraman M, Liang Y, Aldape K, Cha S, Kuo MD. Identification of noninvasive imaging surrogates for brain tumor gene-expression modules. Proc Natl Acad Sci U S A. 2008;105(13):5213–18.
57. Dietrich O, Biffar A, Baur-Melnyk A, Reiser MF. Technical aspects of MR diffusion imaging of the body. Eur J Radiol. 2010;76(3):314–22.
58. Dominietto M, Lehmann S, Keist R, Rudin M. Pattern analysis accounts for heterogeneity observed in MRI studies of tumor angiogenesis. Magn Reson Med. 2013;70(5):1481–90.
59. Dreze M, Charloteaux B, Milstein S, Vidalain PO, Yildirim MA, Zhong Q, Svrzikapa N, Romero V, Laloux G, Brasseur R, et al. 'Edgetic' perturbation of a C. elegans BCL2 ortholog. Nat Methods. 2009;6(11):843–9.
60. Duda RO, Hart PE, Stork DG. Pattern Classification. 2nd ed. New York: John Wiley & Sons; 2001. 654 p.
61. El Naqa I, Grigsby P, Apte A, Kidd E, Donnelly E, Khullar D, Chaudhari S, Yang D, Schmitt M, Laforest R, et al. Exploring feature-based approaches in PET images for predicting cancer treatment outcomes. Pattern Recognit. 2009;42(6):1162–71.
62. Ellingson BM, Kim E, Woodworth DC, Marques H, Boxerman JL, Safriel Y, McKinstry RC, Bokstein F, Jain R, Chi TL, et al. Diffusion MRI quality control and functional diffusion map results in ACRIN 6677/RTOG 0625: a multicenter, randomized, phase II trial of bevacizumab and chemotherapy in recurrent glioblastoma. Int J Oncol. 2015;46(5):1883–92.
63. El-Naqa I, Yang Y, Galatsanos NP, Nishikawa RM, Wernick MN. A similarity learning approach to content-based image retrieval: application to digital mammography. IEEE Trans Med Imaging. 2004;23(10):1233–44.
64. Essig M, Shiroishi MS, Nguyen TB, Saake M, Provenzale JM, Enterline D, Anzalone N, Dorfler A, Rovira A, Wintermark M, et al. Perfusion MRI: the five most frequently asked technical questions. Am J Roentgenol. 2013;200(1):24–34.
65. Evans WE, Relling MV. Pharmacogenomics: translating functional genomics into rational therapeutics. Science. 1999;286(5439):487–91.
66. Fan X, Khaki L, Zhu TS, Soules ME, Talsma CE, Gul N, Koh C, Zhang J, Li YM, Maciaczyk J, et al. NOTCH pathway blockade depletes CD133-positive glioblastoma cells and inhibits growth of tumor neurospheres and xenografts. Stem Cells. 2010;28(1):5–16.
67. Fleishman GM, Gutman BA, Fletcher PT, Thompson PM. Simultaneous longitudinal registration with group-wise similarity prior. Inf Process Med Imaging. 2015;24:746–57.
68. Furnari FB, Fenton T, Bachoo RM, Mukasa A, Stommel JM, Stegh A, Hahn WC, Ligon KL, Louis DN, Brennan C, et al. Malignant astrocytic glioma: genetics, biology, and paths to treatment. Genes Dev. 2007;21(21):2683–710.
69. Gagliardi AR, Wright FC, Davis D, McLeod RS, Urbach DR. Challenges in multidisciplinary cancer care among general surgeons in Canada. BMC Med Inform Decis Mak. 2008;8:59.
70. Gallagher FA. An introduction to functional and molecular imaging with MRI. Clin Radiol. 2010;65(7):557–66.

71. Garcia-Figueiras R, Padhani AR, Beer AJ, Baleato-Gonzalez S, Vilanova JC, Luna A, Oleaga L, Gomez-Caamano A, Koh DM. Imaging of tumor angiogenesis for radiologists – Part 1: biological and technical basis. Curr Probl Diagn Radiol. 2015;44(5):407–24.
72. Garcia-Figueiras R, Padhani AR, Baleato-Gonzalez S. Therapy monitoring with functional and molecular MR imaging. Magn Reson Imaging Clin N Am. 2016;24(1):261–88.
73. Gauberti M, Montagne A, Quenault A, Vivien D. Molecular magnetic resonance imaging of brain-immune interactions. Front Cell Neurosci. 2014;8:389.
74. Gerstner ER, Frosch MP, Batchelor TT. Diffusion magnetic resonance imaging detects pathologically confirmed, nonenhancing tumor progression in a patient with recurrent glioblastoma receiving bevacizumab. J Clin Oncol. 2010;28(6):e91–3.
75. Gilad AA, Winnard Jr PT, van Zijl PC, Bulte JW. Developing MR reporter genes: promises and pitfalls. NMR Biomed. 2007;20(3):275–90.
76. Gillies RJ, Kinahan PE, Hricak H. Radiomics: images are more than pictures, they are data. Radiology. 2016;278(2):563–77.
77. Goldhirsch A, Wood WC, Coates AS, Gelber RD, Thurlimann B, Senn HJ, Panel members. Strategies for subtypes-dealing with the diversity of breast cancer: Highlights of the St. Gallen International Expert Consensus on the Primary Therapy of Early Breast Cancer 2011. Ann Oncol. 2011;22(8):1736–47.
78. Goldman AW, Burmeister Y, Cesnulevicius K, Herbert M, Kane M, Lescheid D, McCaffrey T, Schultz M, Seilheimer B, Smit A, et al. Bioregulatory systems medicine: an innovative approach to integrating the science of molecular networks, inflammation, and systems biology with the patient's autoregulatory capacity? Front Physiol. 2015;6:225.
79. Gong G, He Y, Concha L, Lebel C, Gross DW, Evans AC, Beaulieu C. Mapping anatomical connectivity patterns of human cerebral cortex using in vivo diffusion tensor imaging tractography. Cereb Cortex. 2009;19(3):524–36.
80. Goodman SN. Towards evidence-based medical statistics. 1: The P-value fallacy. Ann Intern Med. 1999;130:995–1004.
81. Grove O, Berglund AE, Schabath MB, Aerts HJ, Dekker A, Wang H, Velazquez ER, Lambin P, Gu Y, Balagurunathan Y, et al. Quantitative computed tomographic descriptors associate tumor shape complexity and intratumor heterogeneity with prognosis in lung adenocarcinoma. PLoS One. 2015;10(3):e0118261.
82. Gu Y, Kumar V, Hall LO, Goldgof DB, Li CY, Korn R, Bendtsen C, Velazquez ER, Dekker A, Aerts H, et al. Automated delineation of lung tumors from CT images using a single click ensemble segmentation approach. Pattern Recognit. 2013;46(3):692–702.
83. Guerrero T, Zhang G, Huang TC, Lin KP. Intrathoracic tumour motion estimation from CT imaging using the 3D optical flow method. Phys Med Biol. 2004;49(17):4147–61.
84. Guo Z, Shu Y, Zhou H, Zhang W, Wang H. Radiogenomics helps to achieve personalized therapy by evaluating patient responses to radiation treatment. Carcinogenesis. 2015;36 (3):307–17.
85. Gupta RK, Cloughesy TF, Sinha U, Garakian J, Lazareff J, Rubino G, Rubino L, Becker DP, Vinters HV, Alger JR. Relationships between choline magnetic resonance spectroscopy, apparent diffusion coefficient and quantitative histopathology in human glioma. J Neurooncol. 2000;50(3):215–26.
86. Gupta S, Takebe N, Lorusso P. Targeting the Hedgehog pathway in cancer. Ther Adv Med Oncol. 2010;2(4):237–50.
87. Gutman DA, Cooper LA, Hwang SN, Holder CA, Gao J, Aurora TD, Dunn Jr WD, Scarpace L, Mikkelsen T, Jain R, et al. MR imaging predictors of molecular profile and survival: multi-institutional study of the TCGA glioblastoma data set. Radiology. 2013;267 (2):560–9.
88. Haralick R, Shanmugam K, Dinstein I. Textural features for image classification. IEEE Trans Syst Man Cybern. 1973;3(6):610–21.

89. Hatanpaa KJ, Burma S, Zhao D, Habib AA. Epidermal growth factor receptor in glioma: signal transduction, neuropathology, imaging, and radioresistance. Neoplasia. 2010;12 (9):675–84.
90. Haynes B. Of studies, syntheses, synopses, summaries, and systems: the “5S” evolution of information services for evidence-based healthcare decisions. Evid Based Nurs. 2007;10 (1):6–7.
91. He L, Greenshields IR. A nonlocal maximum likelihood estimation method for Rician noise reduction in MR images. IEEE Trans Med Imaging. 2009;28(2):165–72.
92. Hegi ME, Diserens AC, Gorlia T, Hamou MF, de Tribolet N, Weller M, Kros JM, Hainfellner JA, Mason W, Mariani L, et al. MGMT gene silencing and benefit from temozolomide in glioblastoma. N Engl J Med. 2005;352(10):997–1003.
93. Heldin CH. Targeting the PDGF signaling pathway in tumor treatment. Cell Commun Signal. 2013;11:97.
94. Heldin CH, Rubin K, Pietras K, Ostman A. High interstitial fluid pressure – an obstacle in cancer therapy. Nat Rev Cancer. 2004;4(10):806–13.
95. Herold CJ, Lewin JS, Wibmer AG, Thrall JH, Krestin GP, Dixon AK, Schoenberg SO, Geckle RJ, Muellner A, Hricak H. Imaging in the Age of precision medicine: summary of the proceedings of the 10th biannual symposium of the international society for strategic studies in radiology. Radiology. 2016;279(1):226–38.
96. Hersh WR, Weiner MG, Embi PJ, Logan JR, Payne PRO, Bernstam EV, Lehmann HP, Hripcsak G, Hartzog TH, Cimino JJ, et al. Caveats for the use of operational electronic health record data in comparative effectiveness research. Med Care. 2013;51(8):S30–7.
97. Heye AK, Culling RD, Hernandez MDV, Thrippleton MJ, Wardlaw JM. Assessment of blood–brain barrier disruption using dynamic contrast-enhanced MRI. A systematic review. Neuroimage-Clin. 2014;6:262–74.
98. Higano S, Yun X, Kumabe T, Watanabe M, Mugikura S, Umetsu A, Sato A, Yamada T, Takahashi S. Malignant astrocytic tumors: clinical importance of apparent diffusion coefficient in prediction of grade and prognosis. Radiology. 2006;241(3):839–46.
99. Hinton GE, Salakhutdinov RR. Reducing the dimensionality of data with neural networks. Science. 2006;313(5786):504–7.
100. Hinton GE, McClelland JL, Rumelhart DE. Distributed representations, in parallel distributed processing: explorations in the microstructure of cognition, In: Rumelhart DE, McClelland JL, editors. Cambridge, MA: MIT Press; 1986.
101. Hobbs SK, Shi G, Homer R, Harsh G, Atlas SW, Bednarski MD. Magnetic resonance image-guided proteomics of human glioblastoma multiforme. J Magn Reson Imaging. 2003;18 (5):530–6.
102. Holodny AI, Nusbaum AO, Festa S, Pronin IN, Lee HJ, Kalnin AJ. Correlation between the degree of contrast enhancement and the volume of peritumoral edema in meningiomas and malignant gliomas. Neuroradiology. 1999;41(11):820–5.
103. Honda T, Kondo T, Murakami S, Saito H, Oshita F, Ito H, Tsuboi M, Nakayama H, Yokose T, Kameda Y, et al. Radiographic and pathological analysis of small lung adenocarcinoma using the new IASLC classification. Clin Radiol. 2013;68(1):e21–6.
104. Horska A, Barker PB. Imaging of brain tumors: MR spectroscopy and metabolic imaging. Neuroimaging Clin N Am. 2010;20(3):293–310.
105. Houhou N, Bresson X, Szlam A, Chan TF, Thiran J-P. Semi-supervised segmentation based on non-local continuous min-cut. Scale Space Variational Methods Comput Vision. 2009;5567:112–23.
106. Hsu W, Taira RK, El-Saden S, Kangarloo H, Bui AA. Context-based electronic health record: toward patient specific healthcare. IEEE Trans Inf Technol Biomed. 2012;16(2):228–34.
107. Hsu W, Han SX, Arnold CW, Bui AA, Enzmann DR. A data-driven approach for quality assessment of radiologic interpretations. J Am Med Inform Assoc. 2016;23(e1):e152–6.

108. Hua X, Hibar DP, Ching CR, Boyle CP, Rajagopalan P, Gutman BA, Leow AD, Toga AW, Jack Jr CR, Harvey D, et al. Unbiased tensor-based morphometry: improved robustness and sample size estimates for Alzheimer's disease clinical trials. Neuroimage. 2013;66:648–61.
109. Hua X, Ching CR, Mezher A, Gutman BA, Hibar DP, Bhatt P, Leow AD, Jack Jr CR, Bernstein MA, Weiner MW, et al. MRI-based brain atrophy rates in ADNI phase 2: acceleration and enrichment considerations for clinical trials. Neurobiol Aging. 2016;37:26–37.
110. Huang RY, Neagu MR, Reardon DA, Wen PY. Pitfalls in the neuroimaging of glioblastoma in the era of antiangiogenic and immuno/targeted therapy – detecting illusive disease, defining response. Front Neurol. 2015;6:33.
111. Hubbard R, Bayarri MJ. Confusion over measures of evidence (p's) versus errors (a's) in classical statistical testing. Am Stat. 2003;57(3):171–8.
112. Ioannidis JPA. Why most published research findings are false. Plos Med. 2005;2 (8):696–701.
113. Jackson A, Kassner A, Annesley-Williams D, Reid H, Zhu XP, Li KL. Abnormalities in the recirculation phase of contrast agent bolus passage in cerebral gliomas: comparison with relative blood volume and tumor grade. AJNR Am J Neuroradiol. 2002;23(1):7–14.
114. Jacobs RE, Cherry SR. Complementary emerging techniques: high-resolution PET and MRI. Curr Opin Neurobiol. 2001;11(5):621–9.
115. Jäger F, Yu D-Z, Frericks B, Wacker F, Hornegger J. A new method for MRI intensity standardization with application to lesion detection in the brain. In: Kobbelt L et al. editors. Vision modeling and visualization 2006. Berlin: Akademische Verlagsgesellschaft Aka GmbH; 2006. p. 269–76.
116. Jaynes ET. Information theory and statistical mechanics. Phys Rev. 1957;106(4):620–30.
117. Jenkinson MD, du Plessis DG, Smith TS, Joyce KA, Warnke PC, Walker C. Histological growth patterns and genotype in oligodendroglial tumours: correlation with MRI features. Brain. 2006;129(Pt 7):1884–91.
118. Joo KM, Jin J, Kim E, Ho Kim K, Kim Y, Gu Kang B, Kang YJ, Lathia JD, Cheong KH, Song PH, et al. MET signaling regulates glioblastoma stem cells. Cancer Res. 2012;72 (15):3828–38.
119. Kalpathy-Cramer J, Gerstner ER, Emblem KE, Andronesi OC, Rosen B. Advanced magnetic resonance imaging of the physical processes in human glioblastoma. Cancer Res. 2014;74 (17):4622–37.
120. Kanewala U, Bieman JM. Testing scientific software: a systematic literature review. Inf Softw Technol. 2014;56(10):1219–32.
121. Kansagra AP, Yu JP, Chatterjee AR, Lenchik L, Chow DS, Prater AB, Yeh J, Doshi AM, Hawkins CM, Heilbrun ME, et al. Big data and the future of radiology informatics. Acad Radiol. 2016;23(1):30–42.
122. Kansal AR, Torquato S, Chiocca EA, Deisboeck TS. Emergence of a subpopulation in a computational model of tumor growth. J Theor Biol. 2000;207(3):431–41.
123. Kim WY, Danias PG, Stuber M, Flamm SD, Plein S, Nagel E, Langerak SE, Weber OM, Pedersen EM, Schmidt M, et al. Coronary magnetic resonance angiography for the detection of coronary stenoses. N Engl J Med. 2001;345(26):1863–9.
124. Klein A, Andersson J, Ardekani BA, Ashburner J, Avants B, Chiang MC, Christensen GE, Collins DL, Gee J, Hellier P, et al. Evaluation of 14 nonlinear deformation algorithms applied to human brain MRI registration. Neuroimage. 2009;46(3):786–802.
125. Koh DM, Collins DJ. Diffusion-weighted MRI in the body: applications and challenges in oncology. AJR Am J Roentgenol. 2007;188(6):1622–35.
126. Kono K, Inoue Y, Nakayama K, Shakudo M, Morino M, Ohata K, Wakasa K, Yamada R. The role of diffusion-weighted imaging in patients with brain tumors. AJNR Am J Neuroradiol. 2001;22(6):1081–8.
127. Koul D. PTEN signaling pathways in glioblastoma. Cancer Biol Ther. 2008;7(9):1321–5.

128. Krizhevsky A, Sutskever I, Hinton GE. Imagenet classification with deep convolutional neural networks. Adv Neural Inf Proces Syst. 2012;2:1097–105.
129. Ku BM, Lee YK, Ryu J, Jeong JY, Choi J, Eun KM, Shin HY, Kim DG, Hwang EM, Yoo JC, et al. CHI3L1 (YKL-40) is expressed in human gliomas and regulates the invasion, growth and survival of glioma cells. Int J Cancer. 2011;128(6):1316–26.
130. Kumar V, Gu Y, Basu S, Berglund A, Eschrich SA, Schabath MB, Forster K, Aerts HJ, Dekker A, Fenstermacher D, et al. Radiomics: the process and the challenges. Magn Reson Imaging. 2012;30(9):1234–48.
131. Kwitt R, Hegenbart S, Rasiwasia N, Vecsei A, Uhl A. Do we need annotation experts? A case study in celiac disease classification. Med Image Comput Comput Assist Interv. 2014;17 (Pt 2):454–61.
132. Lacerda S, Law M. Magnetic resonance perfusion and permeability imaging in brain tumors. Neuroimaging Clin N Am. 2009;19(4):527–57.
133. Lambin P, Rios-Velazquez E, Leijenaar R, Carvalho S, van Stiphout RG, Granton P, Zegers CM, Gillies R, Boellard R, Dekker A, et al. Radiomics: extracting more information from medical images using advanced feature analysis. Eur J Cancer. 2012;48(4):441–6.
134. Landau DA, Carter SL, Stojanov P, McKenna A, Stevenson K, Lawrence MS, Sougnez C, Stewart C, Sivachenko A, Wang L, et al. Evolution and impact of subclonal mutations in chronic lymphocytic leukemia. Cell. 2013;152(4):714–26.
135. Landgrebe TCW, Duin RPW. Approximating the multiclass ROC by pairwise analysis. Pattern Recogn Lett. 2007;28(13):1747–58.
136. LeCun Y, Bottou L, Bengio Y, Haffner P. Gradient-based learning applied to document recognition. Proc IEEE. 1998;86(11):2278–324.
137. LeCun Y, Bengio Y, Hinton G. Deep learning. Nature. 2015;521(7553):436–44.
138. Lee J, Lustig M, Kim DH, Pauly JM. Improved shim method based on the minimization of the maximum off-resonance frequency for balanced steady-state free precession (bSSFP). Magn Reson Med. 2009;61(6):1500–6.
139. Leow AD, Yanovsky I, Chiang MC, Lee AD, Klunder AD, Lu A, Becker JT, Davis SW, Toga AW, Thompson PM. Statistical properties of Jacobian maps and the realization of unbiased large-deformation nonlinear image registration. IEEE Trans Med Imaging. 2007;26 (6):822–32.
140. Liu H, Motoda H. Computational methods of feature selection. Chapman & Hall/CRC data mining and knowledge discovery series. Boca Raton: Chapman & Hall/CRC; 2008. 419 p.
141. Liu H, Setiono R. Feature selection via discretization. IEEE Trans Knowl Data Eng. 1997;9 (4):642–5.
142. Lustgarten JL, Gopalakrishnan V, Grover H, Visweswaran S. Improving classification performance with discretization on biomedical datasets. AMIA Annu Symp Proc. 2008;445–9.
143. Madabhushi A, Udupa JK. Interplay between intensity standardization and inhomogeneity correction in MR image processing. IEEE Trans Med Imaging. 2005;24(5):561–76.
144. Mandelbrot BB. The fractal geometry of nature. Updated and augm. ed. New York: W.H. Freeman; 1983. 468 p.
145. Massoud TF, Gambhir SS. Molecular imaging in living subjects: seeing fundamental biological processes in a new light. Genes Dev. 2003;17(5):545–80.
146. Matiasz NJ, Silva AJ, Hsu W. Synthesizing clinical trials for evidence-based medicine: a representation of empirical and hypothetical causal relations. In: AMIA 2015 joint summits on translational science. Poster Presentation: San Francisco, CA; 2015.
147. McGillicuddy LT, Fromm JA, Hollstein PE, Kubek S, Beroukhim R, De Raedt T, Johnson BW, Williams SM, Nghiemphu P, Liau LM, et al. Proteasomal and genetic inactivation of the NF1 tumor suppressor in gliomagenesis. Cancer Cell. 2009;16(1):44–54.
148. McInerney T, Terzopoulos D. Deformable models in medical image analysis: a survey. Med Image Anal. 1996;1(2):91–108.

149. Nagao M, Murase K. Measurement of heterogeneous distribution on Technegas SPECT images by three-dimensional fractal analysis. Ann Nucl Med. 2002;16(6):369–76.
150. Nagarajan R, Ramadan S, Thomas MA. Detection of amide and aromatic proton resonances of human brain metabolites using localized correlated spectroscopy combined with two different water suppression schemes. Magn Reson Insights. 2010;2010(4):1–9.
151. Nakano T, Asano K, Miura H, Itoh S, Suzuki S. Meningiomas with brain edema: radiological characteristics on MRI and review of the literature. Clin Imaging. 2002;26(4):243–9.
152. National Research Council. (US) Committee on a framework for developing a new taxonomy of disease. Toward precision medicine: building a knowledge network for biomedical research and a new taxonomy of disease. The National Academies Collection: Reports funded by National Institutes of Health. Washington, DC: National Academies Press; 2011.
153. Niemeijer M, van Ginneken B, Staal J, Suttorp-Schulten MSA, Abramoff MD. Automatic detection of red lesions in digital color fundus photographs. IEEE Trans Med Imaging. 2005;24(5):584–92.
154. Nyul LG, Udupa JK. On standardizing the MR image intensity scale. Magn Reson Med. 1999;42(6):1072–81.
155. Ohgaki H, Kleihues P. Genetic pathways to primary and secondary glioblastoma. Am J Pathol. 2007;170(5):1445–53.
156. Ortensi B, Setti M, Osti D, Pelicci G. Cancer stem cell contribution to glioblastoma invasiveness. Stem Cell Res Ther. 2013;4(1):18.
157. Osada R, Funkhouser T, Chazelle B, Dobkin D. Shape distributions. ACM Trans Graph. 2002;21(4):807–32.
158. Ozkan E, West A, Dedelow JA, Chu BF, Zhao W, Yildiz VO, Otterson GA, Shilo K, Ghosh S, King M, et al. CT gray-level texture analysis as a quantitative imaging biomarker of epidermal growth factor receptor mutation status in adenocarcinoma of the lung. AJR Am J Roentgenol. 2015;205(5):1016–25.
159. Padhani AR, Liu G, Koh DM, Chenevert TL, Thoeny HC, Takahara T, Dzik-Jurasz A, Ross BD, Van Cauteren M, Collins D, et al. Diffusion-weighted magnetic resonance imaging as a cancer biomarker: consensus and recommendations. Neoplasia. 2009;11(2):102–25.
160. Pearl J. Probabilistic reasoning in intelligent systems: networks of plausible inference. 2nd ed. Representation and reasoning series. San Francisco: Morgan Kaufmann; 1988.
161. Pedro RWD, Nunes FLS, Machado-Lima A. Using grammars for pattern recognition in images: a systematic review. ACM Comput Surv. 2013;46(2):26.
162. Peng Y, Jiang Y, Yang C, Brown JB, Antic T, Sethi I, Schmid-Tannwald C, Giger ML, Eggener SE, Oto A. Quantitative analysis of multiparametric prostate MR images: differentiation between prostate cancer and normal tissue and correlation with Gleason score--a computer-aided diagnosis development study. Radiology. 2013;267(3):787–96.
163. Pennacchietti S, Michieli P, Galluzzo M, Mazzone M, Giordano S, Comoglio PM. Hypoxia promotes invasive growth by transcriptional activation of the met protooncogene. Cancer Cell. 2003;3(4):347–61.
164. Petrella JR. Use of graph theory to evaluate brain networks: a clinical tool for a small world? Radiology. 2011;259(2):317–20.
165. Pierallini A, Bonamini M, Bozzao A, Pantano P, DiStefano D, Ferone E, Raguso M, Bosman C, Bozzao L. Supratentorial diffuse astrocytic tumours: proposal of an MRI classification. Eur Radiol. 1997;7(3):395–9.
166. Pinker K, Stadlbauer A, Bogner W, Gruber S, Helbich TH. Molecular imaging of cancer: MR spectroscopy and beyond. Eur J Radiol. 2012;81(3):566–77.
167. Pisano ED, Gatsonis C, Hendrick E, Yaffe M, Baum JK, Acharyya S, Conant EF, Fajardo LL, Bassett L, D'Orsi C, et al. Diagnostic performance of digital versus film mammography for breast-cancer screening. N Engl J Med. 2005;353(17):1773–83.
168. Poon M, Hamarneh G, Abugharbieh R. Efficient interactive 3D Livewire segmentation of complex objects with arbitrary topology. Comput Med Imaging Graph. 2008;32(8):639–50.

169. Pope WB, Sayre J, Perlina A, Villablanca JP, Mischel PS, Cloughesy TF. MR imaging correlates of survival in patients with high-grade gliomas. AJNR Am J Neuroradiol. 2005;26(10):2466–74.
170. Prasad M, Sowmya A, Koch I. Efficient feature selection based on independent component analysis. In: Proceedings of the 2004 intelligent sensors, sensor networks & information processing conference; 2004. p. 427–32.
171. Rall LB. Representations of intervals and optimal error-bounds. Math Comput. 1983;41 (163):219–27.
172. Randen T, Husoy JH. FILTERING for texture classification: a comparative study. IEEE Trans Pattern Anal Mach Intell. 1999;21(4):291–310.
173. Raza SM, Fuller GN, Rhee CH, Huang S, Hess K, Zhang W, Sawaya R. Identification of necrosis-associated genes in glioblastoma by cDNA microarray analysis. Clin Cancer Res. 2004;10(1 Pt 1):212–21.
174. Rheinbay E, Suva ML, Gillespie SM, Wakimoto H, Patel AP, Shahid M, Oksuz O, Rabkin SD, Martuza RL, Rivera MN, et al. An aberrant transcription factor network essential for Wnt signaling and stem cell maintenance in glioblastoma. Cell Rep. 2013;3(5):1567–79.
175. Ribeiro MT, Singh S, Guestrin C. "Why should I trust you?": explaining the predictions of any classifier. Computer Science: Learning, Cornell University Library, 2016. Last Accessed: March 23, 2016. Available from http://arxiv.org/abs/1602.04938
176. Rios Piedra EA, Taira RK, El-Saden S, Ellingson B, Bui A, Hsu W. Assessing variability in brain tumor segmentation to improve volumetric accuracy and characterization of change. In: Proceedings of the IEEE international conference on biomedical and health informatics. Las Vegas, NV. 2016.
177. Robbins ME, Brunso-Bechtold JK, Peiffer AM, Tsien CI, Bailey JE, Marks LB. Imaging radiation-induced normal tissue injury. Radiat Res. 2012;177(4):449–66.
178. Robinson PN. Deep phenotyping for precision medicine. Hum Mutat. 2012;33(5):777–80.
179. Rubinov M, Sporns O. Complex network measures of brain connectivity: uses and interpretations. Neuroimage. 2010;52(3):1059–69.
180. Saeys Y, Inza I, Larranaga P. A review of feature selection techniques in bioinformatics. Bioinformatics. 2007;23(19):2507–17.
181. Sahin M, Sur M. Genes, circuits, and precision therapies for autism and related neurodevelopmental disorders. Science. 2015;350(6263):aab3897.
182. Saria S, Goldenberg A. Subtyping: what It is and its role in precision medicine. IEEE Intell Syst. 2015;30(4):70–5.
183. Scheffler K, Hennig J. T1 quantification with inversion recovery TrueFISP. Magn Reson Med. 2001;45(4):720–3.
184. Schneider U, Pedroni E, Lomax A. The calibration of CT Hounsfield units for radiotherapy treatment planning. Phys Med Biol. 1996;41(1):111–24.
185. Shen L, Rangayyan RM, Desautels JL. Detection and classification of mammographic calcifications. Int J Pattern Recognit Artif Intell. 1993;7(6):1403–16.
186. Silva AJ, Muller KR. The need for novel informatics tools for integrating and planning research in molecular and cellular cognition. Learn Mem. 2015;22(9):494–8.
187. Sim I, Carini S, Tu S, Wynden R, Pollock BH, Mollah SA, Gabriel D, Hagler HK, Scheuermann RH, Lehmann HP, et al. The human studies database project: federating human studies design data using the ontology of clinical research. AMIA Jt Summits Transl Sci Proc. 2010;2010:51–5.
188. Simonyan K, Zisserman A. Very deep convolutional networks for large-scale image recognition. Computer science: computer vision and pattern recognition, Cornell University Library, 2015. Last Accessed: 24 March 2016. Available from http://arxiv.org/abs/1409.1556
189. Sinha S, Sinha U, Kangarloo H, Huang HK. Magnetic resonance image synthesis from analytic solutions of spin-echo and radio frequency-spoiled gradient-echo images. Invest Radiol. 1992;27(10):856–64.

190. Smith NB, Webb A. Introduction to medical imaging: physics, engineering and clinical applications. New York: Cambriduge University Press; 2011.
191. Smith JJ, Sorensen AG, Thrall JH. Biomarkers in imaging: realizing radiology's future. Radiology. 2003;227(3):633–8.
192. Sotiras A, Davatzikos C, Paragios N. Deformable medical image registration: a survey. IEEE Trans Med Imaging. 2013;32(7):1153–90.
193. Spence AM, Muzi M, Swanson KR, O'Sullivan F, Rockhill JK, Rajendran JG, Adamsen TCH, Link JM, Swanson PE, Yagle KJ, et al. Regional hypoxia in glioblastoma multiforme quantified with [F-18] fluoromisonidazole positron emission tomography before radiotherapy: correlation with time to progression and survival. Clin Cancer Res. 2008;14(9):2623–30.
194. Sporns O. Networks of the brain. Cambridge, MA: MIT Press; 2011. xi, 412 p., 8 p. of plates.
195. Stebbins GT, Murphy CM. Diffusion tensor imaging in Alzheimer's disease and mild cognitive impairment. Behav Neurol. 2009;21(1):39–49.
196. Suzuki T, Izumoto S, Fujimoto Y, Maruno M, Ito Y, Yoshimine T. Clinicopathological study of cellular proliferation and invasion in gliomatosis cerebri: important role of neural cell adhesion molecule L1 in tumour invasion. J Clin Pathol. 2005;58(2):166–71.
197. Szigeti K, Szabo T, Korom C, Czibak I, Horvath I, Veres DS, Gyongyi Z, Karlinger K, Bergmann R, Pocsik M, et al. Radiomics-based differentiation of lung disease models generated by polluted air based on X-ray computed tomography data. BMC Med Imaging. 2016;16(1):14.
198. Team NLSTR, Aberle DR, Berg CD, Black WC, Church TR, Fagerstrom RM, Galen B, Gareen IF, Gatsonis C, Goldin J, et al. The national lung screening trial: overview and study design. Radiology. 2011;258(1):243–53.
199. Teruel JR, Heldahl MG, Goa PE, Pickles M, Lundgren S, Bathen TF, Gibbs P. Dynamic contrast-enhanced MRI texture analysis for pretreatment prediction of clinical and pathological response to neoadjuvant chemotherapy in patients with locally advanced breast cancer. NMR Biomed. 2014;27(8):887–96.
200. Thomassin-Naggara I, Siles P, Trop I, Chopier J, Darai E, Bazot M, Uzan S. How to measure breast cancer tumor size at MR imaging? Eur J Radiol. 2013;81(12):e790–800.
201. Thompson PM, Toga AW. A framework for computational anatomy. Comput Visual Sci. 2002;5:13–34.
202. Thompson G, Mills SJ, Stivaros SM, Jackson A. Imaging of brain tumors: perfusion/permeability. Neuroimaging Clin N Am. 2010;20(3):337–53.
203. Thompson PM, Stein JL, Medland SE, Hibar DP, Vasquez AA, Renteria ME, Toro R, Jahanshad N, Schumann G, Franke B, et al. The ENIGMA Consortium: large-scale collaborative analyses of neuroimaging and genetic data. Brain Imaging Behav. 2014;8(2):153–82.
204. Thornhill RE, Chen S, Rammo W, Mikulis DJ, Kassner A. Contrast-enhanced MR imaging in acute ischemic stroke: T2* measures of blood–brain barrier permeability and their relationship to T1 estimates and hemorrhagic transformation. AJNR Am J Neuroradiol. 2010;31 (6):1015–22.
205. Thrall JH. Appropriateness and imaging utilization: "computerized provider order entry and decision support". Acad Radiol. 2014;21(9):1083–7.
206. Tien RD, Felsberg GJ, Friedman H, Brown M, MacFall J. MR imaging of high-grade cerebral gliomas: value of diffusion-weighted echoplanar pulse sequences. AJR Am J Roentgenol. 1994;162(3):671–7.
207. Tofts PS, Brix G, Buckley DL, Evelhoch JL, Henderson E, Knopp MV, Larsson HB, Lee TY, Mayr NA, Parker GJ, et al. Estimating kinetic parameters from dynamic contrast-enhanced T (1)-weighted MRI of a diffusable tracer: standardized quantities and symbols. J Magn Reson Imaging. 1999;10(3):223–32.
208. Tong M, Hsu W, Taira RK. A formal representation for numerical data presented in published clinical trial reports. Stud Health Technol Inform. 2013;192:856–60.

209. Tourassi GD, Ike 3rd R, Singh S, Harrawood B. Evaluating the effect of image preprocessing on an information-theoretic CAD system in mammography. Acad Radiol. 2008;15 (5):626–34.
210. Turchi L, Debruyne DN, Almairac F, Virolle V, Fareh M, Neirijnck Y, Burel-Vandenbos F, Paquis P, Junier MP, Van Obberghen-Schilling E, et al. Tumorigenic potential of miR-18A* in glioma initiating cells requires NOTCH-1 signaling. Stem Cells. 2013;31(7):1252–65.
211. Verhaak RG, Hoadley KA, Purdom E, Wang V, Qi Y, Wilkerson MD, Miller CR, Ding L, Golub T, Mesirov JP, et al. Integrated genomic analysis identifies clinically relevant subtypes of glioblastoma characterized by abnormalities in PDGFRA, IDH1, EGFR, and NF1. Cancer Cell. 2010;17(1):98–110.
212. Vlieger EJ, Majoie CB, Leenstra S, den Heeten GJ. Functional magnetic resonance imaging for neurosurgical planning in neurooncology. Eur Radiol. 2004;14(7):1143–53.
213. Wachowicz K. Evaluation of active and passive shimming in magnetic resonance imaging. Res Rep Nucl Med. 2014;4:1–12.
214. Wang Q, Xu X, Zhang M. Normal aging in the basal ganglia evaluated by eigenvalues of diffusion tensor imaging. AJNR Am J Neuroradiol. 2010;31(3):516–20.
215. Weber GL, Parat MO, Binder ZA, Gallia GL, Riggins GJ. Abrogation of PIK3CA or PIK3R1 reduces proliferation, migration, and invasion in glioblastoma multiforme cells. Oncotarget. 2011;2(11):833–49.
216. Weigelt B, Baehner FL, Reis-Filho JS. The contribution of gene expression profiling to breast cancer classification, prognostication and prediction: a retrospective of the last decade. J Pathol. 2010;220(2):263–80.
217. Weiskopf NG, Weng C. Methods and dimensions of electronic health record data quality assessment: enabling reuse for clinical research. J Am Med Inform Assoc. 2013;20 (1):144–51.
218. Wick W, Platten M, Wick A, Hertenstein A, Radbruch A, Bendszus M, Winkler F. Current status and future directions of anti-angiogenic therapy for gliomas. Neuro Oncol. 2016;18 (3):315–28.
219. Wilkinson ID, Romanowski CAJ, Jellinek DA, Morris J, Griffiths PD. Motor functional MRI for pre-operative and intraoperative neurosurgical guidance. Br J Radiol. 2003;76 (902):98–103.
220. Xue S, Qiao J, Pu F, Cameron M, Yang JJ. Design of a novel class of protein-based magnetic resonance imaging contrast agents for the molecular imaging of cancer biomarkers. Wiley Interdiscip Rev Nanomed Nanobiotechnol. 2013;5(2):163–79.
221. Yu K, Lin YQ, Lafferty J. Learning image representations from the pixel level via hierarchical sparse coding. 2011 I.E. conference on computer vision and pattern recognition (Cvpr), 2011. p. 1713–1720.
222. Zacharaki EI, Wang S, Chawla S, Soo Yoo D, Wolf R, Melhem ER, Davatzikos C. Classification of brain tumor type and grade using MRI texture and shape in a machine learning scheme. Magn Reson Med. 2009;62(6):1609–18.
223. Zeiler MD, Fergus R. Visualizing and understanding convolutional networks. In: Computer vision – ECCV 2014: 13th European Conference, Zurich, Switzerland, September 6–12, 2014, Proceedings, part i. New York: Springer; 2014. p. 818–33.
224. Zhang DS, Lu GJ. Review of shape representation and description techniques. Pattern Recogn. 2004;37(1):1–19.
225. Zhao BS, James LP, Moskowitz CS, Guo PZ, Ginsberg MS, Lefkowitz RA, Qin YL, Riely GJ, Kris MG, Schwartz LH. Evaluating variability in tumor measurements from same-day repeat CT scans of patients with non-small cell lung cancer. Radiology. 2009;252(1):263–72.
226. Zhong Q, Simonis N, Li QR, Charloteaux B, Heuze F, Klitgord N, Tam S, Yu H, Venkatesan K, Mou D, et al. Edgetic perturbation models of human inherited disorders. Mol Syst Biol. 2009;5:321.
227. Zhu SC, Mumford D. A stochastic grammar of images. Found Trends Comput Graph Vision. 2006;2(4):259–362.

228. Zhu SC, Wu YN, Mumford D. Minimax entropy principle and its application to texture modeling. Neural Comput. 1997;9(8):1627–60.
229. Zhuge Y, Udupa JK. Intensity standardization simplifies brain MR image segmentation. Comput Vis Image Underst. 2009;113(10):1095–103.
230. Zinn PO, Mahajan B, Sathyan P, Singh SK, Majumder S, Jolesz FA, Colen RR. Radiogenomic mapping of edema/cellular invasion MRI-phenotypes in glioblastoma multiforme. PLoS One. 2011;6(10):e25451.

Chapter 9
LIMS and Clinical Data Management

Yalan Chen, Yuxin Lin, Xuye Yuan, and Bairong Shen

Abstract In order to achieve more accurate disease prevention, diagnosis, and treatment, clinical and genetic data need extensive and systematically associated study. As one way to achieve precision medicine, a laboratory information management system (LIMS) can effectively associate clinical data in a macrocosmic aspect and genomic data in a microcosmic aspect. This chapter summarizes the application of the LIMS in a clinical data management and implementation mode. It also discusses the principles of a LIMS in clinical data management, as well as the opportunities and challenges in the context of medical informatics.

Keywords Big data • Clinical laboratory techniques • Database • Data management

9.1 Introduction

With the rapid development of medical informatics and technological breakthroughs in molecular biology, large amounts of clinical and biomedical data have been accumulated [26]. These data cover multiple levels, including both clinical data in a macrocosmic aspect and genomic data in a microcosmic aspect, which present more approaches and new opportunities to research disease and improve the quality of healthcare [24]. However, most clinical data have no corresponding genomic data, while most genomic data have no precise clinical annotation data. How to comprehensively and systematically utilize these data at all

Y. Chen
Center for Systems Biology, Soochow University, No. 1 Shizi Street, 206, 215006 Suzhou, Jiangsu, China

Department of Medical Informatics, School of Medicine, Nantong University, Nantong, Jiangsu, China

Y. Lin • X. Yuan • B. Shen (✉)
Center for Systems Biology, Soochow University, No. 1 Shizi Street, 206, 215006 Suzhou, Jiangsu, China
e-mail: bairong.shen@suda.edu.cn

B. Shen et al. (eds.), *Translational Biomedical Informatics*, Advances in Experimental Medicine and Biology 939, DOI 10.1007/978-981-10-1503-8_9

levels to achieve the precise early detection, prevention, or treatment of diseases becomes an imminent challenge in the era of precision medicine.

Big data technologies are now increasingly used for biomedical and health-care informatics researches. Breakthroughs in technological omics, especially with next-generation sequencing (NGS), bring avalanches of complicated data which need to undergo effective data management to ensure integrity, security, and maximal knowledge gleaning [14].

An ideal data management system should include strong system compatibility, flexible input formats, diverse data entry mechanisms and views, user-friendliness, attention to standards, hardware and software platform definition, as well as robustness. Relevant solutions elaborated by the scientific community include the laboratory information management system (LIMS), standardization of protocols, and facilitating data sharing and managing.

The traditional LIMS is a type of information system implemented as a software utility specifically designed to improve data acquisition and sample monitoring along laboratory workflows, supporting sample reporting [33]. Most academic laboratories around the world have adopted diverse LIMSs according to their research objectives [2, 33, 38, 47]. Organization and presentation of biodiversity data is greatly facilitated by LIMSs that are specially designed to allow easy data entry and organized data display [51]. One of the most crucial characteristics of a day-to-day LIMS is the collection, storage, and retrieval of information about research subjects and environmental or biomedical samples.

In addition to the routine function of data management, a clinical LIMS should also have powerful capabilities of compatibility, data sharing, and analysis with other clinical information systems, such as hospital information systems (HIS), picture archiving and communication systems, etc. More importantly, as one of the effective ways to achieve precision medicine [37], a clinical LIMS should have the ability to perform systems biology researches of disease, to provide the accurate and comprehensive disease information needed for physicians that will influence clinical decisions. The roles of the LIMS in clinical data management are significant; it can achieve accurate diagnosis, provide convenient and effective access to disease related data, decrease redundancy and costs, and facilitate the integration and collection of data from different types of instruments and systems [28]. Point-of-care testing in the LIMS also allows the rapid and precise analysis of samples to quickly facilitate prompt and accurate clinical decision-making [27].

At present, many commercial and noncommercial LIMSs are available for this purpose in clinical studies. However, many limitations are the bottleneck for the effective utilization of a LIMS in clinical and disease research, such as the effective sharing of genomic and clinical data, standard data exchange format [50], unified work processes, etc. [13].

In this chapter, we summarize the application of a LIMS in clinical data management and implementation mode. We also discuss the principles of the LIMS in clinical data management and the opportunities and challenges in the context of medical informatics.

9.2 Overview of the LIMS in Clinical Data Management

9.2.1 Advances of LIMS in Clinical Data Management

With the gradually deepening study of disease and the rapidly falling cost and availability of high-throughput sequencing and microarray technologies, clinical laboratories have accumulated massive research data including patient demographic data, histopathologic data, imaging data, genomic data, etc. which provide a valuable resource for medical research.

Currently, different LIMSs based on different platforms, commercial and noncommercial, have been developed in clinics to facilitate clinical data management and application, although, largely, available software solutions are limited to a large extent and commercial LIMSs are expensive. The general development of LIMSs in clinical applications mainly experiences the following three stages.

9.2.1.1 Routine Clinical Laboratory Management

At present, a lot of clinical biology laboratories have used multiplex LIMSs to effectively perform data utilizing of those mainly focused on internal data management. Efficient data storage and tracking and monitoring of all phases of laboratory activities can help to improve work efficiency, identify and troubleshoot problems more quickly, and reduce the risk of process failures and their related costs [33].

Some LIMSs are customized to realize special functions such as MOLE: a data management application based on a protein production data model [30]. Many LIMSs have become comprehensive and searchable databases such as PASSIM – an open-source software system for managing information in biomedical studies [48], which contains photographs and micrographs of samples and collection sites, geo-referenced collecting information, taxonomic data, and standardized sequence data. The majority of LIMSs possess user-friendly [40] interfaces in order for research teams to share and explore data generated within different research projects [29].

9.2.1.2 Clinical Study Based on a LIMS

With the promotion of technology and data application, LIMSs gradually play a growing role in clinical researches, from primitive sampling to data analysis. A growing number of LIMSs are developed for special clinical purposes, such as pharmaceutical research [39], drug abuse testing [9], or process development, and continuing to the execution and requisite follow-up of patients on clinical trials [36].

Some academic or clinical researchers have begun to combine the data of the LIMS and HIS to perform data analysis, mining researches, and population investigations [7, 41, 53].

9.2.1.3 Systems Biology Research Based on the LIMS

With the accumulation of clinical data by LIMSs and continually replenished genomic data, the multiple levels of data involved in disease occurrence and development are gradually enriched to form "health big data" [37].

The development of systems biology and translational medicine offers the opportunity for further, systematical investigations of disease. Tracking and monitoring all the data of a single patient in a LIMS can help to identify and troubleshoot problems of individuals more quickly; all this can facilitate the realization of personalized medicine and precise medicine.

Certainly, all these put forward higher requirements and challenges [17] for the function of LIMSs which are discussed in later section.

9.2.2 Resources Available for the LIMS in Clinical Data Management

LIMSs can indeed bring laboratory management to a higher level, but for the meantime, this requires a sufficient investment of money, time, and technical efforts [38]. At present, commercial LIMSs are limited by complexity, insufficient flexibility, high costs, and extended timelines.

Here, based on literature retrieval, we describe a number of the current noncommercial clinical LIMSs. The platform, name, function description, and access link of these LIMSs were extracted.

From the results (Table 9.1), we found that most of the current clinical LIMSs are web based and adopt open-source software, which may be attributed to the flexible features of web-based platforms; they are easy to develop and can be modified according to a research group's needs.

9.2.3 Characters of a Clinical LIMS

The characteristics of LIMS clinical applications can be summarized into three aspects: business drivers, benefits, and requirements, which are represented in Fig. 9.1 [19]. By using these features, the main role and requirements of a LIMS in clinical research can be summarized as follows:

1. To realize the automation and modernization of the clinical laboratory.
2. To meet the analysis requirements of clinical laboratory research work in the hospital and satisfy different laboratories and different testing procedures, to maximize the automation of operation, and to further realize the intelligence.

Table 9.1 LIMS resources available in clinical data management

Type	Name	Function description	Access link
Web based (open source)	AdLIMS [5]	A customized open-source software that allows the bridging of clinical and basic molecular research studies	http://sourceforge.net/projects/adlims/
	MendeLIMS [12]	Management of our clinical genome sequencing studies	http://mendelims.stanford.edu/
	Onco-STS [10]	A sample tracking system for oncogenomic studies	
	SMITH [47]	Handling NGS	
	K-screen [42]	An integrated application environment that supports data analysis, management, and presentation	
	Enzyme tracker [44]	A web-based laboratory information management system for sample tracking, as an open-source and flexible alternative that aims at facilitating entry, mining, and sharing of experimental biological data	http://cubique.fungalgenomics.ca/enzymedb/index.html/
	BonsaiLIMS [4]	A lab information management system for translational medicine	
	User-configurable LIMS [36]	Manage accrual with a healthy blood-donor protocol, as well as manufacturing operations for the production of a master cell bank and several patient-specific stem cell products	
	WIST [15]	Provides common LIMS input components and allows them to be arranged and configured using a flexible language that specifies each component's visual and semantic characteristics	http://vimss.sf.net/
	GNomEx [32]	A tool for generating, analyzing, distributing, and visualizing genomic data	http://sourceforge.net/projects/gnomex/
Excel based	Excel-based LIMS [23]	A simple, flexible, and cost-/time-saving solution for improving workflow efficiencies in early absorption, distribution, metabolism, and excretion screening	
Unknown	Galaxy LIMS	For NGS	http://tron-mainz.de/tron-facilities/computational-medicine/galaxy-lims/

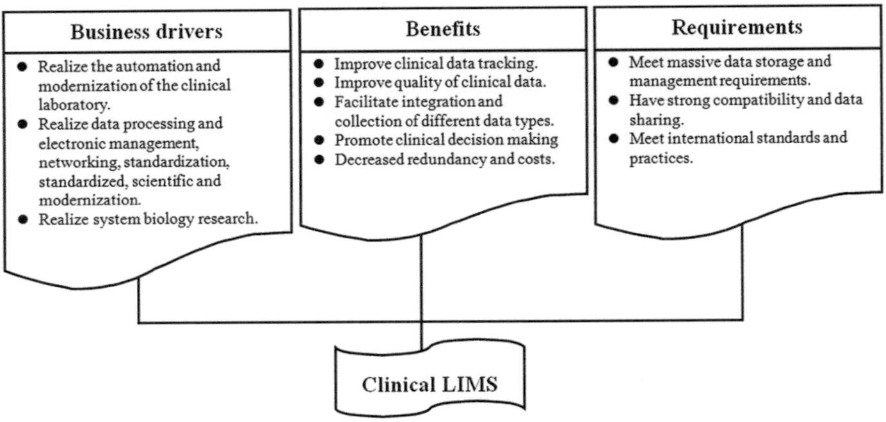

Fig. 9.1 Typical features of the clinical LIMS

3. To allow clinical laboratory work to meet international practices and realize data processing and electronic management, networking, standardization, and modernization.
4. To maximize the realization of an automatic instrument of network operations and automatic data acquisition and fully guarantee the instrument data to be original and make data processing according to special requirements at any time.
5. Adopt a comprehensive configurable method, without the need to adopt the tender to provide development tools or computer language for customization and secondary operation, and can be modified according to standard laboratory analysis and provide simple and flexible expansion and maintenance methods.
6. Unified storage and management of important historical data.
7. Including the spectrum map, LIMS software needs to be able to meet massive data storage and management requirements.
8. Construction according to international standard ISO/IEC17025 and various countries' respective standards, such as the Chinese standard GB/T 15481–2000, and also meet hospital accreditation.

9.3 Application of the LIMS in Clinical Data Management

9.3.1 Application Model of a LIMS in Hospitals

The patient data flow based on a LIMS in the whole hospital system is shown in Fig. 9.2. Once patients are admitted to the hospital, the data is input to the corresponding system and then converted and shared between different systems. The system then analyzes the different levels of data associated with the disease and generates various forms of report to laboratory personnel, medical personnel, and patients and finally reaches efficient and accurate shared decision-making (SDM)[3].

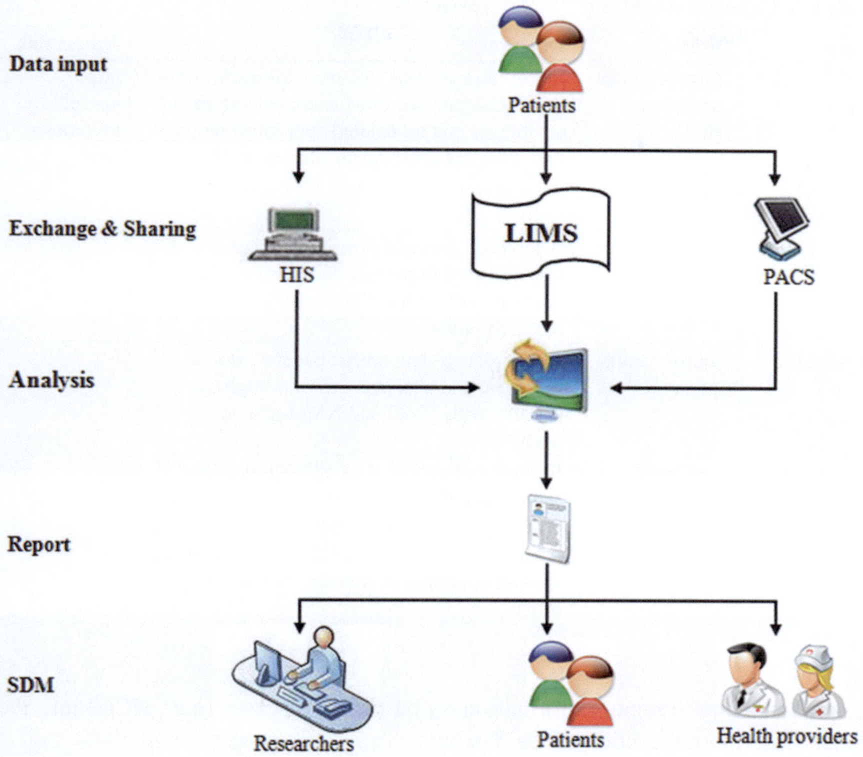

Fig. 9.2 Data flow and application of the clinical LIMS

The LIMS is involved in different levels of function during the whole process, from the data storage, data exchange, sample tracking, data report, etc. It follows that LIMSs play an important role in realizing the precision and personalization of disease treatment.

9.3.2 *Customized LIMS for Specific Disease*

With in-depth research in different fields of medical care, a more and more customized and proprietary LIMS has been used in corresponding fields [44]. Table 9.2 lists the customized LIMSs of several diseases and cancer research including hepatocellular carcinoma [6], neuromyelitis optica (NMO), leukemia, and a customized forensic LIMS that ensures the smooth operation of a death investigation office [20].

Utilizing these customized LIMSs will enable users to extract meaningful results from large datasets while trusting the robustness of their assays [11]. Based on the assigned serial number, physicians and pathologists can analyze and input

Table 9.2 Customized LIMSs for clinical researches

Name	Disease	Description	Access link
YPRC-PDB [6]	Hepatocellular carcinoma (HCC)	Store, retrieve, and analyze various information including two-dimensional electrophoresis images and associated spot information that were obtained during studies of HCC	http://yprcpdb.proteomix.org/
Onco-STS [10]	Oncogenomic	A sample tracking system for oncogenomic studies	
eOncoLIMS [34]	Cancer	A modular data and process management system designed to provide the infrastructure and environment for a collaborative cancer research project	
NMO-DBr [18]	Neuromyelitis optica (NMO)	A database system which collects, stores, retrieves, and analyzes information from patients with NMO and NMO-related disorders	
Leukemia patient LIMS [1]	Leukemia	For entry of patient data, clinical details, sample details, cytogenetics test results, and data mining for various ongoing research areas	
Forensic LIMS [20]	Forensic	A customized forensic LIMS that ensures the smooth operation of a death investigation office	

standardized experimental information on particular diseases (e.g., HCC) into the clinical information section of the database.

9.4 Principles and Standards of the LIMS in Clinical Data Management

Each LIMS according to the various standards sometimes may be considered as onerous and complex; however, these standards force systems to increase accuracy of information, efficiency, and effectiveness of work processes and finally improve patient safety [16, 45].

A lot of standards are involved in the process of the LIMS application, such as database construction, medical information exchange and integration among different systems, laboratory management standards, and so on; partial-related standards are presented as follows:

Medical information exchange and integration is the effective method to solve the interoperability and medical information island and is the basis of medical information sharing.

9.4.1 ISO/IEC 17025

Accreditation criteria for the competence of testing and calibration in laboratories, originally known as the ISO/IEC Guide 25, were initially issued by the International Organization for Standardization in 1999. There are many commonalities with the ISO 9000 standard, but ISO/IEC 17025 is more specific in requirements for competence [21].

Since its initial release, a second release was made in 2005 after it was agreed that it needed to have its quality system words more closely aligned with the 2000 version of ISO 9001. The ISO/IEC 17025 standard itself comprises of five elements which include scope, normative references, terms and definitions, management requirements, and technical requirements [35].

The two main sections in ISO/IEC 17025 are management requirements and technical requirements. Management requirements are primarily related to the operation and effectiveness of the quality management system within laboratories. Technical requirements include factors determining the correctness and reliability of the tests and calibrations performed in the laboratory. The usual contents of the quality manual follow the outline of the ISO/IEC 17025 standard.

9.4.2 Health Level Seven (HL7) Standard

With the development of the technology in medical informatics area, the Health Level Seven (HL7) standard for electronic data exchange in all health-care environments has become a widely used standard. The gateway is used to connect two different medical information systems, a HIS and a LIMS, via HL7 message exchanging [52].

9.4.3 Logical Observation Identifiers Names and Codes (LOINC)

Logical Observation Identifiers Names and Codes (LOINC) is a universal code system for identifying laboratory and clinical researches, which is designed for utilization within messaging standards such as HL7. When LOINC and HL7 are used together, independent systems can electronically exchange test results with one another in an understandable way [49].

This data exchange formula has been used widely around the world. Before health-care organizations can leverage the value of unified data, they must first map their local test codes to codes in LOINC. Unfortunately, this process is complicated and resource intensive.

9.4.4 *Standardized Nomenclature of Medicine Clinical Terms (SNOMED CT)*

Standardized Nomenclature of Medicine Clinical Terms (SNOMED CT) is developed by the International Health Terminology Standards Development Organization. It is the world's largest clinical terminology database and provides broad coverage of clinical medicine, including diseases and phenotypes. SNOMED CT includes pre-coordinated concepts (with their terms) and supports post-coordination, i.e., the principled creation of expressions (logical definitions) for new concepts [8].

The US edition of SNOMED CT dated March 2015 includes about 300,000 active concepts, of which 103,748 correspond to clinical findings.

9.5 Informatics Challenges of a LIMS

The advent of new technologies has led to an increasing growth in both the quality and quantity of health data; modern-automated LIMSs have to operate huge volumes of data. Such growth may entail some challenges, which make adopting appropriate management methods inevitable. In terms of logistics, data capture, data analysis, result visualization, and reporting, new challenges have emerged from such projects [25].

9.5.1 *Reliability and Security*

From the phenotype to genomics data of the patients, with a LIMS covering various data quantity and data type, how to achieve fully effective mining and utilization, to promote individualized and precision medicine, is currently the main direction of medical development. Data structure standardization, quality control, privacy protection, and health data are important problems to be solved in the process [46].

The continuity of data should be fully guaranteed, which not only can achieve efficient sample tracking but also realize the systematic research of disease [26].

Certainly, a series of measures can be taken to improve data reliability and security, guarantee the consistency of the data, and increase laboratory productivity, such as effective data encryption to protect the privacy of the patient; all data for individual samples are linked through unique accession numbers to keep consistency, accession numbers, numerous levels of taxonomy, or collection site provided to users for searching sample information efficiently online [51].

9.5.2 Unification of Standard Data, Workflow, etc.

In terms of unification of standard data, one of the main difficulties to implement a LIMS is the fact that each laboratory has a different routine of experimentation that changes over time; meanwhile, once validated, major modifications to the LIMS, such as revising the user interface, are unlikely [19].

At the present time, few LIMSs use LOINC and SNOMED. Many LIMSs record the order and the result of a laboratory test using local alphanumeric codes (or worse, a combination of local codes and free text) specific to that laboratory or to a particular vendor's LIMS or which may cause new diseases (e.g., severe acute respiratory syndrome or infection due to a new salmonella serotype) that cannot be added rapidly.

Lacking of universally accepted clinical LIMS fitting all requirements may be a great obstacle to medical data sharing and the systematic study of disease [25].

9.5.3 Systems Biology Research Limitations of a LIMS

The organization and systematic analysis of health data are challenging issues today and even more for the rising amount of information in the future. LIMSs that do not accept the currently accepted clinical research on a system, medical data sharing, and diseases will bring great obstacles. Although a large amount of data are generated from high-throughput large-scale techniques, a connection of these mostly heterogeneous data from different analytical platforms and of various experiments is limited.

Data mining procedures and algorithms are often insufficient to extract meaningful results from large datasets and therefore limit the exploitation of generated biological information [31].

How to reduce the gap between clinical data and genetic data and improve the association between phenotyping and genotyping data are all significant issues for the research of diseases by systems biology.

9.5.4 More Communications Between Health-Care Providers and Clinical Laboratory Personnel

Poor communication between health-care providers and clinical laboratory personnel can lead to medical errors. Therefore, researchers and clinicians must collaborate closely to achieve a comprehensive interpretation of heterogeneous biomedical data, especially with respect to clinical diagnosis and treatment [34].

With the development of medical informatics participatory medicine [22, 43], the relationship among doctors, patients, and medical researchers are linked more closely, but how to regard patients as central and how to use disease data to make more accurate decisions are two of the major challenges of precision medicine.

9.6 Future Directions

By increasing the function of the LIMS and better standardization of standards, future LIMSs can achieve data structure standardization and powerful data storage capacity, allow easy data entry and organized data display, streamline all phases of a workflow, and provide powerful compatibility and sharing with other medical systems.

The LIMS function was designed not only to minimize the gaps between different data type and processes but also to perform output function in various report formats to satisfy different users. The procedures for collecting and categorizing samples should be scientific and standardized, which can be verified by physicians, surgeons, and pathologists, and the users can input entire sets of information into LIMS and assign a serial number to each sample. The workflow generator encapsulates a user-friendly visual tool that allows users to design customized workflows [33].

A smart LIMS operating environment in a particular facility strongly influences clinical decisions and developments in medicine. If there are demands for LIMSs, the education and training of the administrator, data analysts, and patients also should be considered in the future.

9.7 Conclusions

Although there are many limitations and obstacles in the way of the clinical LIMS application, we firmly believe that a LIMS can facilitate the "seamless connectivity" between clinical data and genetic data, realize the system researches of disease, improve early disease prevention, also facilitate decision-making, improve quality and productivity of disease care services, promote the development of 4P medicine (predictive medicine, preventive medicine, personalized medicine, and participatory medicine), and, lastly, achieve precision medicine.

References

1. Bakshi SR, Shukla SN, Shah PM. Customized laboratory information management system for a clinical and research leukemia cytogenetics laboratory. J Assoc Genetic Technol. 2009;35:7–8.
2. Barillari C, et al. openBIS ELN-LIMS: an open-source database for academic laboratories. Bioinformatics. 2016;32(4):638–40

3. Barry MJ, Edgman-Levitan S. Shared decision making-pinnacle of patient-centered care. N Engl J Med. 2012;366:780–1.
4. Bath TG, et al. LimsPortal and BonsaiLIMS: development of a lab information management system for translational medicine. Source Code Biol Med. 2011;6:9.
5. Calabria A, et al. adLIMS: a customized open source software that allows bridging clinical and basic molecular research studies. BMC Bioinf. 2015;16 Suppl 9:S5.
6. Cho SY, et al. An integrated proteome database for two-dimensional electrophoresis data analysis and laboratory information management system. Proteomics. 2002;2:1104–13.
7. Davidson DF. A survey of some pre-analytical errors identified from the Biochemistry Department of a Scottish hospital. Scott Med J. 2014;59:91–4.
8. Dhombres F, Bodenreider O. Interoperability between phenotypes in research and healthcare terminologies-Investigating partial mappings between HPO and SNOMED CT. J Biomed Semant. 2016;7:3.
9. Eichhorst JC, et al. Drugs of abuse testing by tandem mass spectrometry: a rapid, simple method to replace immunoassays. Clin Biochem. 2009;42:1531–42.
10. Gavrielides M, et al. Onco-STS: a web-based laboratory information management system for sample and analysis tracking in oncogenomic experiments. Source Code Biol Med. 2014;9:0.
11. Grandjean G, Graham R, Bartholomeusz G. Essential attributes identified in the design of a Laboratory Information Management System for a high throughput siRNA screening laboratory. Comb Chem High Throughput Screen. 2011;14:766–71.
12. Grimes SM, Ji HP. MendeLIMS: a web-based laboratory information management system for clinical genome sequencing. BMC Bioinf. 2014;15:290.
13. Hakkinen J, Levander F. Laboratory data and sample management for proteomics. Methods Mol Biol. 2011;696:79–92.
14. Harel A, et al. Omics data management and annotation. Methods Mol Biol. 2011;719:71–96.
15. Huang YW, Arkin AP, Chandonia JM. WIST: toolkit for rapid, customized LIMS development. Bioinformatics. 2011;27:437–8.
16. Isfahani SS, et al. The evaluation of hospital laboratory information management systems based on the standards of the American National Standard Institute. J Educ Health Promot. 2014;3:61.
17. Kuhn S, Schlorer NE. Facilitating quality control for spectra assignments of small organic molecules: nmrshiftdb2-a free in-house NMR database with integrated LIMS for academic service laboratories. Magn Reson Chem. 2015;53:582–9.
18. Lana-Peixoto MA, et al. NMO-DBr: the Brazilian Neuromyelitis Optica Database System. Arq Neuropsiquiatr. 2011;69:687–92.
19. Lemmon VP, et al. Challenges in small screening laboratories: implementing an on-demand laboratory information management system. Comb Chem High Throughput Screen. 2011;14:742–8.
20. Levy BP. Implementation and user satisfaction with forensic laboratory information systems in death investigation offices. Am J Forensic Med Pathol. 2013;34:63–7.
21. Lopez MA, et al. The challenge of ciemat internal dosimetry service for accreditation according to Iso/Iec 17025 standard, for in vivo and in vitro monitoring and dose assessment of internal exposures. Radiat Protect Dosimetry. 2015.
22. Lopez-Campos G, Ofoghi B, Martin-Sanchez F. Enabling self-monitoring data exchange in participatory medicine. Stud Health Technol Inform. 2015;216:1102.
23. Lu X. Development of an excel-based laboratory information management system for improving workflow efficiencies in early ADME screening. Bioanalysis. 2016;8:99–110.
24. Luo J, et al. Big data application in biomedical research and health care: a literature review. Biomed Inform Insights. 2016;8:1–10.
25. Maier H, et al. Principles and application of LIMS in mouse clinics. Mamm Genome. 2015;26:467–81.
26. Melo A, et al. SIGLa: an adaptable LIMS for multiple laboratories. BMC Genomics. 2010;11 Suppl 5:S8.

27. Mirzazadeh M, et al Point-of-care testing of electrolytes and calcium using blood gas analysers: it is time we trusted the results. Emerg Med J. 2016;33(3):181–6
28. Mohammadzadeh N, Safdari R. The intelligent clinical laboratory as a tool to increase cancer care management productivity. Asian Pac J Cancer Prev. 2014;15:2935–7.
29. Monnier S, et al. T.I.M.S: TaqMan Information Management System, tools to organize data flow in a genotyping laboratory. BMC Bioinf. 2005;6:246.
30. Morris C, et al. MOLE: a data management application based on a protein production data model. Proteins. 2005;58:285–9.
31. Nebrich G, et al. PROTEOMER: a workflow-optimized laboratory information management system for 2-D electrophoresis-centered proteomics. Proteomics. 2009;9:1795–808.
32. Nix DA, et al. Next generation tools for genomic data generation, distribution, and visualization. BMC Bioinf. 2010;11:455.
33. Palla P, et al. QTREDS: a Ruby on Rails-based platform for omics laboratories. BMC Bioinf. 2014;15 Suppl 1:S13.
34. Quo CF, Wu B, Wang MD. Development of a laboratory information system for cancer collaboration projects. Conference proceedings: annual international conference of the IEEE engineering in medicine and biology society. IEEE engineering in medicine and biology society. Annual Conference. 2005:3:2859–2862.
35. Romero AM, et al. Ciemat external dosimetry service: iso/iec 17025 accreditation and 3 y of operational experience as an accredited laboratory. Radiat Prot Dosimetry. 2015.
36. Russom D, et al. Implementation of a configurable laboratory information management system for use in cellular process development and manufacturing. Cytotherapy. 2012;14:114–21.
37. Shin Y, et al. First step to big data research in hospital. Stud Health Technol Inform. 2015;216:924.
38. Singh KS, et al. SaDA: from sampling to data analysis-an extensible open source infrastructure for rapid, robust and automated management and analysis of modern ecological high-throughput microarray data. Int J Environ Res Public Health. 2015;12:6352–66.
39. Song Y, et al. A high efficiency, high quality and low cost internal regulated bioanalytical laboratory to support drug development needs. Bioanalysis. 2014;6:1295–309.
40. Stocker G, et al. iLAP: a workflow-driven software for experimental protocol development, data acquisition and analysis. BMC Bioinf. 2009;10:390.
41. Swanson CM, et al. Update on inpatient glycemic control in hospitals in the United States. Endocr Pract. 2011;17:853–61.
42. Tai D, Chaguturu R, Fang J. K-Screen: a free application for high throughput screening data analysis, visualization, and laboratory information management. Comb Chem High Throughput Screen. 2011;14:757–65.
43. Townsend A, et al. eHealth, participatory medicine, and ethical care: a focus group study of patients' and health care providers' use of health-related internet information. J Med Internet Res. 2015;17, e155.
44. Triplet T, Butler G. The EnzymeTracker: an open-source laboratory information management system for sample tracking. BMC Bioinf. 2012;13:15.
45. Turner E, Bolton J. Required steps for the validation of a Laboratory Information Management System. Qual Assur. 2001;9:217–24.
46. Ulma W, Schlabach DM. Technical considerations in remote LIMS access via the World Wide Web. J Autom Methods Manage Chem. 2005;2005:217–22.
47. Venco F, et al. SMITH: a LIMS for handling next-generation sequencing workflows. BMC Bioinf. 2014;15 Suppl 14:S3.
48. Viksna J, et al. PASSIM-an open source software system for managing information in biomedical studies. BMC Bioinf. 2007;8:52.
49. Vreeman DJ, Hook J, Dixon BE. Learning from the crowd while mapping to LOINC. J Am Med Inform Assoc. 2015;22:1205–11.
50. Walzer M, et al. qcML: an exchange format for quality control metrics from mass spectrometry experiments. Mol Cell Proteomics. 2014;13:1905–13.

51. Wang N, et al. The Hawaiian Algal database: a laboratory LIMS and online resource for biodiversity data. BMC Plant Biol. 2009;9:117.
52. Zhang Q, Gao S. [Design and realization of HL7 gateway], Sheng wu yi xue gong cheng xue za zhi. J Biomed Eng = Shengwu yixue gongchengxue zazhi. 2003;20:111–5.
53. Zhi YJ, et al. [Impact analysis of parenterally administered shuxuetong on abnormal changes of BUN index based on hospital information system data]. Zhongguo Zhong yao za zhi = Zhongguo zhongyao zazhi = China J Chin materia medica. 2013;38:3048–52.

Chapter 10
Biobanks and Their Clinical Application and Informatics Challenges

Lan Yang, Yalan Chen, Chunjiang Yu, and Bairong Shen

Abstract Biobanks are one of the most important biomedical research resources and contribute to the development of biomarker detection, molecular diagnosis, translational medicine, and multidisciplinary disease research, as well as studies of interactions between genetic and environmental or lifestyle factors. Aiming for the wide clinical application of biobanks, biobanking efforts have recently switched from a focus on accumulating samples to both formalizing and sustaining collections in light of the rapid progress in the fields of personalized medicine and bioinformatics analysis. With the emergence of novel molecular diagnostic technologies, although the bioinformatics platform of biobanks ensures reliable bioinformatics analysis of patient samples, there are a series of challenges facing biobanks in terms of the overall harmonization of policies, integrated processes, and local informatics solutions across the network. Further, there is a controversy regarding the increased role of ethical boards, governance, and accreditation bodies in ensuring that collected samples have sufficient informatics capabilities to be used in biobanks. In this volume, we present a selection of current issues on the inevitable challenges of the clinical application of biobanks in informatics.

Keywords Biomedical • Harmonization • Personalized medicine • Standardization

L. Yang • B. Shen (✉)
Center for Systems Biology, Soochow University, No. 1 Shizi Street, 206, 215006 Suzhou, Jiangsu, China
e-mail: bairong.shen@suda.edu.cn

Y. Chen
Center for Systems Biology, Soochow University, No. 1 Shizi Street, 206, 215006 Suzhou, Jiangsu, China

Department of Medical Informatics, School of Medicine, Nantong University, Nantong, Jiangsu, China

C. Yu
Suzhou Industrial Park Institute of Services Outsourcing, No. 99 Ruoshui Road, Suzhou Industrial Park, Suzhou 215123, Jiangsu, China

B. Shen et al. (eds.), *Translational Biomedical Informatics*, Advances in Experimental Medicine and Biology 939, DOI 10.1007/978-981-10-1503-8_10

10.1 Introduction

10.1.1 *Definition of the Biobank*

Biobanks are storage banks for samples of human origin for use in national and international research in the field of biomedicine [38]. Their most significant feature is the storing of specimens for research purposes. Biobanks collect, store, and distribute biospecimens such as blood, urine, tissue, and DNA/RNA and involve not only the specimens, disease history, and characteristics but also their associations with lifestyle, environmental factors, and comorbid health burden [42]. Biobanks are the best large-scale instruments to link molecular and clinical findings to patients' prognosis. They vary in size, ranging from small disease-specific biobanks to ones for general health care.

All biomedical research requires collecting data and biological samples from people affected by a disease, as well as from people not affected by the disease, to analyze and draw conclusions for improving knowledge and advancing the diagnosis and/or treatment of diseases under research. Nowadays, biobanks are redefining many aspects of research by allowing ongoing access to research populations, exploring methods of consent and governance, and creating new models for conducting translational research [45].

10.1.2 *Classification of Biobanks*

Revisions in the purpose, operations, and clientele of biobanks have caused a review of biobank classification systems, which previously grouped human biobanks according to their research purpose [34].

Human-driven biobanks include three major types [35]:

(A) Population banks. Their primary goal is to obtain biomarkers of susceptibility and population identity, and their operational substrate is germinal-line DNA acquired from a large number of healthy donors that is representative of a discrete country/region or ethnic cohort.
(B) Disease-oriented banks for epidemiology. Their activity is focused on biomarkers of exposure using a huge number of samples, usually following a healthy exposed cohort/case–control design, and studying germinal-line DNA or serum markers and a great amount of certain designed and collected data.
(C) Disease-oriented general biobanks (i.e., tumor banks). Their goals correspond to biomarkers of diseases through prospective and/or retrospective collections of tumor and non-tumor samples and their derivates (DNA/RNA/proteins), usually in association with clinical data and sometimes with clinical trials.

According to the CTRNet system [44, 55, 56], mono-user biobanks aim to facilitate one research project, oligo-user biobanks serve several research groups, and poly-user biobanks support unspecified research projects undertaken by external researchers. Biobanking, the large-scale, systematic collection of data and tissue for open-ended research purposes, is on the rise, particularly in clinical research.

10.2 Clinical Application of Biobanks

10.2.1 The Importance of the Establishment of Biobanks

We are increasingly becoming aware of the importance of biomedical research to improve human health. Among the many existing biomedical and health platforms, biobanks are one of the most attractive options and contribute to building bridges between basic, translational, and clinical research and practice. The key purpose of biobanking in translational medicine and other medical research is to provide biological samples that are integrated with clinical information.

The main mission for which biobanks have been established is to empower those biomedical studies considered particularly relevant, focusing on the analysis and improvement of knowledge of conditions or diseases including cancer, infections, "rare" diseases, etc. [10]. Further, biobanks are comprehensively applied in many types of research such as in epidemiological studies [26], health-related quality of life deficits [42], central nervous system disease [20], lipid-related diseases [58], etc. Therefore, large samples are required to obtain certain demographic characteristics of not only the samples and social and food habits but also the environmental characteristics of the patients' living environment, lifestyle of patients and risk factors associated with the diseases of patients.

By sharing clinical experiences, patient treatment principles, and biobank strategies, clinical teams in Japan and Sweden, respectively, are aiming to develop predictive and drug-related protein biomarkers [32]. For instance, pre-therapeutic histological distinction between large-cell neuroendocrine cancer (LCNEC) of the lung and small-cell lung carcinoma has been problematic so far, leading to adverse clinical outcomes. Thus, the establishment of protein targets characteristic of those in the LCNEC biobank would be quite helpful for optimizing decision-making in therapeutic strategies by diagnosing individual patients.

The storage of human biological material and associated information in biobanks not only enables biological research but also the development and utilization of new diagnostic and therapeutic techniques in previously stored specimens. In Denmark, opportunities for exploiting leftover dried blood spot cards from neonatal screening for genomic research have been considered [47] in recent years.

The emergence of clinical biobanking is associated with general shifts in biomedical research toward molecular-level investigations to understand and intervene in the mechanisms of disease, particularly with the uptake of genomics in clinical research and medicine. Clinical samples can also be archived into repositories for future studies to investigate the root causes of disease using genetic, genomic, proteomic, and metabonomic approaches. For example, Middha and colleagues examined the genotype–phenotype correlation from whole-exome sequencing (WES) studies in a series of individuals representing a broad range of phenotypes based on a set of samples from the Mayo Clinic Biobank [28].

Using different approaches, such as genomics and proteomics, researchers can identify biomarkers of these processes and help design new target molecules for the development of drugs and therapeutic alternatives. In the case of rheumatology research, the possibility offered by biobanks in terms of samples from each patient in different formats (solid, liquid) and at different times of the disease progression (diagnosis, progression, pre- or posttreatment, etc.) may help to elucidate the mechanisms involved in various pathologies associated with this area [10]. As for proteomics in biomedical research, it can help with the identification of specific biomarkers for diagnosis, classification, and prediction and may contribute to defining new therapeutic targets [41].

Research using samples from large biobanks is essential for understanding not only genetic risks for common diseases caused by gene variants but also uncommon environmental and other risk exposures that impact health. Furthermore, studies utilizing biobank samples are useful in developing personalized therapeutics, targeting biomarkers in disease progression and prognosis, and implementing personalized medicine projects. The application areas of biobanks in human research can be seen in Fig. 10.1.

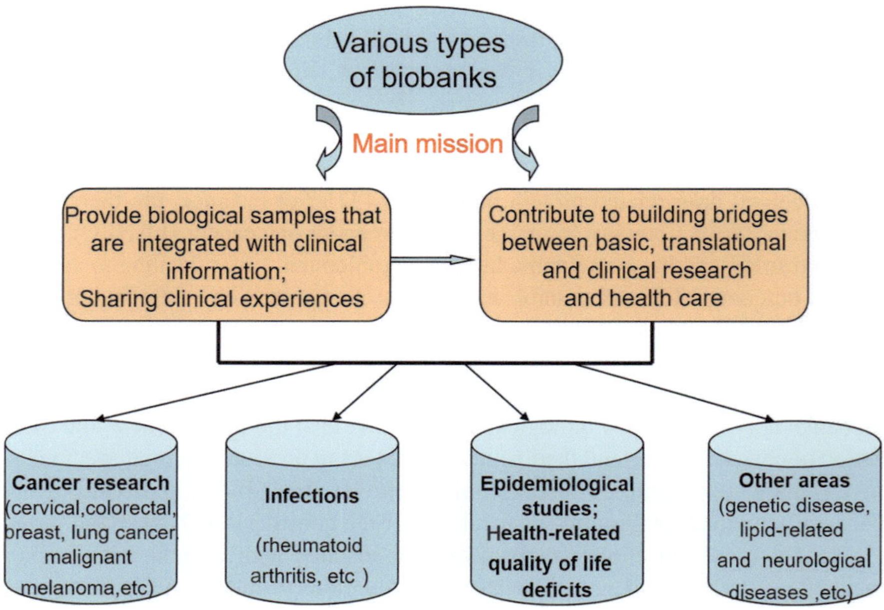

Fig. 10.1 Clinical applications of biobanks

10.2.2 The Prerequisite of Biobanks for Clinical Applications

Facilitating the integration required to achieve proper clinical applications of biobanking involves large amounts of work standardizing and harmonizing the data and tissue provision, as well as establishing quality control, certification of workflow, essential and procedural benchmarks for data and tissue collection and management, and evidence-based data models [29, 38].

The prerequisite for maintaining a tissue sample repository is the establishment of quality standards to ensure that samples are equivalent in terms of form (fresh, frozen, wax block, sections on slides, etc.), physical integrity (processing and storage), and availability for use (identity labels, aliquot size). Standard operating procedures should define the exact steps performed for each process in the sample life cycle to minimize both pre-analytic variability and storage artifacts [27].

However, the mass storage of biological samples raises complex technical issues that affect sample collection, transportation, identification, traceability, storage at different temperatures, the recovery of the stored samples, the processing of the data, etc. Thus, several essential processes must be developed, such as the unification of the protocols, establishment of an appropriate methodology for coding and identification of samples, rigorous informed consent, and hiring of well-qualified staff [10]. Further, a biobank information management system (BIMS) is necessary to manage the data and samples in biobanks. Worldwide standardization of samples, preparation, storage, and analysis will help to ensure regulated utilization of the sample sets in future health-care studies, serving in the development of new medicines and diagnostics as well as novel methodologies and innovative technology platforms. Improving the ways we collect and utilize samples will also depend on integrated systems that allow for search routines on sample collections, types, and datasets that have been generated from biobanks [27]. Additionally, data on recurrences and survival status related to treatment therapies and genetic profiling will provide much information on the progression and outcome of the disease. In order to exploit the potential of genomics and other molecular analytical techniques, there has been increasing emphasis on the differentiation and stratification of target objectives and populations; however, there are many challenges in establishing statistically significant associations between diseases and disease markers, which require data from ever larger target populations of both healthy individuals and patients [4]. With these efforts, the combination and harmonization of clinical examinations, pathological diagnoses, follow-up cases, and clinical samples could also create shortcuts for quick and efficient analysis.

In this era of increased interest in personalized medicine, the linkage of molecular data and genetic profiling with demographic, pathologic, and clinical records will greatly enhance future studies of disease etiology and risk factors. As long-term storage of biological materials and data is a critical component of any epidemiological or clinical study, when designing biobanks, informatics play a vital role in the handling of samples and data in a timely fashion [25]. However, a series of challenges have emerged because of the large amount of biomedical informatics data from various biobanks as well as the high complexity of the data analysis and integration. Policies for enrolling participants, returning research results, and obtaining samples and data will have far-reaching implications on the types of research that can be carried out on each biobank. Population-based biobanking setup specifically for research purposes has received considerable attention in studies devoted to ethical, legal, and social aspects of biobanking [21, 46].

10.3 Informatics Challenges Faced by Biobanks

10.3.1 Properly Designed Projects and Approaches Demanded in Biobanks

As biobank samples are increasingly used for translational research and clinical implementation projects, issues that must be addressed by each biobank include questions regarding the appropriate means to have ongoing engagement with participants, best consenting methods, the return of personal results, and other policy issues. With the rapid development of bioinformatics, which we often emphasize together with the objective of "personalized medicine," current approaches to biomedical research therefore involve not only individualized, stratified, and differentiated forms of intervention but also new forms of population-level surveillance [37]. The scientific goal is to initiate and stimulate scientific research of high quality and of international importance. However, during the clinical application of biobanks, there are inevitably new informatics challenges associated with ethical, legal, and privacy issues.

Recent research has reported that the most important factor for achieving this goal is the implementation of a solid national research infrastructure with standardized data collection and a solid long-term storage strategy [19]. Therefore, firstly, we should design appropriate projects that can help us to further understand how biomedical findings might correlate with various medical diseases. If the clinical data are linked to the biomaterial, it can be used to conduct more advanced research. This may ultimately lead to better health care for patients with the studied medical condition, as the treatment can be more personalized. For example, the Netherlands has the Esophageal and Gastric Cancer Pearl, a nationwide clinical biobanking project. In this project, all participating researchers are permitted to submit their study proposal to the scientific committee of the Esophageal and Gastric Cancer Pearl to conduct research using the clinical data and biomaterial gathered in the database, thus providing opportunities for future studies to gain more insight into the etiology, treatment, and prognosis of esophageal and gastric cancer [19]. In order to determine prognostic biomarkers and therapeutic targets for esophageal adenocarcinoma including whole-genome sequencing, Oesophageal Cancer

Clinical and Molecular Stratification study group in the UK has also started a similar program for tissue and data collection from patients with esophageal cancer across multiple specialist centers [35, 57].

Secondly, because the integrity of collection protocols in biobanking is essential for a high-quality sample preparation process, with multisource well-informed consent [7], which can improve patient autonomy in conscious decision-making, an effective sample collection process such as Community Networks Program Centers ([51]) and an electronic specimen collection protocol schema (eSCPS) [12] can improve the biospecimen collection as well as involvement in prevention, screening, and therapeutic trials.

The study of the molecular basis of cancer also needs many, often thousands of, biospecimens in order to find answers to questions related to environmental exposure to genetic predispositions (which are typically beyond the control of the individuals) [18]. Besides, these centers try to increase the community's capacity to conduct cancer education and, via training activities, to enhance the probability of successful research. Accumulating biospecimens and recruiting individuals from diverse groups to prevention and treatment trials will help allow the contribution of personalized medicine to all, regardless of their health status.

With the quickly expanding biospecimen needs and limited health-care budgets, biobanks may need to be selective as to what is stored. Thirdly, we still require effective methods such as the open electronic health record (EHR) archetype approach [48] and the specific data model [11] to help model the data in the database as well as enable data analysis compliance with the data privacy regulations of the existing BIMS. Although establishing exact mapping between the fields in the database and the elements of the existing archetypes that have been designed for clinical practice can be challenging and time-consuming and involves resolving many common system integration conflicts, we can provide a proper harmonization procedure which should be developed that allows large-scale collaboration and contributes to the harmonization of the clinical and test data collection acquired in various biobank resources [54]. The main projects and approaches used in current biobanks are listed in Table 10.1.

10.3.2 Bioinformatics Technologies for Data Management

As concomitant analysis of patients' personal and clinical data, such as family history, smoking and drinking habits, race, medical history, etc., can also be performed to identify disease-related factors, the key data management challenges faced by disease research projects include the high complexity and heterogeneity of the data types involved and the variability among experimental platforms. The data obtained from these different sources can be analyzed jointly to determine which environmental or behavioral factors play meaningful roles in cancer genesis [16]. Additionally, the high volume of data and the distributed nature of the sources make traditional approaches to data management impractical, and new solutions are therefore required.

Table 10.1 Projects and approaches currently used in biobanks

Number	Tool name used in biobanks (types)	Function	Application field	Reference
1	The Esophageal and Gastric Cancer Pearl (a nationwide clinical biobanking project in the Netherlands)	Provide opportunity for future studies to gain more insight in the etiology, treatment, and prognosis of esophageal and gastric cancer	Esophageal and gastric cancer	[19]
2	A data management procedure	Offers an easy way for the transformation of a nonautomated biobank from the small-scale, early stage to the large-scale, highly automated level	Different types of diseases	[17]
3	Endometriosis Phenome and Biobanking Harmonization Project	Harmonizes the collection of nonsurgical clinical and epidemiologic data relevant to endometriosis research, allowing large-scale collaboration	Endometriosis	[54]
4	Community Networks Program Centers	Successfully engages the groups to participate in the biospecimen collection and in prevention, screening, and therapeutic trials	Cancer research	[52]
5	An electronic specimen collection protocol schema (eSCPS)	Improves the integrity and facilitates the exchange of specimen collection protocols in the existing open-source BIMS	Prostate cancer	[12]
6	Linkage of data from diverse data sources (LDS)—a data model	Provides an effective approach to distribute clinical and repository data from different data sources to enable data analysis compliance with data privacy regulations	Prostate cancer	[11]
7	A multisource informed consent procedure	Allows a high rate of understanding and the study participations' awareness of their roles as research stakeholders in a cancer biobank	Cancer	[7]

(continued)

Table 10.1 (continued)

Number	Tool name used in biobanks (types)	Function	Application field	Reference
8	The ONCO-I2b2 project	Integrates biobank information and clinical data to support translational research in oncology	Oncology	[43]
9	Open EHR archetype approach	Development of an interoperable electronic biomedical research record (eBMRR) to support biomedical knowledge discovery	Prostate cancer	[48]
10	The Clinical Information Integration System (CIIS)	Systematically collects and manages various human-origin biomedical resources and the donors' clinical information with their signed consent	Cancer, digestive organ disease, pulmonary disease, other diseases, and cohort projects	[22]

With the emergence of new sequencing techniques, it is necessary for researchers to develop more effective bioinformatics techniques to manage and analyze the large amounts of data in biobanks, with the aim of being able to understand the mechanisms that contribute to the development of diseases. Ferretti and colleagues developed a web-based computer system—called BioBankWarden (BBW)—that enables researchers to store, retrieve, and integrate data on biomolecular researchers trying to integrate information across sample collections. BBW allows for the creation and association of different projects involving different groups of users and has a requisition module to manage the output of material based on roles and permissions for exchanges among collaborative groups. The researchers who participate in the projects contribute by detailing the system requirements and evaluating the prototypes. Along with some databases such as Institut Curie [31] and Breast Cancer Campaign Tissue Bank (BCCTBbp) [9], we also need valuable software such as the Sample avAILability (SAIL) system, which first created harmonized variables and annotated and made searchable information on the number of specimens available in individual biobanks in various phenotypic categories [49]. Furthermore, an integrated platform connecting databases, registries, biobanks, and clinical bioinformatics or a sharing service platform is another requisite for data analysis ([8]; [52]). As data must be linked at both the individual patient and whole cohort levels to enable researchers to gain a complete view of their disease and the patient population of interest, data access and authorization procedures are required to allow researchers in multiple institutions to securely compare results and gain new insights. The databases, software, and platforms that can be used on data containing disease associations between different phenotypes and mechanisms of diseases are listed in Table 10.2.

Table 10.2 Databases, software, and platforms used in biobanks

Number	Tool name used in biobanks (types)	Function	Application field	Reference
1	Institut Curie (database)	Responsible for the overall coordination and management of data	Cervical cancer	[31]
2	BCCTBbp (database)	Improved the ability of the biobanks to participate and share their samples and data within the network	Breast cancer	[9]
3	The Sample avAILability (SAIL) (software package)	Data linking, harmonization, submission of samples and phenotype information	Prostate cancer quality registry; SUMMIT for GWAS genotyping and omics analysis	[50]
4	BioBankWarden (BBW, database)	Be used to store and retrieve specific information from different clinical fields linked to biomaterials	Provides different datasets for each clinical area according to the user's needs	[16]
5	A sharing service platform	Integrates clinical practice and biological information that can be used in diverse medical and pharmaceutical research studies	Diverse medical and pharmaceutical research studies	[8]
6	The Human Tissue and Cell Research (HTCR) web application	To develop an information system supporting acquisition, processing, and storage of remnant biomaterial from surgical treatment, as well as its allocation to research projects	Support of research based on human specimens	[30]
7	RD-Connect platform	An integrated platform connecting databases, registries, biobanks, and clinical bioinformatics	For rare disease research	[52]
8	"Compass" approach (self-organizing maps and association mining)	Can generate a highly condensed and structured output for efficient manual screening of potentially interesting rules through the use of Associative Variable Groups	Prostate cancer and breast cancer	[23]

(continued)

Table 10.2 (continued)

Number	Tool name used in biobanks (types)	Function	Application field	Reference
9	The Federated Utah Research and Translational Health electronic Repository (FURTHeR)	Combine electronic health records (EHR) and biospecimen data by both institutions to demonstrate the robustness of the infrastructure	More than 20 diseases (not detailed)	[24]
10	Dataset called MIABIS (Minimum Information About BIobank data Sharing)	Facilitates data discovery through harmonization of data elements describing a biobank at the aggregate level	Not mentioned	[33]

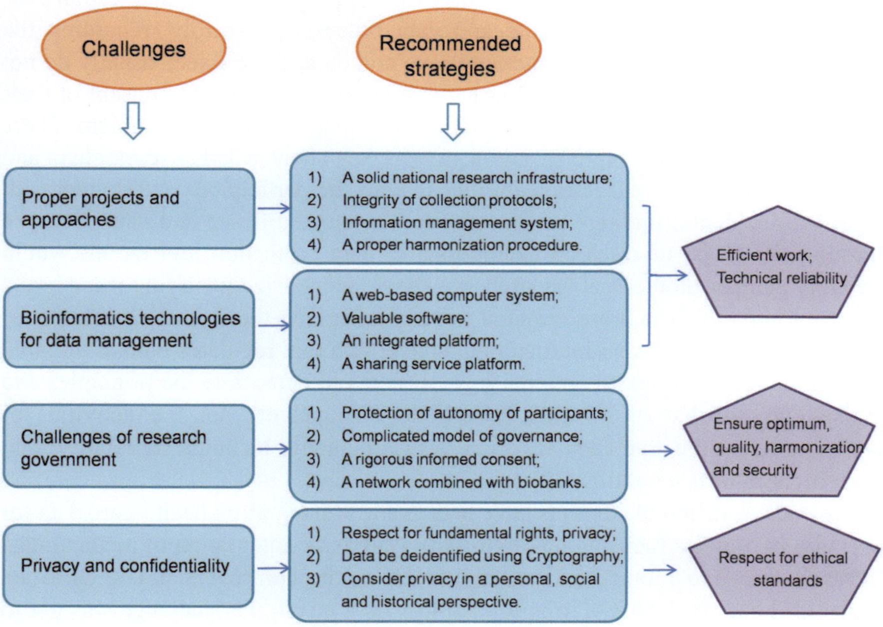

Fig. 10.2 Informatics challenges and recommended strategies for the development of biobanks

Furthermore, running such a comprehensive clinical bank requires massive coordination among the biobanks, specific clinical departments, pathology department, biochemistry lab, and hospital administration department. Powerful technologies critical to personalized medicine and targeted therapeutics require the analysis of carefully validated, procured, stored, and managed biospecimens. Informatics challenges that are faced in the development of biobanks and the recommended strategies can be seen in Fig. 10.2.

10.3.3 Clinical Biobanking Challenges Research Governance

Due to the heterogeneity of the specimens in biobanks, the uncertain and heterogeneous landscape also demands careful consideration and planning by biobank managers to maintain high-quality practices in the acquisition, storage, and release of specimens, all the while striving to protect the rights of the subjects. Clinical biobanking also poses challenges to clinical research governance. Based on a recent study [1], such governance is aimed at enacting and enforcing the distinctions and boundaries between research and care. In the authors' opinion, the first issue raised by the bio-objectification of clinical biobanking relates to the core principle underpinning the ethics of human subject research: the protection of the autonomy of the research participants. Secondly, such models of governance are complicated by the extent to which clinical biobanking initiatives are organized as complex arrangements. Further, these models often involve overlapping organizational responsibilities for various aspects of tissue and data processing. Thirdly, regarding the practices of the residual use of human tissue and data, these issues usually do not adapt to the remit of most clinical research legislations. Informed consent of their participants of various scope and specificity may help to solve this problem. There has been debate among ethicists and legal scholars about whether researchers and biobanks have duties and responsibilities toward the participants and donors with regard to incidental findings generated from the banked tissue and data. Because limiting the scope of research exemptions in data protection legislations would severely hamper biomedical research, we should emphasize justifying these broad forms of consent, which are oriented toward enhancing the collective benefits of research. Since there is no international convention that regulates human research biobanks, we must look to national legal systems to determine the principles and legislations capable of forming a regulatory framework for the activities of biobanks to ensure that they function with efficiency, technical reliability, and respect for ethical standards [15].

Nowadays, different attempts have been made at integrating health-care data for research in projects focusing on large-scale systematic integration of medical data infrastructure into biomedical research, such as the controversial UK care.data project [6] and the ONCO-I2b2 project [43]. As a result, the challenge was not to set up a new biobank system but to provide a mechanism that could facilitate searching across many existing database systems and structures [36]. As research becomes more globalized, the systems should also be able to record sample data in a global database. Thus, an international network will be necessary to realize this goal. While informatics capabilities of biobanks remain consistent with what is represented within the ISBER (International Society for Biological and Environmental Repositories) informatics survey, forging virtual networks of biobanks with meaningful annotations will be extremely challenging [13]. Biobanks and biobank networks are established as the optimal methods to store large amounts of human biological samples to ensure their optimum quality, harmonization, and security, as well as the ethical and legal requirements guaranteeing the rights of participants [10]. In contrast to the more straightforward technical and management issues, ethical and regulatory practices often involve issues that are more controversial and difficult to standardize [53].

10.3.4 Considerations About Privacy and Confidentiality

Biological issues from the interpretation of genetic data can provide genetic information. Genetic information retrieves an individual's biological inheritance in terms of individual, family, and lineage history. For this reason, research projects that involve biological materials stored in biobanks always have ethical and legal considerations. Privacy and confidentiality are the main issues we should be concerned with: privacy as a human right and confidentiality as a professional prima facie duty [40]. Although the boundaries of privacy vary in different countries because of cultural differences, confidentiality remains a core professional duty in all countries, even after a patient dies [39]. There are many other infra-constitutional laws, such as the new Civil Code [2] and the Biosecurity Law [3], as well as other administrative guidelines and regulations [14] that codes for fundamental rights and privacy.

Research should not infringe the rules concerning personal rights; further, they should protect the patient's right of privacy. Besides, in terms of confidentiality, health research investigators have the same duty to prevent personally identifiable data from disclosure. Data handling is one of the most important aspects in the activity of biobanks. During processing, personal data must be identified and stored using cryptography, which helps to protect privacy. Sometimes, however, this cannot be accomplished in some exceptional situations. For instance, if the research subject gives specific permission for disclosure, the researcher does not have to observe the duty of confidentiality toward the data obtained during research activities. Furthermore, if the data concerns an endemic or highly contagious disease, we may inevitably need to disclose data as long as it is carried out in a sincere, conscientious, and responsible manner, because of consideration of public order, or even by the force of the law [15]. Thus, the problem lies between private life and public interest. Either the issue primarily concerns an aspect of someone's private life, in which case it must be kept within the private sphere, or it is something that deserves broad visibility, owing to the acknowledged presence of a public interest [5].

Another important aspect of the management of information in the activity of biobanks that affects the privacy of research subjects is the sharing of information among researchers and the creation of research networks that are increasingly connected globally. The cultural borders of the concept of privacy must be considered as a new challenge to research projects that involve the utilization of biological samples stored in biobanks. The challenge of privacy in the context of genetics research must be emphasized according to the criteria established by the legal system so as to preserve justice. We should consider privacy from the personal, social, and historical perspectives in the clinical application of biobanks.

10.4 Conclusions

In this chapter, we discussed the clinical application of biobanks and bioinformatics challenges faced by biobanks during the practice process. Biobanks are an important resource for medical research. Biological material from large numbers of specimens can yield valuable information that could improve our understanding of mechanisms and genetic–environmental interactions and the genesis and development of the early onset disorders of diseases. Although every biobank has internal standards for record-keeping, quality assurance, and medical procedures, considering the complexity of different types of data, efforts are required to standardize sample quality, form, processes, and an effective integration of multidimensional data across the network, or at least harmonization among various ethical, social, and legal issues, as well as common practices for the management of biobanks. As long as we recognize the importance of ethical, legal, and social issues in human tissue research in both society and the research community specifically and take proper measures to deal with informatics challenges during the realization process of biobanks, we will gain the full benefits of biobanks and improve modern scientific biomedical research.

References

1. Boeckhout M, Douglas CM. Governing the research-care divide in clinical biobanking: Dutch perspectives. Life Sci Soc Policy. 2015;11:7. doi:10.1186/s40504-015-0025-z.
2. Brasil. Código Civil: Lei 10. 406. Brasilia. 2002. Available at:http://www.planalto.gov.br/ccivil_03/Leis/2002/L10406.htm. Accessed 10 Jan 2015.
3. Brasil. Lei de Biossegurança: Lei n 11.105. Brasília. 2005. Available at: http://www.planalto.gov.br/ccivil_03/_ato2004-2006/2005/lei/l11105.htm. Accessed 10 Jan 2015.
4. Burton PR, Hansell AL, Fortier I, Manolio TA, Khoury MJ, Little J, Elliott P. Size matters: just how big is BIG?: Quantifying realistic sample size requirements for human genome epidemiology. Int J Epidemiol. 2009;38(1):263–73. doi:10.1093/ije/dyn147.
5. Cachapuz M. Intimidade e vida privada no novo código civil brasileiro: uma leitura orientada no discurso jurídico [Intimidade and vida privada according to the new Brazilian Civil Code: an interpretation based on legal discourse]. Fabris Porto Alegre. 2006.
6. Carter P, Laurie GT, Dixon-Woods M. The social licence for research: why care.data ran into trouble. J Med Ethics. 2015;41(5):404–9. doi:10.1136/medethics-2014-102374.
7. Cervo S, Rovina J, Talamini R, Perin T, Canzonieri V, De Paoli P, Steffan A. An effective multisource informed consent procedure for research and clinical practice: an observational study of patient understanding and awareness of their roles as research stakeholders in a cancer biobank. BMC Med Ethics. 2013;14:30. doi:10.1186/1472-6939-14-30.

8. Cui W, Zheng P, Yang J, Zhao R, Gao J, Yu G. Integrating clinical and biological information in a shanghai biobank: an introduction to the sample repository and information sharing platform project. Biopreserv Biobank. 2015;13(1):37–42. doi:10.1089/bio.2014.0091.
9. Cutts RJ, Guerra-Assuncao JA, Gadaleta E, Dayem Ullah AZ, Chelala C. BCCTBbp: the Breast Cancer Campaign Tissue Bank bioinformatics portal. Nucleic Acids Res. 2015;43 (Database issue):D831–6. doi:10.1093/nar/gku984.
10. Domenech Garcia N, Cal Purrinos N. Biobanks and their importance in the clinical and scientific fields related to Spanish biomedical research. Reumatol Clin. 2014;10(5):304–8. doi:10.1016/j.reuma.2014.02.011.
11. Eminaga O, Ozgur E, Semjonow A, Herden J, Akbarov I, Tok A, . . . Wille S. Linkage of data from diverse data sources (LDS): a data combination model provides clinical data of corresponding specimens in biobanking information system. J Med Syst. 2013;37(5):9975. doi: 10.1007/s10916-013-9975-y.
12. Eminaga O, Semjonow A, Oezguer E, Herden J, Akbarov I, Tok A, . . . Wille S. An electronic specimen collection protocol schema (eSCPS). Document architecture for specimen management and the exchange of specimen collection protocols between biobanking information systems. Methods Inf Med. 2014;53(1);29–38. doi: 10.3414/ME13-01-0035.
13. Fearn P, Michels C, Meagher K, Cada M. 2012 international society for biological and environmental repositories informatics working group: survey results and conclusions. Biopreserv Biobank. 2013;11(1):64–6. doi:10.1089/bio.2012.1115.
14. Fernandes M, Ashton-Prolla P, Matte U, Meurer L, Osvaldt A, Bittelbrunn A, . . . Goldim J. A normativa do Hospital de Clínicas de Porto Alegre para o armazenamento e utilização de materiais biológicohumanos e informações associadas em pesquisa: uma proposta interdisciplinar [Rules and standards used by the Hospital de Clínicas Biobank in Porto Alegre concerning the storage and use of human biospecimens and related research data: an interdisciplinary proposal]. Rev HCPA. 2010;(30):169–79.
15. Fernandes M, Ashton-Prolla P, de Moraes L, Matte Ú, Goldim J, Martins-Costa J. Genetic information and biobanking: a Brazilian perspective on biological and biographical issues. J Commun Genet. 2015;(6):295–9.
16. Ferretti Y, Miyoshi NS, Silva Jr WA, Felipe JC. BioBankWarden: a web-based system to support translational cancer research by managing clinical and biomaterial data. Comput Biol Med. 2015. doi:10.1016/j.compbiomed.2015.04.008.
17. Gao Z, Gu Y, Lv Z, Yu G, Zhou J. Practical electronic information system and printed recording promote management accuracy in an early-stage small-scale non-automatic biobank. Biopreserv Biobank. 2015;13(1):61–6. doi:10.1089/bio.2014.0102.
18. Giuliano AR, Mokuau N, Hughes C, Tortolero-Luna G, Risendal B, Ho RCS, . . . McCaskill-Stevens WJ. Participation of minorities in cancer research: the influence of structural, cultural, and linguistic factors. Ann Epidemiol. 2000;10(8 Suppl):S22–34.
19. Haverkamp L, Parry K, van Berge Henegouwen MI, van Laarhoven HW, Bonenkamp JJ, Bisseling TM, . . . Ruurda JP. Esophageal and Gastric Cancer Pearl: a nationwide clinical biobanking project in the Netherlands. Dis Esophagus. 2015. doi:10.1111/dote.12347.
20. Jurga M, Forraz N, Basford C, Atzeni G, Trevelyan AJ, Habibollah S, . . . McGuckin CP. Neurogenic properties and a clinical relevance of multipotent stem cells derived from cord blood samples stored in the biobanks. Stem Cells Dev. 2012;21(6):923–36. doi:10.1089/scd.2011.0224.
21. Kaye Jane, Mark S. Principles and practice in biobank governance. Ashgate Publishing Ltd. 2009.
22. Kim H, Yi BK, Kim IK, Kwak YS. Integrating clinical information in National Biobank of Korea. J Med Syst. 2011;35(4):647–56. doi:10.1007/s10916-009-9402-6.
23. Krysiak-Baltyn K, Nordahl Petersen T, Audouze K, Jorgensen N, Angquist L, Brunak S. Compass: a hybrid method for clinical and biobank data mining. J Biomed Inform. 2014;47:160–70. doi:10.1016/j.jbi.2013.10.007.

24. Lasalle B, Varner M, Botkin J, Jackson M, Stark L, Cessna M, … Mitchell J. (2013). Biobanking informatics infrastructure to support clinical and translational research. AMIA Jt Summits Transl Sci Proc. 2013;2013:132–5.
25. Litton JE. Biobank informatics: connecting genotypes and phenotypes. Methods Mol Biol. 2011;675:343–61. doi:10.1007/978-1-59745-423-0_21.
26. Macfarlane GJ, Beasley M, Smith BH, Jones GT, Macfarlane TV. Can large surveys conducted on highly selected populations provide valid information on the epidemiology of common health conditions? An analysis of UK Biobank data on musculoskeletal pain. Br J Pain. 2015;9 (4):203–12. doi:10.1177/2049463715569806.
27. Marko-Varga G, Vegvari A, Welinder C, Lindberg H, Rezeli M, Edula G, … Fehniger TE. Standardization and utilization of biobank resources in clinical protein science with examples of emerging applications. J Proteome Res. 2012;11(11):5124–34. doi: 10.1021/pr300185k.
28. Middha S, Lindor NM, McDonnell SK, Olson JE, Johnson KJ, Wieben ED, … Thibodeau SN. How well do whole exome sequencing results correlate with medical findings? A study of 89 Mayo Clinic Biobank samples. Front Genet. 2015;6:244. doi: 10.3389/fgene.2015.00244.
29. Mook, L. Parelsnoer op weg naar een gezamenlijke nationale biobank infrastructuur. Tijdschrift voor Gezondheidswetenschappen. 2011.
30. Muller TH, Thasler R. Separation of personal data in a biobank information system. Stud Health Technol Inform. 2014;205:388–92.
31. Ngo C, Samuels S, Bagrintseva K, Slocker A, Hupe P, Kenter G, … http://www.raids-fp7.eu, RAc. From prospective biobanking to precision medicine: BIO-RAIDs – an EU study protocol in cervical cancer. BMC Cancer. 2015;15:842. doi: 10.1186/s12885-015-1801-0.
32. Nishimura T, Kawamura T, Sugihara Y, Bando Y, Sakamoto S, Nomura M, … Marko-Varga G. Clinical initiatives linking Japanese and Swedish healthcare resources on cancer studies utilizing Biobank Repositories. Clin Transl Med. 2014;3(1):61. doi: 10.1186/s40169-014-0038-x.
33. Norlin L, Fransson MN, Eriksson M, Merino-Martinez R, Anderberg M, Kurtovic S, Litton JE. A minimum data Set for sharing biobank samples, information, and data: MIABIS. Biopreserv Biobank. 2012;10(4):343–8. doi:10.1089/bio.2012.0003.
34. Otlowski M, Nicol D, Stranger M. Biobanks information paper. National Health and Medical Research Council, Canberra. 2015. . Available from http://www.nhmrc.gov.au/files_nhmrc/file/your_health/egenetics/practioners/biobanks_information_paper.pdf. Accessed 2010.
35. Peters CJ, Rees JR, Hardwick RH, Hardwick JS, Vowler SL, Ong CA, … Molecular Stratification Study G. A 4-gene signature predicts survival of patients with resected adenocarcinoma of the esophagus, junction, and gastric cardia. Gastroenterology. 2010;139 (6):1995–2004 e15. doi: 10.1053/j.gastro.2010.05.080.
36. Quinlan PR, Groves M, Jordan LB, Stobart H, Purdie CA, Thompson AM. The informatics challenges facing biobanks: a perspective from a United Kingdom biobanking network. Biopreserv Biobank. 2015;13(5):363–70. doi:10.1089/bio.2014.0099.
37. Raman S, Richard T. Life, science, and biopower. Technol Hum Values. 2010;35:711–34.
38. Riegman PH, Morente MM, Betsou F, de Blasio P, Geary P, Marble Arch International Working Group on Biobanking for Biomedical, R. Biobanking for better healthcare. Mol Oncol. 2008;2(3):213–22. doi:10.1016/j.molonc.2008.07.004.
39. Rosler, H. Dignitarian posthumous personality rights—an analysis of U.S. and German Constitutional and Tort Law. Berkeley J Int Law. 2008;(26):153.
40. Ross W. The right and the good. Oxford: Oxford; 2002.
41. Ruiz-Romero C, Blanco FJ. Proteomics role in the search for improved diagnosis, prognosis and treatment of osteoarthritis. Osteoarthr Cartil. 2010;18(4):500–9. doi:10.1016/j.joca.2009.11.012.
42. Ryu E, Takahashi PY, Olson JE, Hathcock MA, Novotny PJ, Pathak J, … Sloan JA. Quantifying the importance of disease burden on perceived general health and depressive

symptoms in patients within the Mayo Clinic Biobank. Health Qual Life Outcomes. 2015;13:95. doi: 10.1186/s12955-015-0285-6.

43. Segagni D, Tibollo V, Dagliati A, Perinati L, Zambelli A, Priori S, Bellazzi R. The ONCO-I2b2 project: integrating biobank information and clinical data to support translational research in oncology. Stud Health Technol Inform. 2011;169:887–91.
44. Simeon-Dubach D, Watson P. Biobanking 3.0: evidence based and customer focused biobanking. Clin Biochem. 2014;47(4–5):300–8. doi:10.1016/j.clinbiochem.2013.12.018.
45. Smith ME, Aufox S. Biobanking: the melding of research with clinical care. Curr Genet Med Rep. 2013;1(2):122–8. doi:10.1007/s40142-013-0014-6.
46. Solbakk JH, Holm S, Hofmann B. The ethics of research biobanking. New York: Springer Science & Business Media; 2009.
47. Sorensen KM, Jespersgaard C, Vuust J, Hougaard D, Norgaard-Pedersen B, Andersen PS. Whole genome amplification on DNA from filter paper blood spot samples: an evaluation of selected systems. Genet Test. 2007;11(1):65–71. doi:10.1089/gte.2006.0503.
48. Spath MB, Grimson J. Applying the archetype approach to the database of a biobank information management system. Int J Med Inform. 2011;80(3):205–26. doi:10.1016/j.ijmedinf.2010.11.002.
49. Spjuth O, Krestyaninova M, Hastings J, Shen HY, Heikkinen J, Waldenberger M, … Harris JR. Harmonising and linking biomedical and clinical data across disparate data archives to enable integrative cross-biobank research. Eur J Hum Genet. 2015. doi: 10.1038/ejhg.2015.165.
50. Spjuth O, Krestyaninova M, Hastings J, Shen HY, Heikkinen J, Waldenberger M, … Harris JR. Harmonising and linking biomedical and clinical data across disparate data archives to enable integrative cross-biobank research. Eur J Hum Genet. 2016;24(4):521–8. doi: 10.1038/ejhg.2015.165.
51. Thompson B, Hebert JR. Involving disparate populations in clinical trials and biobanking protocols: experiences from the community network program centers. Cancer Epidemiol Biomarkers Prev. 2014;23(3):370–3. doi:10.1158/1055-9965.EPI-14-0118.
52. Thompson R, Johnston L, Taruscio D, Monaco L, Beroud C, Gut IG, … Lochmuller H. - RD-Connect: an integrated platform connecting databases, registries, biobanks and clinical bioinformatics for rare disease research. J Gen Intern Med. 2014;29 Suppl 3:S780–7. doi: 10.1007/s11606-014-2908-8.
53. Vaught J, Lockhart NC. The evolution of biobanking best practices. Clin Chim Acta. 2012;413(19–20):1569–75. doi:10.1016/j.cca.2012.04.030.
54. Vitonis AF, Vincent K, Rahmioglu N, Fassbender A, Buck Louis GM, Hummelshoj L, … Group WEW. World endometriosis research foundation endometriosis phenome and biobanking harmonization project: II. Clinical and covariate phenotype data collection in endometriosis research. Fertil Steril. 2014;102(5):1223–32. doi: 10.1016/j.fertnstert.2014.07.1244.
55. Watson PH, Barnes RO. A proposed schema for classifying human research biobanks. Biopreserv Biobank. 2011;9(4):327–33. doi:10.1089/bio.2011.0020.
56. Watson PH, Nussbeck SY, Carter C, O'Donoghue S, Cheah S, Matzke LA, … Schacter B. A framework for biobank sustainability. Biopreserv Biobank. 2014;12(1):60–8. doi: 10.1089/bio.2013.0064.
57. Weaver JM, Ross-Innes CS, Shannon N, Lynch AG, Forshew T, Barbera M, … Consortium O. Ordering of mutations in preinvasive disease stages of esophageal carcinogenesis. Nat Genet. 2014;46(8):837–43. doi:10.1038/ng.3013.
58. Wilke RA. High-density lipoprotein (HDL) cholesterol: leveraging practice-based biobank cohorts to characterize clinical and genetic predictors of treatment outcome. Pharmacogenomics J. 2011;11(3):162–73. doi:10.1038/tpj.2010.86.

Chapter 11
XML, Ontologies, and Their Clinical Applications

Chunjiang Yu and Bairong Shen

Abstract The development of information technology has resulted in its penetration into every area of clinical research. Various clinical systems have been developed, which produce increasing volumes of clinical data. However, saving, exchanging, querying, and exploiting these data are challenging issues. The development of Extensible Markup Language (XML) has allowed the generation of flexible information formats to facilitate the electronic sharing of structured data via networks, and it has been used widely for clinical data processing. In particular, XML is very useful in the fields of data standardization, data exchange, and data integration. Moreover, ontologies have been attracting increased attention in various clinical fields in recent years. An ontology is the basic level of a knowledge representation scheme, and various ontology repositories have been developed, such as Gene Ontology and BioPortal. The creation of these standardized repositories greatly facilitates clinical research in related fields. In this chapter, we discuss the basic concepts of XML and ontologies, as well as their clinical applications.

Keywords Data exchange • Data integration • Knowledge representation • Ontology • XML

11.1 Introduction

At present, various clinical systems are available such as clinical information systems, clinical decision support systems (CDSSs), clinical practice guideline (CPG) system, electronic health record (EHR) systems, case-based reasoning (CBR)-driven medical diagnostic systems, clinical trial management systems (CTMSs), unified medical language systems, laboratory information management systems, and electronic patient

C. Yu
Suzhou Industrial Park Institute of Services Outsourcing, No. 99 Ruoshui Road, Suzhou Industrial Park, Suzhou 215123, Jiangsu, China

B. Shen (✉)
Center for Systems Biology, Soochow University, No. 1 Shizi Street, 206, 215006 Suzhou, Jiangsu, China
e-mail: bairong.shen@suda.edu.cn

B. Shen et al. (eds.), *Translational Biomedical Informatics*, Advances in Experimental Medicine and Biology 939, DOI 10.1007/978-981-10-1503-8_11

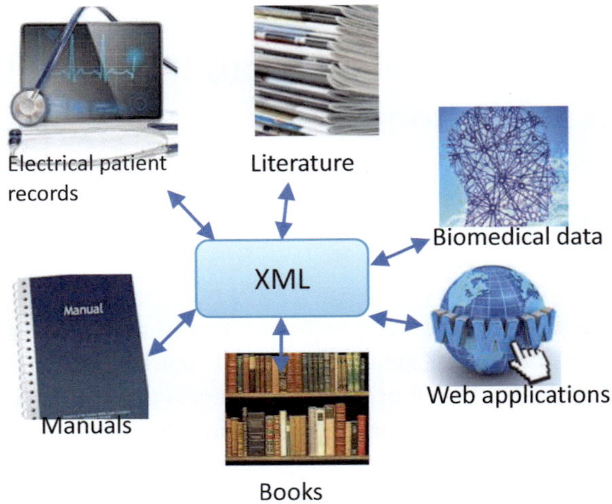

Fig. 11.1 XML provides a common mechanism for data interchange

care systems. These systems are used to collect, save, query, and utilize data. Data may be collected in various formats, including electronic patient records (EPRs), biomedical data, literature, manuals, and books, where these data may be scattered among a variety of information sources, e.g., information systems, on the Internet, and text data. After the data has been collected, they must be consolidated, organized, and processed into a unified data format (Fig. 11.1).

The design goals of Extensible Markup Language (XML) emphasize the simplicity, generality, and usability of the Internet. XML data are known as self-describing, which means that the structure of the data is embedded with the data. The XML format can be used by anyone to share information in a consistent manner. XML is defined by a number of related specifications, including those for XML, XML Pointer Language, XML Linking Language, Extensible Stylesheet Language (XSL), XSL Transformation (XSLT), and document-type definitions.

During data processing, XML can organize data in different formats using the unified XML format. For example, CBR-driven medical diagnostic systems require a large number of clinical cases [1]. In general, clinical cases are distributed in the clinical system of each hospital, and the clinical systems used by hospitals may differ, with various data storage formats. XML can be used to convert clinical cases in different formats into a unified format. In some scenarios, we may need to exchange heterogeneous data within or between systems, e.g., data obtained from experiments using human tissue specimens have little actual value unless they can be combined with medical data [5]. In the past, research data were correlated with medical data by manually retrieving pathology reports, patient charts, radiology reports, and the results of special procedures. However, manually annotating research data are very difficult when experiments involve thousands of tissue specimens, which involve the export of large and complex data collections.

The same word may have different meanings, e.g., in the UK, football refers to "association football," whereas "soccer" is used in the USA, but it also could refer to "rugby" football, while football generally means "American football" in the

USA. In addition, the same thing can be represented using different words. Thus, an ontology is a good solution if we want to share a common understanding of the structure of information among people or software agents, as well as reusing domain knowledge, making domain assumptions explicit, separating domain knowledge from operational knowledge, and analyzing domain knowledge. An ontology is the basic level of a knowledge representation scheme, which defines a common vocabulary for researchers who need to share information within a domain. An ontology includes machine-interpretable definitions of basic concepts in the domain and the relations among them.

There is no common definition of the term "ontology" itself, but the definitions can be categorized into three rough groups:

1. Ontology is a philosophical term that means "theory of existence."
2. Ontology is an explicit specification of conceptualization.
3. Ontology is a body of knowledge that describes a domain.

Ontology is important for enabling the sharing and reuse of knowledge. The backbone of ontology is often a taxonomy, which is a classification of things in a hierarchical form.

Many disciplines have now developed standardized ontologies, which can be used by domain experts to share and annotate information in their fields, e.g., large standardized and structured vocabularies have been produced in medicine. General-purpose ontologies are also emerging such as the Basic Formal Ontology and Open Biomedical Ontology Foundry.

When researchers conduct domain-specific research, they need to develop an ontology for their specific domain by defining a common vocabulary in order to share information. XML can facilitate data-level integration, but it is difficult to use XML for semantic-level data integration. An ontology can integrate data at the semantic level, where the integration of multiple databases involves creating an ontology for each database, before integrating multiple ontologies into a unified ontology. Users can search data using the unified ontology in order to execute a semantic query [31]. CPGs are increasingly important for clinical applications, and they have become instruments for supporting patient care [25]. CPGs are generally provided in the form of free text. However, it is known that nationally or internationally produced guidelines, particularly those that do not involve medical processes at the time of the consultation, do not consider local factors and they have no consistent implementation strategy, which limits their impact on changing the behavior of physicians or patterns of care [15]. We can perform semantic-sensitive searches by creating a CPG ontology because semantic-based searches outperform free-text queries. In general, the precision is greater when more ontological elements are used in the query [26]. The widespread adoption of CDSSs in clinical practice has been hampered mainly by the difficulty of expressing domain knowledge and patient data in a unified formalism [50], but the CPG ontology can be integrated into CDSSs. Thus, Zhang et al. proposed a semantic-based approach by integrating Health Level Seven (HL7) Reference Information Model (RIM) and an ontology to obtain a unified representation of healthcare domain knowledge and patient data to support practical clinical decision-making applications [50]. Clinical trials are designed to assess whether a new intervention is better than the current alternatives [30] and they play important roles in the development of therapeutic methods, drug development, and verification. Ontologies are very helpful for supporting clinical trial data reuse and semantic queries.

11.2 Clinical Applications of XML

11.2.1 Application of XML to Clinical Data Storage

11.2.1.1 Using XML to Store Data

XML data are known as self-describing or self-defining, which means that the structure of the data is embedded with the data. Thus, the power of XML is related to its simplicity, where large chunks of information can be consolidated into an XML document, and the meaningful pieces structure and organize the information (Fig. 11.2).

GeneClinics is an online genetic information resource that comprises descriptions of specific inherited disorders. GeneClinics acquires content from authors via templates and converts it into an XML document that reflects the underlying database schema, which is then loaded into a database [46]. A drug formulary comprises drug data and treatment guidelines, where the guidelines are a textual description, with related information such as drug substances and drug brand names. If a clinical user wants to access the information, this will require some effort, but XML can be used to restructure the text description [37]. The XML Transaction Architecture for the Management of Internet Objects (TAMINO) database management system (DBMS) is designed for handling XML documents. Queries can be processed rapidly

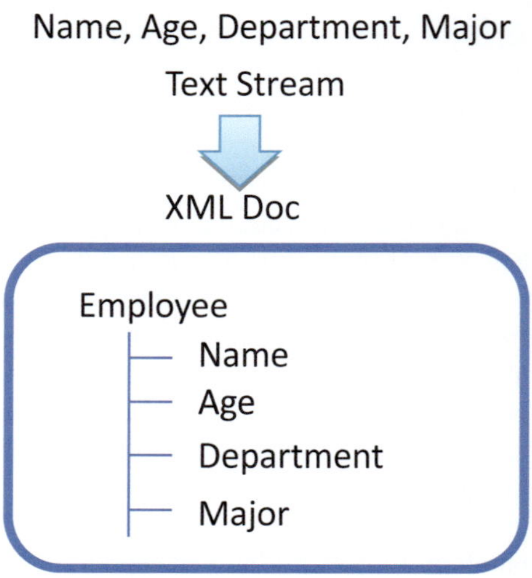

Fig. 11.2 XML for a simple text

against text using purely native XML TAMINO DBMS to store electronic medical records (EMRs), where the queries are compared with annotations added to the text to select documents from many different areas of interest [20]. A clinical path is a method for managing care and checklists for a certain disease, which provides a useful tool for hospital management. Clinical paths can help hospitals to reduce variation in the care of patients. At present, clinical path is defined by each separate hospital and there is no standard format. Benchmark testing between the clinical paths used by different hospitals is important for evaluating medical practices in order to develop more effective improved practices. Description rules for medications have been introduced in XML, which can be used to compare the different medications in the clinical paths prescribed in multiple hospitals [28].

11.2.1.2 Creation of an XML Vocabulary to Store Domain-Specific Documents

Generic XML document vocabularies, such as DocBook XML for software documentation, XHTML for Web documents, and the HL7 Clinical Document Architecture narrative block, cannot meet the requirements of domain-specific documents. Thus, to format domain-specific documents, we need to create domain-specific XML markup vocabularies. Compared with a document formatted using a general-purpose XML markup vocabulary, a document formatted by a domain-specific markup vocabulary is easier to edit and read, with a simpler structure, and it is easier to traverse during search and retrieval. The Pittsburgh Biomedical Informatics Training Program created the Clinical Laboratory Procedure Markup Language (CLP-ML) for formatting CLP manuals. When used with appropriate software, CLP-ML can support electronic authoring, reviewing, distributing, and searching of CLPs from a central repository, thereby decreasing the procedure maintenance effort and increasing the utility of procedure information [34].

11.2.1.3 Formatting Data into XML for Querying

Clinical data are found in various forms in documents, e.g., discharge letters, reports, forms, textbooks, articles, and guidelines, but these unstructured or semi-structured documents make it difficult to find useful information when a reference is required. This lack of structure limits the automatic identification and extraction of the information contained in these resources. For example, most clinical guidelines are text based, so it will take a long time to go through these documents when medical staff need to refer to clinical guidelines in clinical practice. If we store these types of documents in a structured manner using XML and establish the corresponding query system, then medical staff can readily access selected and specific information at the point of care. Thus, XML empowers applications for in-context searching as well as allows the same content to be represented in different ways [15].

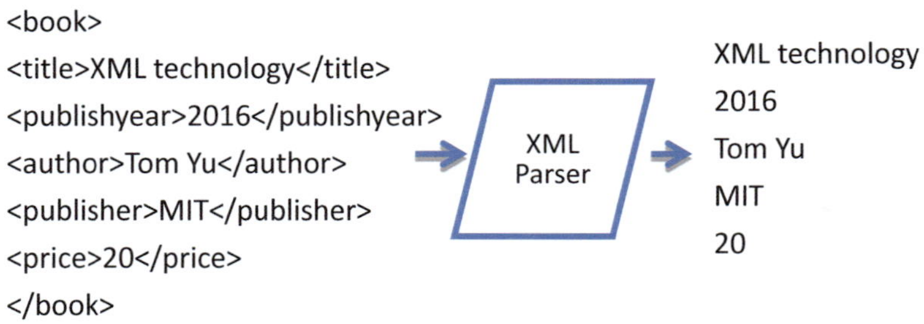

Fig. 11.3 XML parsers extract information from XML

11.2.1.4 Data Retrieval from XML Documents

The content of XML documents can be customized using XML technology, where XSL is used to refer to a family of languages for transforming and rendering XML documents. Using XSLT technology, an XML document can be transformed into an HTML document or another XML document. Cascading Style Sheets (CSS) is a style sheet language used for describing the presentation of a document written in a markup language. CSS was primarily designed to allow the separation of the document content from document presentation. Using these methods, we can extract the XML content to display it in different forms (Fig. 11.3).

At present, large volumes of clinical data are available, such as XML-based EPRs, and useful information is present in these data. The extraction of useful information from EPRs can help to create a corresponding system for use by physicians or researchers, including extracting clinical cases from XML-based EPRs for medical CBR systems. The development and usage of CBR-driven medical diagnostic systems require a large volume of clinical cases to illustrate the problem-solving methods of medical experts [22]. In addition, clinical cases need to be updated continually, but collecting CBR-compliant cases is quite challenging. Adding clinical cases manually places a heavy workload on physicians who may be unfamiliar with computer systems, thereby preventing the implementation of this task. Abidi et al. proposed a novel medical knowledge acquisition approach that exploits routinely generated XML-based EMRs as an alternative source for CBR-compliant cases [1].

11.2.2 Application of XML to Clinical Data Exchange

EMRs have been under development for many years and they are already at a certain scale. In recent years, the application of EMRs has developed rapidly, and the DICOM3, HL7, and other data exchange standards have been created. However, there is no standard unified technology or implementation framework to meet different demands and address the main problems. Therefore, it is difficult to exchange and share data between various systems.

XML data are known as self-describing or self-defining, which means that the structure of the data is embedded with the data, which facilitates data exchange between systems that can "understand" these features. The data in a system can be imported into an XML document with a specific structure. Systems that need to use the data can parse the XML document and extract useful information to allow data exchange. XML provides an ideal solution to address the massive flow of information in EMRs with various complex data types, as well as data exchange and sharing between heterogeneous systems.

11.2.2.1 Exchanging Data Within Systems

Clinical systems usually contain multiple function modules, where each module executes an independent function. We need to define a data exchange format if we want to transfer information between modules, which can be achieved using XML. For example, a CPG system may include a query interface module, logic inference module, and recommendation display module. The query interface module accepts query criteria from users, and XML formats the query criteria before transfer to the logic inference module. The logic inference module parses the query criteria and uses the knowledge base for logic inference to obtain a recommendation. The logic inference module uses XML to format the recommendation before its transfer to the recommendation display module. The recommendation display module shows recommendations to users [2]. During the software design process, the program is modularized according to the requirements of software engineering, thereby improving the efficiency of the program and clarifying the structure of the program. Each module is independent, which ensures that the system has a low degree of coupling and it is highly coherent, as well as easy to manage and maintain.

11.2.2.2 Exchanging Data Between Systems

Clinical information systems often integrate multiple systems to provide services, e.g., EHR systems, clinical patient record systems, CDSSs, and patient accounting systems, where each system has a relatively independent function. However, these systems need to exchange data to provide better services. Thus, a CDSS needs data from clinical patient record systems, and a patient accounting system needs data from clinical patient record systems. If a unified data format is not employed, we cannot exchange data effectively between these systems. XML can allow the exchange of information between systems, but we need to predefine the unified data format before exchanging data. A system that provides data to other systems requires a data output interface, so the systems that want to import data can call the interface. For example, a patient accounting system can call the clinical patient record system's interface to obtain data related to therapy, medication, and an operation to calculate the overall cost when a patient is discharged from hospital [11]. Thus, clinical patient record systems require the implementation of an interface, which accepts data from the caller, retrieves the corresponding data from the system, formats the data into an XML file, and sends it back to the caller.

11.2.2.3 Exchanging Data Using XML Topic Maps

Most clinical data comprise narrative text, and they are often not accessible or searchable at the point of care. Text matching methods will fail to represent implicit relationships between data, e.g., the relationship between HIV and AIDS [38]. Thus, XML can be used to produce a topic map, which provides a flexible data model for representing arbitrary relationships between resources. When searching for information, instead of performing simple text matching, the search content is processed semantically using the XML topic map, thereby allowing the search engine to improve the search accuracy and hit rate.

11.2.2.4 Exchanging Data in Different Formats

XML documents can be converted into different data formats. By employing XML with other W3C standards, informaticists can design systems for the storage and retrieval of structured knowledge, as well as for the rapid transformation of this knowledge into many different usable formats. A single XML document can serve multiple purposes, including content review, application data, printed material, and executable logic. Using XSLT, the content of an XML document can be retrieved according to different demands and provided to the user in different formats. For example, certain professionals might use tables to display data, whereas nonprofessionals could employ a user-friendly interface to present the data. An XML document can also be converted into a PDF file for printing or retrieving data in a system-specific format to import into a system, e.g., import into a CDSS [18].

11.2.3 Application of XML to Clinical Data Integration

Increasingly heterogeneous data sources are being produced due to the development and popularization of networks and distributed application. Thus, the integration of heterogeneous data is becoming more important, but the problem of integrating data from a wide range of sources with high heterogeneity is difficult to solve.

The early data integration systems generally adopted a relational model or object model as common data models before the appearance of XML, but XML provides the opportunity to integrate heterogeneous data from different sources. Thus, researchers have implemented data integration based on XML technology.

11.2.3.1 Integrating Different Types of Clinical Data

Clinical information systems comprise multiple subsystems, each of which defines its own format according to the data employed. These data need to be integrated with each other to make them more useful. For example, large amounts of human tissue samples have little value if they cannot be integrated with pathological and clinical information. In the past, research data were correlated with medical data by manually reading, retrieving, abstracting, and assembling patient charts, pathology reports, radiology reports, and the results of special tests and procedures. However, the manual annotation of research data is impractical when experiments involve hundreds or thousands of tissue samples in large and complex data collections [5]. Defining general data standards in XML can facilitate the integration of biomedical data.

11.2.3.2 Integrating the Same Types of Clinical Data

Large amounts of homogeneous data are distributed in many hospitals and research centers, which would be more valuable if they could be aggregated according to some framework. There are many solutions to this problem, such as retrieving distributed data in XML files and importing them into a central database, thereby allowing data use simply by querying the central database. Another solution to the distribution of data in multiple systems is producing an integration framework to access the data. The integration framework usually includes the data source layer, XML middle layer, and application layer. The data source layer is the lowest level, which provides data to the system from different data sources such as databases, documents, and multimedia. The XML middle layer provides data transfer tools or modules, where this layer can access the data in the database, transfer database data to XML files, or transfer XML files to database data. The application layer allows the manipulation of data in XML files and transmission to the XML middle layer, or data can be obtained from the XML middle layer and transmitted to the application layer.

In clinical settings, some patients need joint treatments [24]. In order to share patient profiles, hospitals need to share distributed EPRs so different hospitals can view the EPRs for the same patient. Thus, different doctors can view the contents of the patient profiles. XML is suitable for addressing this problem because it is system independent and it is easy to retrieve content from XML documents. In order to support clinical research to improve the treatment of HIV, access to multisite clinical data regarding the treatment and outcomes of HIV-infected patients in routine care is required. Thus, a relational XML schema has been developed to extend the existing observational research repository and to integrate real-time clinical information from EMRs at six centers for AIDS research into a repository [6].

11.3 Clinical Applications of Ontologies

11.3.1 Creation of Clinical Ontologies

The power of ontologies is based on their ability to represent knowledge explicitly, to encode semantics, and to facilitate a shared understanding of formal knowledge representations within and between humans and machines. Formally, an ontology comprises entities, relationships, properties, instances, functions, constraints, rules, and other inference procedures.

11.3.1.1 Creation of Ontology Principles

Base on their experience, researchers have proposed various principles and methods for constructing ontologies. In 1995, Gruber proposed five principles of ontology construction: clarity, coherence, extendibility, minimal encoding bias, and minimal ontological commitment.

Gómez-Pérez supplemented this list with principles that have proved useful in the development of ontologies: the ontological distinction principle (i.e., classes in an ontology should be disjoint), diversification (multiple inheritance) of hierarchies, modularity, minimization of the semantic distance between sibling concepts, and standardization of names.

There are no unified standards for guiding principles, procedures, or methods for evaluating ontology construction. In each domain, researchers summarize their experience based on practice as a method for guidance. However, during the process of building a domain-specific ontology, it is generally accepted that experts in the field should participate.

11.3.1.2 Creating Ontology Methods

1. Basic ideas for ontology creation:

   (a) Ontology creation employs domain-specific resources, including unstructured text, semi-structured Web pages, XML documents, lexicon, and structured relational databases. The most commonly used method is creating an ontology from the bottom up with the help of domain experts.

   (b) Transforming an existing thesaurus or taxonomy into an ontology. An ontology is an effective expansion of a thesaurus, which can be considered a simplified ontology. The existing thesaurus concept and the conceptual relationships of a thesaurus can be used to create an ontology.

   (c) Integrating with an existing ontology. A general ontology or reference ontology can be created after merging with an existing ontology and organizing it in an effective manner.

Table 11.1 Commonly used methods for building ontologies

Ontology creation method	Developer	Application domain	Creation tool
Seven-Step method	Stanford University School of Medicine	Widely used in the field of subject knowledge	Protégé
METHONTOLOGY method	Technical University of Madrid	Creating a chemical ontology	WebODE
IDEF5 method	KBSI company of the USA	Describing and producing enterprise ontologies	
TOVE method	Gruninger, Fox, et al.	Business process and activity model ontology	
Skeleton method	Uschold and King	Enterprise modeling ontology for definition and terminology collection between commercial enterprises	

2. Approaches for building ontologies

The most commonly used methods for building ontologies include the Seven-Step method, METHONTOLOGY method, IDEF5 method, TOVE method, and skeleton method (Table 11.1).

11.3.1.3 Tools for Creating Ontologies

Research into ontology has led to the increased development of ontologies by various research communities. Ontology development is an enormous knowledge engineering task. However, various problems are encountered when creating ontology using the methods mentioned above, such as consistency checking and ontology presentation. Thus, there is a need for tools to facilitate the task of ontology development, and some ontology creation tools have emerged. Many research organizations have attempted to develop an ontology creation environment for specific fields to support several stages of ontology development. Using these tools, ontology creators can focus on the organization of the ontology contents without understanding the details of the ontology description language and the description method, which greatly facilitates the creation of ontologies. The currently available ontology creation tools include Protégé, OntoEdit, WebOnto, WebODE, KAON, etc.

Protégé is a tool developed for knowledge acquisition by Stanford University, which is mainly applied to knowledge acquisition as well as the combination and arrangement of an existing ontology. Protégé is open source and it can be downloaded for free. It supports the creation and editing of one or more ontologies in a single workspace via a completely customizable user interface, where its refactor operations include ontology merging, moving axioms between ontologies, and renaming multiple entities.

Protégé has the following features: W3C standards compliance, semantic Web technology support (e.g., OWL, RDF, and SPARQL), customizable user interface, visualization support, ontology refactoring support, direct interface to reasoners, and a highly pluggable architecture.

Protégé is one of the most popular ontology editing tools because of its ease to use, continuous upgrades, free availability, and powerful functional scalability.

A lot of clinical ontologies were created by researchers (Table 11.2).

11.3.2 Application of Ontologies to Clinical Data Integration

It is increasingly important for investigators to efficiently and effectively access, interpret, and analyze data from diverse biological, literature, and annotation sources in a unified manner. Ontology-based data integration is a good solution for addressing the heterogeneity and semantic conflicts involved in heterogeneous data integration.

Clinical and biological research often requires different systems to handle various tasks, thereby producing different types of data. Researchers can obtain more useful data, expand the study samples, and achieve excellent outcomes via data integration. However, semantic conflicts will be encountered during heterogeneous data integration, such as conflicts caused by using different terms in heterogeneous systems to express the same entity or the same term in heterogeneous systems to denote a different entity, where an example is the use of the word football, as mentioned above. A common strategy for addressing semantic conflicts is the use of an ontology with explicitly defined schema terms. This approach is called ontology-based data integration. An ontology also allows the users to query different database systems together by merging them at a semantic level.

An ontology has many advantages during data integration. First, an ontology provides a rich and predefined lexicon, which can be used as the stable concept interface for a data source and it is independent of the data mode. Second, the knowledge in an ontology representation is sufficient to convert all relevant information sources. Third, an ontology supports consistent management and non-consistent data recognition.

At the computer technology center of the University of Bremen in Germany, Wache and colleagues investigated the existing ontology-based integrated systems and research in Europe and the USA, where they analyzed 25 ontology-based integrated systems, and three ontology-based integration approaches were summarized: single ontology integration method, multiple ontology integration method, and hybrid ontology integration method [49]:

1. The single ontology integration method is also known as the global ontology integration method, which uses a global ontology to describe all of the data sources in an integrated system. Thus, all of the information sources establish semantic relationships with the same shared lexicon. In addition, all user queries are processed with this ontology (Fig. 11.4).

Table 11.2 Clinical ontologies

Ontology name	Domain	Function
Adverse Drug Event Ontology [3]	Clinical surveillance	To address the problem of adverse drug event identification, a gap assessment was completed by creating a comprehensive ontology using a minimal clinical data set framework, which incorporated existing identification approaches, clinical literature, and a large set of inpatient clinical data
Disease Ontology [36]	Biological and clinical human disease-related data	The Disease Ontology is populated with consensus-driven disease data descriptors, which incorporate disease terms utilized by genomic and genetic projects, as well as resources employed in studies to understand the genetics of human disease using model organisms
Clinical Data Element Ontology [19]	Data elements	This ontology organizes clinical data elements originating from different medical data repositories into a single unified conceptual structure. This allows the highly selective search and retrieval of relevant data elements from multiple medical data repositories, thereby enabling clinical documentation and clinical research data aggregation
Bacterial Clinical Infectious Diseases Ontology [10]	Clinical infectious disease treatment	This ontology defines a controlled terminology for clinical infectious diseases and domain knowledge that is widely used in hospital settings for making clinical infectious disease treatment decisions
Clinical Measurement Ontology, Measurement Method Ontology, and Experimental Condition Ontology [41]	Rat Genome Database	These ontologies were developed for the Rat Genome Database to standardize quantitative rat phenotype data in order to integrate results from multiple studies into the PhenoMiner database and data mining tool
Core Clinical Protocol Ontology [40]	Clinical guidelines	This ontology includes definitions for clinical guideline recommendations and the process of recommendation
Epilepsy and Seizure Ontology [35]	Epilepsy and seizure	This ontology uses a four-dimensional epilepsy classification system, which integrates the latest International League Against

(continued)

Table 11.2 (continued)

Ontology name	Domain	Function
		Epilepsy terminology recommendations and National Institute of Neurological Disorders and Stroke common data elements
Biomedical Resource Ontology [47]	Biomedical resources	This ontology enables semantic annotation and the discovery of biomedical resources
Haghighi-Koeda Mood Disorder Ontology [12]	Mood disorder	This ontology includes both medical and psychological approaches to mood disorders in order to promote the exchange of information between psychiatrists and psychologists
Clinical Bioinformatics Ontology [16]	Clinical bioinformatics	This ontology is a semantic network for describing clinically significant genomics concepts

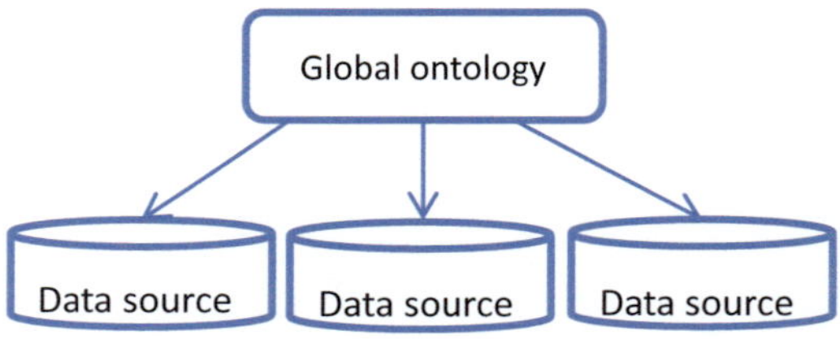

Fig. 11.4 Single ontology integration method

If the granularity of the local view of a data source exhibits high disparity, it is not easy to formulate a unified global ontology, and this increased the difficulty of integration. In addition, the single ontology integration method is readily affected by changes in the data source, e.g., adding a new data source or removing an old data source. The global ontology also needs to make corresponding changes after changes to the data sources. This is an inherent problem with the single ontology integration method.

The process of integrating different databases using an ontology is generally divided into three layers: the presentation layer, the database and ontology mapping layer, and the data layer (Fig. 11.5). The user sends a database query request through the presentation layer. The mapping layer analyzes the query request. The database and ontology mapping relationship converts the user query request into a query against the database. The result of the query against the database is returned to the mapping layer. The mapping layer maps the data onto the semantic ontology and it is returned to the user. In the presentation layer, the user can query data via the semantic contents without considering the storage format of the underlying data [23].

The schemas employed for databases and ontologies are similar in some respects, but they differ in many other. A database schema defines the structure

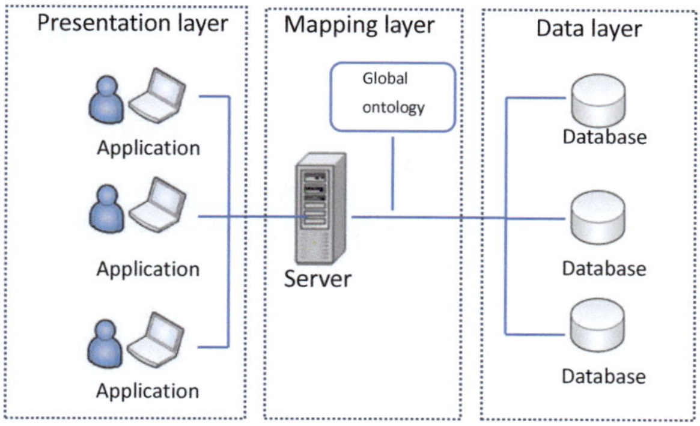

Fig. 11.5 Database integration model using an ontology

of a database, whereas an ontology describes the knowledge for a subject area. Therefore different applications obtain various schemas for databases. An ontology is a body of knowledge that describes a domain, particularly a commonsense domain, so it is independent of specific applications.

The mapping layer needs to map table fields from several databases onto the ontology concepts. There are three types of mapping: one-to-one mapping, many-to-one mapping, and one-to-many mapping. The problem of semantic heterogeneity is encountered during mapping, e.g., the same patient might be in a different table name in two systems, or two systems may define the same table field, but the meaning of the fields could be different. This semantic heterogeneity problem can be addressed by mapping between database fields and ontology concepts.

If fields with the same meaning in different databases use different standards, we must create a mapping between these standards. For example, prostate cancer data uses different standards to determine how much cancer is in the body and where it is located. Staging describes the severity of an individual's cancer based on the magnitude of the original tumor as well as the extent of the cancer's spread throughout the body. For example, the American Joint Committee on Cancer uses the tumor-node-metastasis (TNM) staging system and the International Federation of Gynecology and Obstetrics (FIGO) staging system.

2. The multiple ontology integration method is used to overcome the inherent shortcomings of the single ontology integration method, as shown in Fig. 11.6. In the multiple ontology integration method, each data source is described by its own ontology, and it is not affected by the semantics of other data sources. It is necessary to create a mapping with the local ontology during data source integration. During mapping, a query of one data source is transferred to other data sources to allow multiple data source integration.

The biggest problem with the multiple ontology integration method is establishing mapping relationships between multiple local ontologies. In general, an additional representation is used to define the mapping between ontologies, which is one of the main difficulties with ontology-based data integration.

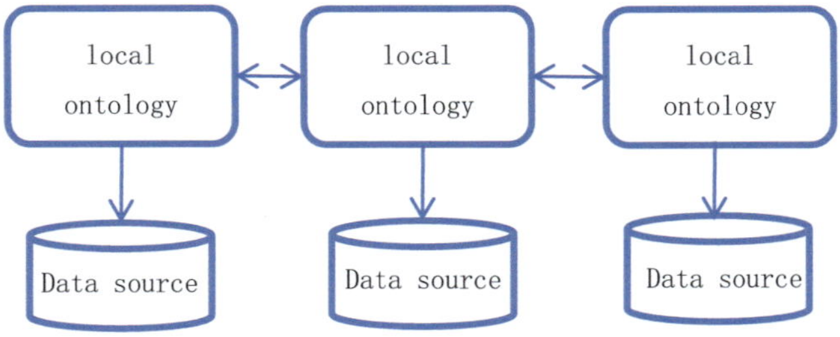

Fig. 11.6 Multiple ontology integration method

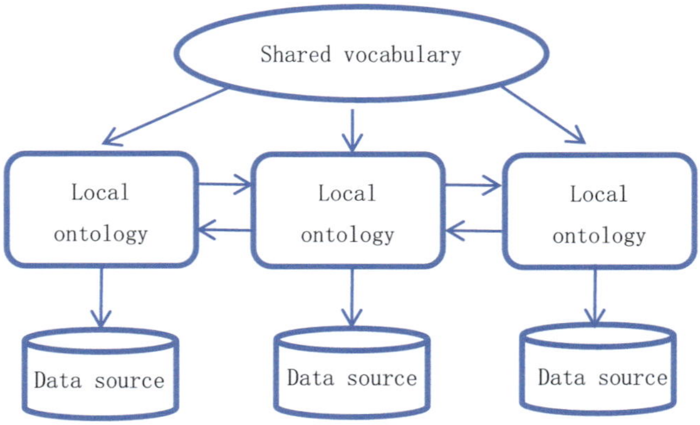

Fig. 11.7 Hybrid ontology integration method

3. The hybrid ontology integration method overcomes the shortcomings of the multiple ontology integration method by establishing mappings between different ontologies (Fig. 11.7). Similar to the multiple ontology integration method, during the process of hybrid ontology integration, each data source uses its own local ontology to describe the semantics in order to ensure the autonomy of the local data sources. In addition, the hybrid ontology integration method applies the idea of single ontology integration, where a shared vocabulary is established above the local ontology to make comparisons between local ontologies. This shared vocabulary contains the basic terminologies from the local ontology. Thus, a query based on the shared vocabulary can easily be converted into a local query.

Nevertheless, hybrid ontology integration methods still need to solve the problems of ontology mapping. Thus, establishing a mapping between the local ontology and shared vocabulary is an important task for the hybrid ontology integration method.

Pérez-ReyIn et al. proposed the ONTOFUSION system using the hybrid ontology integration method as an ontology-based system for biomedical database integration. In the ONTOFUSION system, the physical database layer contains private databases, public databases, and biomedical ontology databases. During the mapping process, the physical schema of each database is mapped onto a virtual schema. Each physical database has a corresponding virtual schema. During the unification process, the local ontologies are unified to create unified virtual schemas, which can be accessed by users in order to retrieve data from various sources at the same time. To create the unified virtual schema, the biomedical ontology is referenced as a knowledge base. The unified virtual schemas are ontologies that reflect the conceptual structure of the information stored in various databases. The user can retrieve data through the unified virtual schemas, where the unified virtual schemas retrieve data from virtual schemas, and the virtual schemas retrieve data from physical databases [31] (Fig. 11.8).

11.3.3 Application of Ontologies to CPGs

11.3.3.1 CPG Concept

CPGs are defined by the Institute of Medicine as "systematically developed statements to assist practitioner and patient decisions about appropriate health care for specific clinical circumstances" [27].

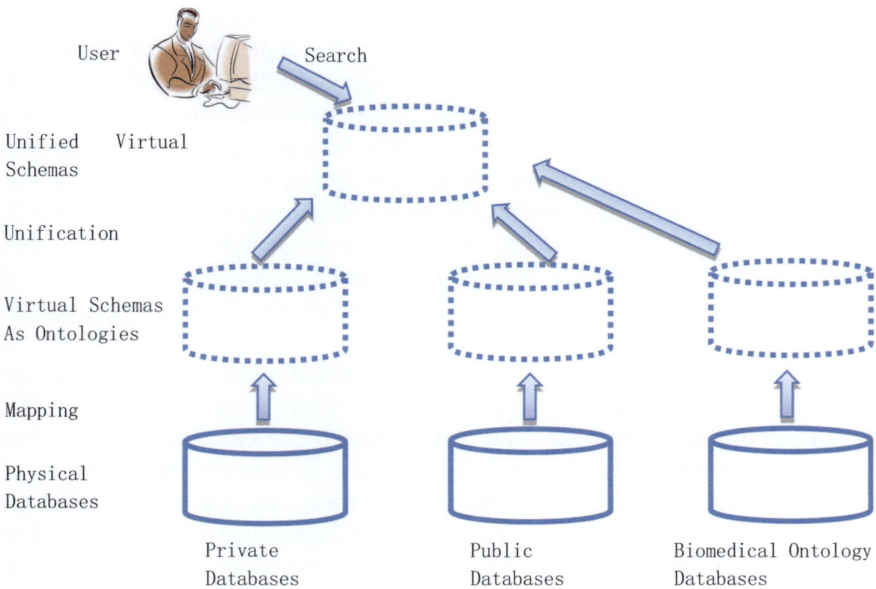

Fig. 11.8 ONTOFUSION mapping and unification

CPGs are generally specified by a CPG committee, where the statements contain recommendations based on evidence from a rigorous systematic review and a synthesis of the published medical literature. CPGs are evidence-based documents that allow healthcare professionals and providers to use existing knowledge in the diagnosis and management of patients. Every year, many CPGs are specified by various organizations. For example, the National Guideline Clearinghouse (NGC) is a public resource for evidence-based CPGs. The mission of the NGC is to provide physicians and others with an accessible mechanism for obtaining objective detailed information about CPGs and to further their dissemination, implementation, and use.

The application of CPGs can have many benefits for medical staff, such as providing a standard operating process and reducing the incidence of errors. CPGs also have many constraints because they are based on existing evidence and they are not specific to a particular hospital. In addition, CPGs identify and describe generally recommended courses of intervention without considering the specific environment. Thus, when hospitals use CPGs, they need to be customized first. CPGs are usually represented in free text, so it is difficult to use them effectively at the point of care. Furthermore, several CPGs may be available for the same disease, and conflicts may exist between the different CPGs, which prevent their widespread application.

11.3.3.2 Application of Ontologies to CPGs

Several CPGs may be available for the same disease because different organizations have developed their own CPGs. For example, the European Association of Urology developed "Guidelines on Prostate Cancer" and the National Comprehensive Cancer Network (NCCN) of America developed the "NCCN Clinical Practice Guidelines in Oncology—Prostate Cancer." Many inconsistencies are found if we analyze different versions of CPGs. Even within the same state, different versions developed by different organizations for the same disease can be very different. For example, Galopin et al. used ontological modeling to evaluate the consistency of adult hypertension CPGs [8], where they found inconsistencies in CPGs covering the same topic. The analysis of different profiles and their associated recommended actions showed that the recommended actions were potentially inconsistent for the majority of profiles.

Ontological modeling is generally used to model CPGs first in order to facilitate the use of CPGs. The Guideline Elements Model (GEM) is an XML-based guideline document model, which can store and organize the heterogeneous information contained in practical guidelines. GEM is intended to facilitate the translation of natural language guideline documents into a format that can be processed by computers. The GEM Cutter tool can annotate CPGs with GEM elements. The computerization of paper-based CPGs will identify many problems such as ambiguities. Several methods can resolve these ambiguities such as consulting with cancer oncologists, reviewing the available literature, and applying personal clinical experience. The CPG ontology is derived from the contents of the knowledge components in the GEM representation of CPGs. The developed CPG ontology can be applied to CDSS. Thus, Abidi et al. performed a project to computerize and deploy breast cancer follow-up CPGs in a breast cancer follow-up decision support system for use by family physicians in a primary care setting [2].

Most guidelines are text based, and they are published primarily in medical journals or posted on the Internet. However, it can be difficult and time consuming to browse the Internet to find the correct guidelines for an existing diagnosis and adequate recommendations for a specific clinical problem [15]. For example, the NGC initiative by the Agency for Healthcare Research and Quality contains CPGs from different Web portals in a uniform internal structure, which are indexed by MeSH concepts. The Guidelines International Network includes CPGs from a wide range of countries, and its members can browse and perform full-text searches. Guidelines Finder3 contains CPGs from the UK, and it allows simple text-based search to access free-text CPGs. However, most digital libraries mainly provide unstructured free-text CPGs and the most basic search techniques using terms and keywords. These basic search techniques cannot deliver the required query efficiency. Thus, using an ontology can facilitate the context-sensitive search and retrieval of CPGs. For most recall levels, context-sensitive search methods outperform traditional full-text search [27].

11.3.4 Application of Ontologies to Clinical Trials

11.3.4.1 Clinical Trial Concept

Clinical trials are defined by the WHO International Clinical Trials Registry Platform as: "For the purposes of registration, a clinical trial is any research study that prospectively assigns human participants or groups of humans to one or more health-related interventions to evaluate the effects on health outcomes" (http://www.who.int/ictrp/en/). When a new product or approach is being studied, it is usually unclear whether it will be helpful, harmful, or no different compared with available alternatives. Thus, it is necessary to determine the safety and efficacy of the intervention by measuring certain outcomes in participants. There is a growing international recognition of the need to record the existence of clinical trials so they can be publicly accessed, which improves research transparency and ultimately strengthens the validity and value of the scientific evidence base. There has been a push by governments and international organizations, especially since 2005, to make clinical trial information more widely available as well as to standardize registries and registration processes.

At present, there are several clinical trials registry platforms in use throughout the world. ClinicalTrials.gov is a Web-based resource, which provides patients, their family members, healthcare professionals, researchers, and the public with readily accessible information on publicly and privately supported clinical studies of a wide range of diseases and conditions. ClinicalTrials.gov currently lists 209,390 studies located in all 50 states of the USA and 192 countries (https://clinicaltrials.gov/).

11.3.4.2 Application of Ontologies to Clinical Trials

In recent years, it has been recognized that ontologies can play important roles in clinical trials. However, most clinical trials fail to recruit participants on schedule, and it is difficult to find eligible patients automatically based on EHR systems. This process is time consuming and inefficient, and it also requires specialized training. Patrao et al. developed an ontology-based information retrieval system for clinical trial recruitment [30], where this system uses EHR data, represents medical knowledge with ontologies, integrates several databases, and allows searches for structured data and free text. The preliminary quality assessments obtained excellent recall rates.

CTMSs promise to help researchers in hospitals and biotechnology companies to better manage the tremendous amounts of data that are generated when conducting clinical trials. It is still usual to collect data at each trial site on paper-based case report forms and to input them into CTMSs. In general, these case report forms are designed according to specific clinical trial requirements. Thus, it is difficult to compare and exchange data between different clinical trials because there is no unified standard. Stenzhorn et al. developed an ontology-based trial management application (ObTiMA), which allows data reusability according to shared concepts defined in an ontology that covers the entire cancer care and research spectrum [44].

The management of clinical trials involves the use of several software applications, which generate large volumes of data during the course of a trial. It is important to solve the problem of semantically integrating heterogeneous applications to facilitate the efficient management of trials and subsequent analyses of clinical trial data. Shankar et al. devised an ontology-based architecture to support interoperability among clinical trial software applications, where this approach focused on a suite of clinical trial ontologies that define the vocabulary and semantics required to represent information about clinical trials [39].

A clinical study is conducted according to a research plan known as a clinical trial protocol, which is a document that describes how a clinical trial will be conducted (the objective, design, methodology, statistical considerations, and organization of the clinical trial), thereby ensuring the safety of the trial subjects and the integrity of the data collected. The number of clinical trials is increasing, and international multicenter clinical trials are being conducted more frequently. However, no standards are available for the structure of trial protocols or reusable concepts for use in a clinical trial context. Heller et al. developed a medical and trial-specific term data dictionary for clinical trial protocols to improve quality assurance in clinical trial protocols. The data dictionary is based on domain-specific ontologies and the top-level ontology of the General Ontological Language [13].

11.3.5 Application of Ontologies to CDSSs

11.3.5.1 CDSS Concept

A CDSS is a health information technology system designed to provide physicians and other health professionals with clinical decision support (CDS). The following working definition was proposed by Hayward at the Centre for Health Evidence: "Clinical decision support systems link health observations with health knowledge to influence health choices by clinicians for improved health care." The American Medical Informatics Association defines CDS as: "Clinical decision support provides clinicians, staff, patients or other individuals with knowledge and person-specific information, intelligently filtered or presented at appropriate times, to enhance health and better health care." There are two main types of CDSSs: a CDSS that uses a knowledge base, applies rules to patient data using an inference engine, and displays the results to the end user and systems without a knowledge base, which rely on machine learning to analyze clinical data.

CDS interventions can be applied throughout the medication management cycle to optimize medication safety and other pertinent outcomes. A useful framework for achieving success in this regard is the "CDS Five Rights" approach. The CDS Five Rights framework asserts that to improve targeted healthcare decisions with well-developed CDS interventions, the interventions must provide the right information (evidence-based guidance in response to a clinical need) to the right people (the entire care team, including the patient) through the right channels (e.g., EHR, mobile device, or patient portal) with the right intervention formats (e.g., order sets, flow sheets, dashboards, or patient lists) at the right points in the workflow (for decision making or action).

11.3.5.2 Application of Ontologies to CDSSs

The widespread adoption of CDSSs in clinical practice has been hampered mainly by the difficulty of expressing domain knowledge and patient data in a unified formalism. However, integrating domain knowledge and patient data by using an ontology to develop a CDSS can yield high accuracy and a better acceptance rate in practical applications. Evaluation results demonstrate the technical feasibility and application prospects of the ontology approach [50]. There is a large amount of clinical evidence in medical records. In the medical domain, knowledge representation is a key issue because it should be used effectively in reasoning as part of the CDSS. Ontologies describe and organize domain knowledge in a manner that machines can read and humans can understand, so decision support may be promoted by computer-based systems or applications [4].

In clinical treatment, CPGs can guide medical staff to make decisions. CPG ontologies can be integrated into CDSSs for disease treatment. Thus, Omaish et al. conducted the ontology-based computerization of acute coronary syndrome clinical guidelines to develop a CDSS for acute coronary syndrome [29]. CPGs generally focus on a specific medical disorder, but actual patients often present multiple pathologies, and the management of multiple morbidities can be a major challenge for clinicians. Using an ontology to integrate multiple relevant CPGs can solve this problem. Galopin et al. developed a framework where ontological reasoning was used to enrich the patient description at different levels of abstraction, thereby increasing the number of appropriate recommendations [9]. CPG computerization involves modeling and the conversion of paper-based CPGs into an electronic and executable format, which can be accessed by physicians as well as be embedded within clinical decision support systems at the point of care. Several CPG modeling formalisms are available such as GEM. Abidi et al. used GEM to model breast cancer follow-up CPGs and the Jena inference engine to develop a CDSS [2].

11.4 XML and Ontology Tools

11.4.1 XML Tools

Many XML tools are available for editing XML files as shown in Table 11.3.

Many XML tools have been created during the development of clinical domain projects, as shown in Table 11.4.

11.4.2 Ontology Tools

Many ontology tools are available for editing ontology files, as shown in Table 11.5.

Many ontology tools have been created during the development of clinical domain projects, as shown in Table 11.6.

Table 11.3 XML tools

No.	Name	Function	Free	URL
1	XMLSpy	XMLSpy is the industry's best-selling XML editor and development environment for all XML-related technologies. It offers the world's leading schema designer; code generation; file converters; debuggers; profilers; full database integration; support for XSD, XSLT, XPath, XQuery, WSDL, SOAP, XBRL, and Office Open XML; and many more features	No	http://www.altova.com/xml_tools.html
2	MapForce	MapForce is an award-winning any-to-any graphical data mapping, conversion, and integration tool for mapping data between any combination of XML, SQL database, EDI, XBRL, flat file, Excel, JSON, and/or Web service structures, which transforms data instantly or autogenerates royalty-free data integration code for executing recurrent conversions	No	http://www.altova.com/xml_tools.html
3	StyleVision	StyleVision is an award-winning tool for designing compelling presentation layouts and document formats from input sources including XML, SQL databases, and XBRL. StyleVision makes the full presentation and format conversion power of XSLT readily available in a graphical design tool for HTML, Word, PDF, and Authentic® output	No	http://www.altova.com/xml_tools.html
4	DiffDog	DiffDog is a powerful, XML-aware diff/merge utility for files, directories, and database schemas and tables. DiffDog makes it easy to compare and merge text or source code files, synchronize directories, and compare database schemas and tables. DiffDog also provides advanced XML-aware differencing and editing capabilities	No	http://www.altova.com/xml_tools.html
5	SchemaAgent	SchemaAgent is a visionary tool for analyzing and managing relationships between XSD, XML, XSLT, and WSDL files across a project, an intranet, or even an enterprise. It allows you to view and manage XML file relationships easily via its graphical design view and drag and drop to automatically configure imports, includes, and/or redefines (IIRs)	No	http://www.altova.com/xml_tools.html

(continued)

Table 11.3 (continued)

No.	Name	Function	Free	URL
6	Authentic	Authentic is a powerful and dynamic electronic form editor for enterprise XML solutions. By employing Authentic, business users can work with intuitive and dynamic forms to view and edit information stored in XML documents and SQL databases, as well as XBRL information management systems, without being exposed to the underlying technology	No	http://www.altova.com/xml_tools.html
7	Microsoft XML Core Services (MSXML)	MSXML allows customers to build high-performance XML-based applications, which provide a high degree of interoperability with other applications that adhere to the XML 1.0 standard. MSXML provides developer support for the following core services: Document Object Model, Helper APIs, XML Schema Definition, Simple API for XML, Schema Object Model, and XML Digital Signatures	Yes	https://msdn.microsoft.com/en-us/library/cc507432(v=vs.85).aspx
8	EditiX	EditiX is a powerful and easy to use XML editor, Visual Schema Editor, XQuery Editor, and XSLT debugger for Windows, Linux, and Mac OS X, which was designed to help Web authors and application programmers to take advantage of the latest XML and XML-related technologies, such as XSLT/FO, DocBook, and XSD Schema	No	http://www.editix.com
9	XmlPad	XmlPad is a professional editor for XML processing documents, which allows the presentation of data in a tabular format. It includes a text editor with syntax highlighting, string numeration, collapsing, and element autocompletion options	Free	http://www.wmhelp.com/xmlpad3.htm
10	XML Notepad	XML Notepad is an open-source XML editor written by Chris Lovett and published by Microsoft. This editor features incremental search in both tree and text views, drag/drop support, IntelliSense, find/replace with regular expressions and XPath expressions, and support for XInclude	Free	https://xmlnotepad.codeplex.com

Table 11.4 Domain-specific XML tools

No.	Name	Function
1	CLP-ML [34]	CLP-ML was designed to support electronic authoring, reviewing, distribution, and searching of CLPs from a central repository, thereby decreasing the procedure maintenance effort and increasing the utility of procedure information
2	cMDX Editor [7]	cMDX can represent pathological findings related to the prostate in schematic styles. cMDX documents can be converted into different data formats such as text, graphics, and PDFs
3	DoT-U2 [32]	DoT-U2 is an XML-based knowledge-supported checklist software system for documenting newborn clinical screening examinations. Physicians can enter findings in a tree-structured protocol with management of the logical dependencies
4	CLinical Accounting InforMation (CLAIM) intranet health clinic [11]	CLAIM is an XML-based data exchange standard for connecting EMR systems to patient accounting systems

Table 11.5 Ontology tools

No.	Name	Function	Free	URL
1	Protégé	Protégé is a free, open-source visual ontology editor and knowledge-based framework The Protégé platform supports two main ways of modeling ontologies via the Protégé-Frames and Protégé-OWL editors. Protégé ontologies can be exported into a variety of formats, including RDF (S), OWL, and XML schema	Yes	http://protege.stanford.edu
2	SemanticWorks	Altova's SemanticWorks is a visual Semantic Web editor, which features a graphical RDF and RDFS editor and a graphical OWL editor. It supports OWL Lite, OWL Full, and OWL DL dialects	No	http://www.lesliesikos.com/altova-semanticworks-visual-rdf-and-owl-editor
3	BioMixer	BioMixer is a Web-based environment for visualizing and exploring biomedical ontologies. It is the underlying technology for the visualization components found in BioPortal, the world's most comprehensive repository of biomedical ontologies	Yes	http://thechiselgroup.org/biomixer

(continued)

Table 11.5 (continued)

No.	Name	Function	Free	URL
4	OntoBuilder	The OntoBuilder project supports the extraction of ontologies from Web search interfaces, which range from simple search engine forms to multiple-page, complex reservation systems. OntoBuilder enables fully automatic ontology matching	Yes	http://ontobuilder.bitbucket.org
5	NeOn toolkit	The NeOn toolkit is a state-of-the-art, open-source multi-platform ontology engineering environment, which provides comprehensive support for the ontology engineering life cycle	Yes	http://neon-toolkit.org/wiki/Main_Page

Table 11.6 Domain-specific ontology tools

No.	Name	Function
1	OntoStudyEdit [49]	OntoStudyEdit is a software tool for ontology-based representation and management of metadata in clinical and epidemiological research
2	Recruit [30]	An ontology-based information retrieval system for clinical trial recruitment
3	MorphoCol [43]	An ontology-based knowledge base for the characterization of clinically significant bacterial colony morphologies
4	Onto Clinical Research Forms (OntoCRF) [21]	OntoCRF is a framework for the definition, modeling, and instantiation of clinical data repositories
5	Duke Enterprise Data Unified Content Explorer (DEDUCE) [33]	DEDUCE is a self-service query tool developed to provide clinicians and researchers with access to data within the Duke Medicine Enterprise Data Warehouse
6	Semantator [42]	Semantator is a semiautomatic tool for document annotation with Semantic Web ontologies
7	OnWARD [48]	OnWARD is an ontology-driven, secure, rapidly deployed, Web-based framework to support data capture for large-scale multicenter clinical research
8	TrialWiz [14]	TrialWiz is an authoring tool for encoding a clinical trial knowledge base. TrialWiz manages the complexity of the protocol-encoding process and improves the efficiency of knowledge acquisition

11.5 Conclusion

The application of information technology in clinical areas has facilitated the development of medical and biotechnology technology, as well as gene technology, thereby generating increasing numbers of clinical systems and volumes of data. The growing volume and diversity of health and biomedical data indicate that the era of Big Data has arrived for healthcare. In the Big Data era, the management and exploitation of data will bring new challenges for modern medicine. The organization and management of these data sources will present problems involving data exchange and data integration. XML techniques facilitate data management from the data level, while ontologies promote data management from the semantic level, and these methods have been remarkably effective in clinical applications in recent years. The large volumes of available medical data contain useful information, and mining these data can generate information with clinical applications, thereby facilitating the future development of medical science.

References

1. Abidi SS, Manickam S. Leveraging XML-based electronic medical records to extract experiential clinical knowledge. An automated approach to generate cases for medical case-based reasoning systems. Int J Med Inform. 2002;68:187–203.
2. Abidi SR, et al. Ontology-based modeling of clinical practice guidelines: a clinical decision support system for breast cancer follow-up interventions at primary care settings. Stud Health Technol Inform. 2007;129:845–9.
3. Adam TJ, Wang J. Adverse drug event ontology: gap analysis for clinical surveillance application. AMIA Jt Summits Transl Sci Proc. 2015;2015:16–20.
4. Bau CT, Chen RC, Huang CY. Construction of a clinical decision support system for undergoing surgery based on domain ontology and rules reasoning. Telemed J E Health. 2014;20:460–72.
5. Berman JJ, Bhatia K. Biomedical data integration: using XML to link clinical and research data sets. Expert Rev Mol Diagn. 2005;5:329–36.
6. Drozd DR, et al. Developing a relational XML schema for sharing HIV clinical data. AMIA Annu Symp Proc. 2005;2005:943.
7. Eminaga O, et al. Clinical map document based on XML (cMDX): document architecture with mapping feature for reporting and analysing prostate cancer in radical prostatectomy specimens. BMC Med Inform Decis Mak. 2010;10:71.
8. Galopin A, et al. Using an ontological modeling to evaluate the consistency of clinical practice guidelines: application to the comparison of three guidelines on the management of adult hypertension. Stud Health Technol Inform. 2014;205:38–42.
9. Galopin A, et al. An ontology-based clinical decision support system for the management of patients with multiple chronic disorders. Stud Health Technol Inform. 2015;216:275–9.
10. Gordon CL, et al. Design and evaluation of a bacterial clinical infectious diseases ontology. AMIA Annu Symp Proc. 2013;2013:502–11.
11. Guo J, et al. CLAIM (CLinical Accounting InforMation)—an XML-based data exchange standard for connecting electronic medical record systems to patient accounting systems. J Med Syst. 2005;29:413–23.

12. Haghighi M, et al. Development of clinical ontology for mood disorder with combination of psychomedical information. J Med Dent Sci. 2009;56:1–15.
13. Heller B, et al. Standardized terminology for clinical trial protocols based on top-level ontological categories. Stud Health Technol Inform. 2004;101:46–60.
14. Hirose Y, Yamamoto R, Ueda S. The nodes focusing tool for clinical course data of hypergraph structure in the ontological framework CSX output from POMR-based EMR system. Stud Health Technol Inform. 2007;129:741–5.
15. Hoelzer S, et al. Value of XML in the implementation of clinical practice guidelines—the issue of content retrieval and presentation. Med Inform Internet Med. 2001;26:131–46.
16. Hoffman M, Arnoldi C, Chuang I. The clinical bioinformatics ontology: a curated semantic network utilizing RefSeq information. Pac Symp Biocomput. 2005;2005:139–50.
17. http://www.who.int/ictrp/en/WHO International clinical trials registry platform
18. Hulse NC, et al. Application of an XML-based document framework to knowledge content authoring and clinical information system development. AMIA Annu Symp Proc. 2003;2003:870.
19. Jeong S, et al. Clinical data element ontology for unified indexing and retrieval of data elements across multiple metadata registries. Healthc Inform Res. 2014;20:295–303.
20. Johnson SB, et al. A native XML database design for clinical document research. AMIA Annu Symp Proc. 2003;2003:883.
21. Lozano-Rubi R, Pastor X. OWLing clinical data repositories with the ontology web language. JMIR Med Inform. 2014;2:e14.
22. Manickam S, Abidi SS. Extracting clinical cases from XML-based electronic patient records for use in web-based medical case based reasoning systems. Stud Health Technol Inform. 2001;84:643–7.
23. Min H, et al. Integration of prostate cancer clinical data using an ontology. J Biomed Inform. 2009;42:1035–45.
24. Mludek V, et al. Integration of clinical practice guidelines into a distributed regional electronic patient record for tumour-patients using XML: a means for standardization of the treatment processes. Stud Health Technol Inform. 2001;84:658–62.
25. Mludek V, et al. Supporting clinical practice guidelines: lifecycle of guidelines for oncology within an XML-based guideline framework. Stud Health Technol Inform. 2002;90:466–70.
26. Moskovitch R, et al. A comparative evaluation of full-text, concept-based, and context-sensitive search. J Am Med Inform Assoc. 2007;14:164–74.
27. Moskovitch R, Shahar Y. Vaidurya: a multiple-ontology, concept-based, context-sensitive clinical-guideline search engine. J Biomed Inform. 2009;42:11–21.
28. Okada O, Ohboshi N, Yoshihara H. Clinical path modeling in XML for a web-based benchmark test system for medication. J Med Syst. 2005;29:539–53.
29. Omaish M, Abidi S, Abidi SS. Ontology-based computerization of acute coronary syndrome clinical guideline for decision support in the emergency department. Stud Health Technol Inform. 2012;180:437–41.
30. Patrao DF, et al. Recruit—an ontology based information retrieval system for clinical trials recruitment. Stud Health Technol Inform. 2015;216:534–8.
31. Perez-Rey D, et al. ONTOFUSION: ontology-based integration of genomic and clinical databases. Comput Biol Med. 2006;36:712–30.
32. Philipp F, et al. Introducing DoT-U2—an XML-based knowledge supported checklist software for documentation of a newborn clinical screening examination. Comput Methods Prog Biomed. 2005;77:115–20.
33. Roth C, et al. DEDUCE clinical text: an ontology-based module to support self-service clinical notes exploration and cohort development. AMIA Jt Summits Transl Sci Proc. 2013;2013:227.
34. Saadawi GM, Harrison Jr JH. Definition of an XML markup language for clinical laboratory procedures and comparison with generic XML markup. Clin Chem. 2006;52:1943–51.
35. Sahoo SS, et al. Epilepsy and seizure ontology: towards an epilepsy informatics infrastructure for clinical research and patient care. J Am Med Inform Assoc. 2014;21:82–9.

36. Schriml LM, Mitraka E. The Disease Ontology: fostering interoperability between biological and clinical human disease-related data. Mamm Genome. 2015;26:584–9.
37. Schweiger R, et al. XML structured clinical information: a practical example. Stud Health Technol Inform. 2000;77:822–6.
38. Schweiger R, et al. Linking clinical data using XML topic maps. Artif Intell Med. 2003;28:105–15.
39. Shankar RD, et al. An ontology-based architecture for integration of clinical trials management applications. AMIA Annu Symp Proc. 2007;2007:661–5.
40. Slaughter L, et al. The core clinical protocol ontology (C2PO): a realist ontology for representing the recommendations within clinical guidelines. Stud Health Technol Inform. 2013;192:997.
41. Smith JR, et al. The clinical measurement, measurement method and experimental condition ontologies: expansion, improvements and new applications. J Biomed Semantics. 2013;4:26.
42. Song D, Chute CG, Tao C. Semantator: annotating clinical narratives with semantic web ontologies. AMIA Jt Summits Transl Sci Proc. 2012;2012:20–9.
43. Sousa AM, Pereira MO, Lourenco A. MorphoCol: an ontology-based knowledgebase for the characterisation of clinically significant bacterial colony morphologies. J Biomed Inform. 2015;55:55–63.
44. Stenzhorn H, et al. The ObTiMA system—ontology-based managing of clinical trials. Stud Health Technol Inform. 2010;160:1090–4.
45. Tao C, Solbrig HR, Chute CG. CNTRO 2.0: a harmonized semantic web ontology for temporal relation inferencing in clinical narratives. AMIA Jt Summits Transl Sci Proc. 2011;2011:64–8.
46. Tarczy-Hornoch P, et al. GeneClinics: a hybrid text/data electronic publishing model using XML applied to clinical genetic testing. J Am Med Inform Assoc. 2000;7:267–76.
47. Tenenbaum JD, et al. The Biomedical Resource Ontology (BRO) to enable resource discovery in clinical and translational research. J Biomed Inform. 2011;44:137–45.
48. Tran VA, et al. OnWARD: ontology-driven web-based framework for multi-center clinical studies. J Biomed Inform. 2011;44 Suppl 1:S48–53.
49. Uciteli A, Herre H. OntoStudyEdit: a new approach for ontology-based representation and management of metadata in clinical and epidemiological research. J Biomed Semantics. 2015;6:41.
50. Zhang YF, et al. Integrating HL7 RIM and ontology for unified knowledge and data representation in clinical decision support systems. Comput Methods Prog Biomed. 2016;123:94–108.

Chapter 12
Bayesian Computation Methods for Inferring Regulatory Network Models Using Biomedical Data

Tianhai Tian

Abstract The rapid advancement of high-throughput technologies provides huge amounts of information for gene expression and protein activity in the genome-wide scale. The availability of genomics, transcriptomics, proteomics, and metabolomics dataset gives an unprecedented opportunity to study detailed molecular regulations that is very important to precision medicine. However, it is still a significant challenge to design effective and efficient method to infer the network structure and dynamic property of regulatory networks. In recent years a number of computing methods have been designed to explore the regulatory mechanisms as well as estimate unknown model parameters. Among them, the Bayesian inference method can combine both prior knowledge and experimental data to generate updated information regarding the regulatory mechanisms. This chapter gives a brief review for Bayesian statistical methods that are used to infer the network structure and estimate model parameters based on experimental data.

Keywords Bayesian inference • Approximate Bayesian computation • Genetic regulation • Reverse engineering

12.1 Introduction

Precision medicine involves using detailed, patient-specified molecular information to diagnose and categorize disease, then guide treatment to improve clinic outcome [42]. To achieve these goals, precision medicine aims to develop computational models that integrate data and knowledge from both clinic and basic research to gain a mechanistic understanding of disease [14]. Compared with bioinformatics approaches, computational models are able to predict mode of action and responses

T. Tian (✉)
School of Mathematical Science, Monash University, Clayton, VIC 3800, Australia
e-mail: tianhai.tian@monash.edu

B. Shen et al. (eds.), *Translational Biomedical Informatics*, Advances in Experimental Medicine and Biology 939, DOI 10.1007/978-981-10-1503-8_12

to treatments not only at the molecular level but across all levels of biological organizations as well, including molecular level (gene network, cell-signaling pathway, and metabolic network), cell population level, tissue level, and even whole organism levels. However, a significant challenge facing precision medicine is the incorporation of models at different levels into a single framework by the integration of heterogeneous datasets [58].

With the rapid advancement of high-throughput technologies such as microarray, RNA sequencing, and mass spectrometry (MS)-based proteomics, enormous amounts of information are available for gene expression and kinase activity in the genome-wide scale [9, 44, 55]. These datasets give opportunities to develop mathematical models to explore the regulatory mechanisms and study system dynamics of molecular networks. Although the datasets contain enormous amounts of information, it is still a challenge to develop effective methods to extract useful knowledge from observations [41]. Currently regulatory networks can be constructed by using one of the 2 days, namely, approaches for identifying large-scale molecular interaction networks from "omics" datasets and methods for examining detailed mechanisms through functional properties of the interactions between network components [6]. The latter approach not only captures the dynamic behavior of molecular networks but also provides more detailed regulatory information than the former one [15]. However, the limitation of this method is that it can deal with small-scale network model only.

Since many unknown parameters need to be estimated in mechanism models, it is particularly important to design effective and efficient methods for parameter inference, which is also referred to as model calibration, model fitting, or parameter estimation, in order to produce small simulation errors against experimental data. There are two types of inference methods, namely, optimization methods and Bayesian statistical methods. Aiming at minimizing an objective function is the optimization method search in a directed manner within the parameter space. The inferred set of parameters produces the best fit between simulations and experimental data [28]. A variety of different approaches have been developed. They all share two main ingredients: a cost function for specifying the distance between simulated data and experimental data and an optimization algorithm for searching for parameters that optimize the cost function. When the cost function landscape is complex, which is often the case for systems biology models with high-dimensional parameter spaces, these methods are unlikely to find the global optimum. To tackle this issue, global optimization methods try to explore complex surfaces as widely as possible; among these, genetic algorithms are particularly well known and have been applied to ordinary various models [49]. Comparison studies have been conducted by applying several global optimization algorithms on the test models [16].

Compared with optimization methods, Bayesian inference is able to infer the whole probability distribution of parameters by updating prior probability estimates using Bayes' rule. In addition, Bayesian methods are more robust in dealing with stochastic models and/or experimental data with noise [19, 56]. The recent advances in approximate Bayesian computation (ABC) provide more effective methods without any restriction on the requirement of likelihood function. In recent years Bayesian methods have been used successfully in a diverse range of fields and provide promise to the application in precision medicine [47]. This chapter provides a brief review for Bayesian computation and also prospective for future application in systems biology and precision medicine.

12.2 High-Throughput Biomedical Data

Current high-throughput techniques allow simultaneous examination of thousands of genes, transcripts, proteins, and metabolites. In addition, various bioinformatics tools have been designed to extract insightful information from omics datasets. With the help of omics technologies, it is possible to investigate cellular states as well as biological correlations in cell lines, primary tissues, and the whole organism.

12.2.1 Genomics

The genome is the total DNA of a cell or organism, and genomics is the systematic study of an organism's genome. Genomics applies recombinant DNA, DNA-sequencing methods, and bioinformatics tools to sequence, assemble, and analyze the function and structure of genomes. Research areas of genomics include functional genomics for describing the dynamic aspects such as gene transcription, translation, and protein–protein interactions, structural genomics for studying the three-dimensional structure of every protein encoded by a given genome, epigenomics for investigating the complete set of epigenetic modifications on the genetic material of a cell, and metagenomics for exploring genetic material recovered directly from environmental samples. The variations in DNA sequences between people are of particular interest when linked with diseases with a genetic determination. These studies have a role in pharmacogenomics in exploring individual patient responses to drugs.

12.2.2 Transcriptomics

Transcriptome is the total mRNA in a cell or organism and the template for protein synthesis in the process of translation. Compared with the static information of DNA sequence, transcriptomics measures transcriptome that reflects the genes that are actively expressed at any given moment. A key omics technique is the gene expression microarrays that measure packaged mRNA (mRNA with the introns spliced out) as a summary of gene activity. During the past 10 years, advances in microarray technology have generated a huge amount of gene expression data and resulted in progress in genomics and transcriptomics. Microarray gene expression datasets are major resources for the reverse engineering of genetic regulatory networks [30]. However, gene expression microarrays only measure changes in mRNA abundance, not protein, and thus there is a lack of consensus around the interpretation of microarray data.

12.2.3 Proteomics

The proteome is the entire complement of proteins. Unlike the genome that is more or less constant, the proteome differs from cell to cell, as well as varies over time and distinct requirements that a cell or organism undergoes. Proteomics represents the large-scale study of proteins, especially their structure and function. It aims to characterize information flow within the cell and the organism, through protein pathways and networks, with the eventual aim of understanding the functional relevance of proteins [9]. In recent years, mass spectrometry-based phosphoproteome has generated huge amount of quantitative data of kinase activities in different types of cells and under various experimental conditions. The generated proteomic datasets are precious information for the development of mathematical models for large-scale cell-signaling pathway [50].

12.2.4 Metabolomics

Metabolome refers to the complete set of small-molecule metabolites (such as metabolic intermediates, hormones and other signaling molecules, and secondary metabolites) to be found within a biological sample, such as a single organism. Metabolomics can generally be defined as the study of global metabolite profiles in a system (cell, tissue, or organism) under a given set of conditions [18]. Unlike other omics studies that have much unknown molecular knowledge, metabolomics has a number of theoretical advantages over the other omic approaches. The metabolome is the final downstream product of gene transcription and, therefore, changes in the metabolome are amplified relative to changes in the transcriptome and the proteome. Although the metabolome contains the smallest domain (5000 metabolites), it is more diverse, containing many different biological molecules, making it more physically and chemically complex than the other omics studies. In 2015, real-time metabolome profiling was demonstrated for the first time [29].

12.3 Bayesian Inference Methods

This section gives a brief overview over methods from Bayesian statistics for estimating parameters in systems biology models. In statistical inference the central concept is the likelihood that measures the probability of a given parameter set θ for realizing experimental data D, given by

$$L(\theta) = P(D|\theta)$$

Here $P(D|\theta)$ is a probability density function if it is considered as a function of data D; or a likelihood function if it is a function of θ. The purpose of inference is, for the given observation data D, to find the optimal parameter θ that the probability $P(\theta|D)$ reaches the maximum. According to the Bayesian theorem, this probability can be written as

$$\begin{aligned} P(\theta|D) &= \frac{P(D|\theta)P(\theta)}{P(D)} \\ &\propto P(D|\theta)P(\theta) \end{aligned}$$

where $P(\theta)$ is the prior distribution of parameter, which summarizes our prior knowledge about the parameter under investigation, and $P(\theta|D)$ is the posterior probability distribution over the model parameter. If likelihood function $P(D|\theta)$ is available, the classic Metropolis–Hastings algorithm is applied to find a Markov chain of the parameter. If the target distribution function is $f(D|\theta)$ and the prior distribution of parameter is denoted as $f(\theta)$, the Metropolis–Hastings algorithm is given below.

Algorithm 1 (Metropolis–Hastings Algorithm)

1. Initialization: Choose a starting value θ_0 for which $P(\theta|D) > 0$.
2. For $i = 1, 2, \cdots, N$

   (a) Choose candidate θ^* from a proposal distribution $g(\theta^*|\theta_{i-1})$.

   (b) Compute the ratio

$$r(\theta^*|\theta_{i-1}) = \frac{f(D|\theta^*)f(\theta^*)/g(\theta^*|\theta_{i-1})}{f(D|\theta_{i-1})f(\theta_{i-1})/g(\theta_{i-1}|\theta^*)}$$

   (c) Get samples from the uniformly distributed random variable $s \sim U(0, 1)$, and set

$$\theta_i = \begin{cases} \theta^*, & \text{if } s < r(\theta^*, \theta_{i-1}) \\ \theta_{i-1}, & \text{else} \end{cases}$$

Note that in the classic Metropolis algorithm, the proposal distribution is symmetric, namely, $g(\theta^*|\theta_{i-1}) = g(\theta_{i-1}|\theta^*)$, while in the Metropolis–Hastings algorithm, the symmetric property is not required. The Bayesian "equivalent" to the cost function in optimization methods is the Bayesian posterior distribution. In the Bayesian setting, Markov chain Monte Carlo (MCMC) methods have been used to establish a framework for Bayesian parameter estimation and evidential model ranking over models of biochemical systems [54]. In addition, Kalman filtering has been applied to the estimation of both parameters and hidden variables of nonlinear state-space models [39]. To obtain confidence intervals for a point estimate in a frequentist setting, a range of techniques can be applied that include variance–covariance matrix-based techniques, profile likelihood, and bootstrap methods [46]. However, the limitation of the Metropolis–Hastings algorithm is the requirement for the density function $P(D|\theta)$. For complex models in systems biology, it is difficult to have such analytical density function. To tackle this challenge, ABC was proposed to avoid directly calculation of the likelihood function.

12.4 Approximate Bayesian Computation (ABC)

ABC was initially designed for studying the plausible mutation schemes and estimating the variance in mutation size. Instead of calculating likelihood function $P(D/\theta)$, ABC generates a simulation of the model that is regarded as an artificial dataset Y. The method then relies on some metric to determine the distance between simulated data Y and experimental data D. The widely used metric includes the sum of squared error or more generally the weighted square error [33]. For stochastic models the normal kernel density function is widely used to measure the probability of transitional density [21]. Similar to the optimization methods, a key issue is how to find point estimates of parameters by minimizing the metric using a given technique. However, unlike the optimization methods, the goal of ABC is not to find a set of estimate for parameters with minimum metric. Instead, we hope to obtain the posterior distribution of the estimated parameters. There are three major types of ABC methods: the rejection method, ABC-MCMC (Markov chain Monte Carlo) method, and ABC-sequential Monte Carlo (SMC) method [53]. In the simplest form, for the given experimental data D and model with unknown parameters θ, an error threshold value ε, and a prior distribution $\pi(\theta)$, the rejection method is implemented as follows:

Algorithm 2 (The Rejection Method)

Step 1: Sample a candidate parameter from the prior distribution $\theta_i^* \sim \pi(\theta)$.
Step 2: Generate a simulated dataset Y from the model with parameter θ_i^*.
Step 3: Compare the distance between the simulated dataset Y and experimental data D using a distance metric $\rho(D, Y)$.

Step 4: If $\rho(D,Y) < \varepsilon$, accept the sample $\theta_i = \theta_i^*$ and set step index to $i+1$. Otherwise, reject sample θ_i^*. Go to Step 1.

The output of the ABC rejection algorithm is samples of parameters from a distribution $P(\theta|\rho(D,Y) < \varepsilon)$. If error threshold value ε (tolerance level) is sufficiently small, then the distribution $P(\theta|\rho(D,Y) < \varepsilon)$ will be a good approximation to the posterior distribution $P(\theta|D)$. The key difference of this method from the optimization methods is that we accept a number of samples that satisfy the error tolerance criterion, rather than only the best estimate that has the smallest error. The tolerance ε in Algorithm 2 may determine the efficiency of ABC. The ABC rejection algorithm is one of the basic ABC algorithms that may result in long computing time when a badly prior distribution that is far away from posterior distribution is chosen. In addition, there is no learning process in this algorithm; and thus no information could be obtained from the previous accepted samples of parameters. When the search space is complex, the convergence rate may be very slow.

12.4.1 ABC-MCMC Algorithm

Following the principle of the rejection method, a number of algorithms have been proposed to improve the efficiency. One of them is the combination of the Markov chain Monte Carlo (MCMC) sampling technique with approximation computation. ABC-MCMC introduces a concept of acceptance probability during the decision-making step which saves computing time. MCMC sampling is a process that filters proposed values for θ to arrive at a sample of values drawn from the desired posterior distribution. There are a number of MCMC samplers. Based on the most popular Metropolis–Hastings algorithm, the ABC-MCMC algorithm accepts a sample θ_i^* based on

$$\alpha = \begin{cases} \min\left(1, \dfrac{\pi(\theta_i^*)q(\theta_{i-1}|\theta_i^*)}{\pi(\theta_{i-1})q(\theta_i^*|\theta_{i-1})}\right) & \text{if } \rho(D,Y) < \varepsilon \\ 0 & \text{else} \end{cases}$$

where $\pi(\theta)$ is the prior distribution for θ, and q is the proposed distribution. Similar to the Metropolis–Hastings algorithm, we draw a sample from the uniform distribution and accept the sample θ^* if this sample is less than the value of α [31].

The convergence property of the generated chain $(\theta_1, \theta_2, \ldots, \theta_n)$ is important because MCMC algorithm may suffer if the proposal distribution is poorly chosen [10]. A potential issue is that the chain may be likely to get "stuck" in low probability region of the posterior, and we may never be able to get a good approximation [13]. The ABC-MCMC algorithm is particularly susceptible to this because the proposed sample θ^* must meet two criteria, namely, the generated data should be close to the experimental data and the standard Metropolis–Hastings sample criteria. Thus the rejection rate of the ABC-MCMC may be extremely high.

12.4.2 Sequential Monte Carlo Sampling Algorithm

To tackle the challenges in ABC-MCMC, the idea of particle filtering and sequential Monte Carlo sampling has been introduced. Sequential Monte Carlo sampling differs from the MCMC approach by using the technique of particle filtering. Rather than drawing one candidate sample θ^* at a step, this algorithm considers a pool with a large number of samples $\left(\theta_1^*, \ldots, \theta_N^*\right)$ simultaneously and treats each parameter vector as a particle. The sequential Monte Carlo-based ABC algorithm [45, 52] starts from sampling a pool of N particles for parameter vector θ through prior distribution $\pi(\theta)$. The sampled particle candidates $(\theta_1, \ldots, \theta_N)$ will be chosen randomly from the prior distribution, and their distance measures satisfy the tolerance level. Then we will assign each particle a corresponding weight w_i to be considered as the sampling probability. In the first step, the weight of each particle is assumed to be the same, namely, $1/N$.

In the subsequent iterations, samples of parameter will be drawn from the particles in the previous iteration according to the weight w_i assigned to the particles. A perturbation is generated from the transition kernel $q\left(\cdot\,|\theta^*\right)$ to generate sample candidate $\theta^{**} = \theta^* + \Delta$. The filtering process is then applied to let the generated sample still meet the required tolerance level. A new weight will be assigned to the accepted particle based on the prior distribution, error of simulation, and weights of the previous iteration as well as the transitional kernel. For discrete chemical reaction systems, we proposed to use the following formula to calculate the weight of the ith particle in kth iteration

$$w_i^k = \frac{\pi\left(\theta_i^k\right) b_k\left(\theta_i^k\right)}{\sum_{j=1}^{N} w_j^{k-1} q\left(\theta_i^k | \theta_j^{k-1}\right)}$$

where $b_k(\theta)$ is the match of simulation data generated by using parameter θ to experimental data [60].

A number of effective methods have been designed by using different transitional kernel and/or different approaches to assign weights to each particle, including the ABC-PRC (partial rejection control) algorithm [45], ABC-PMC (population Monte Carlo) sampling [4], and ABC-SMC (sequential Monte Carlo) sampling [52]. Among them, the ABC-SMC algorithm is particularly useful when the transition kernel in ABC-PMC cannot have infinite support (e.g., cannot be Gaussian). For models in systems biology, this property is important because the estimated rate constants in chemical reactions are positive. In addition, we may have prior knowledge about the range of rate constant. Another advantage of the SMC technique is that the tolerance level may change over iterations, and the tolerance level can be determined adaptively [25].

12.4.3 Implementation of ABC-SMC

Implementation of ABC requires the choice of three major factors, namely, the summary statistics, tolerance level, and distance measure. For high-dimensional data, it is very important to identify an informative and low-dimensional set of summaries. Techniques and methods for selecting ABC summary statistics fall into three major categories, namely, subset selection, projection technique, and regularization technique [5]. The subset selection approach selects a best subset from a number of candidates that are evaluated and ranked using various information-based criteria. Projection techniques reduce the dimension by projecting the original dataset into a lower-dimensional space. The regularization technique aims to reduce over-fitting in a model by penalizing model complexity. Although summary statistics is an important issue in discussing high-dimensional data, this issue is less important when we study small-scale regulatory networks.

The threshold value for accepting samples is important for the efficiency of ABC. A large threshold value may lead to posterior distribution with larger error, but it would be difficult to search satisfactory samples if the threshold value is too small. This result is illustrated in Fig. 12.1. Instead of choosing threshold value carefully, it can be selected adaptively. For example, assuming the simulation errors to experimental data in kth iteration are $\left(e_1^k, e_2^k, \ldots, e_N^k\right)$ for the N particles. We can use the median value of these values or the mean of these values as the threshold value of the $(k+1)$th iteration [25]. In addition, the number of samples in each generation and number of generations can also be determined by accepted samples [11]. This approach is very effective for deterministic models such as ordinary differential equation (ODE) models for genetic regulation. However, for stochastic models such as chemical reaction systems, our recent simulation results suggest that more work is still needed for the adaptive techniques when inferring stochastic models.

The distance function is the key measure to assess the quality of parameter samples. For dynamic models such as ODE models or chemical reaction systems, experimental data normally are observed molecular activities/concentrations $(D_1, \ldots D_M)$ in a number of observation time points. The distance function will measure the error of simulation $(Y_1, \ldots Y_M)$ to experimental data. The widely used error function is the mean square root function, the absolute error function, or the weighted error function

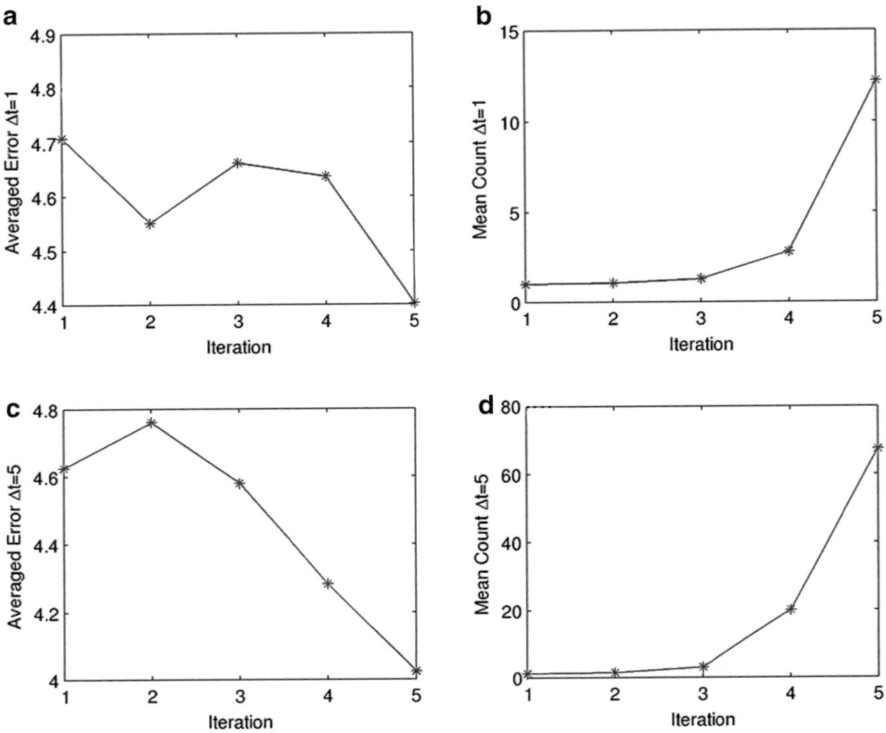

Fig. 12.1 The averaged error of estimated parameters and mean count of sample tests for obtaining an accepted particle with step size Δt of 1 and 5 of the data using decreased tolerance levels over five iterations of ABC for inferring parameters in a stochastic model of a gene network [59]. A smaller tolerance level normally leads to smaller simulation value but larger number of sample tests

$$\text{Error} = \sum_{i=1}^{N} \frac{1}{w_i} |Y_i - D_i|^p$$

where $p = 1$ and 2 are for the absolute error and mean square error, respectively, and w_i is the weight. If $w_i = 1$ ($w_i = D_i$), it is the absolute error (relative error).

In addition to the values of simulation, the derivative and second-order derivative have also been used in the error function, which gives smoother approximation to experimental data. We have further developed a continuous distance function using a spline function to generate the continuous simulation and dataset over the whole time period. Compared with the discrete criteria using data at observation time points only, the continuous criteria lead to more accurate estimates [12]. In addition, the model using estimated parameter from continuous approaches has better robustness property than that using parameter from discrete approaches.

For stochastic models, the transitional probability density function $f\left(D_{i+1}\middle|D_i,\theta\right)$ measures the probability of realizing the observation D_{i+1} with the given D_i and parameter θ. A good estimate should make the joint transitional density

$$f(D_0,\theta)\sum_{i=0}^{N-1} f\left(D_{i+1}\middle|D_i,\theta\right)$$

to reach the maximum. An equivalent measure is the negative log-likelihood function

$$L(\vartheta) = -\log\bigl(f(D_0,\vartheta) - \sum_{i=1}^{N-1}\log\bigl(f\left(D_{i+1}\middle|D_i,\vartheta\right)$$

Note that, if we choose the density function as an exponential function

$$f(D_{i+1}|D_i,\theta) = \exp\left(-(Y_{i+1} - D_{i+1})^2\right),$$

the negative log-likelihood function actually is the mean square error. Since the closed-form expression of the transitional density is usually unavailable, we use a nonparametric kernel density function

$$f(D_{i+1},t_{i+1}) = \frac{1}{MB}\sum_{j=1}^{M} K\left(\frac{Y_{ij} - D_i}{B}\right)$$

to evaluate the transitional density based on M stochastic simulations [21]. Here $K(\cdot)$ is a nonnegative kernel function enclosing unit probability mass. The normal kernel is one of the widely used kernel functions for inferring parameters in stochastic models. In addition, based on the discrete nature of chemical reactions, the frequency distribution of simulation molecular numbers was proposed to evaluate the transitional density [51]. Numerical results showed that the frequency distribution gives more stable estimations of the transitional density than the normal kernel density functions for discrete chemical reaction systems.

We have proposed two ABC algorithms using simulated likelihood density, which have been applied to estimate unknown rate constants in chemical reaction systems. Compared with other distance measures, the simulated likelihood density function provides more accurate estimates. Figure 12.2 suggests that the probabilistic distribution starts from nearly a uniform distribution in the second iteration (Fig. 12.2a) and gradually converges to a normalized-like distribution (Fig. 12.2d) with a mean value that is close to the exact rate constant.

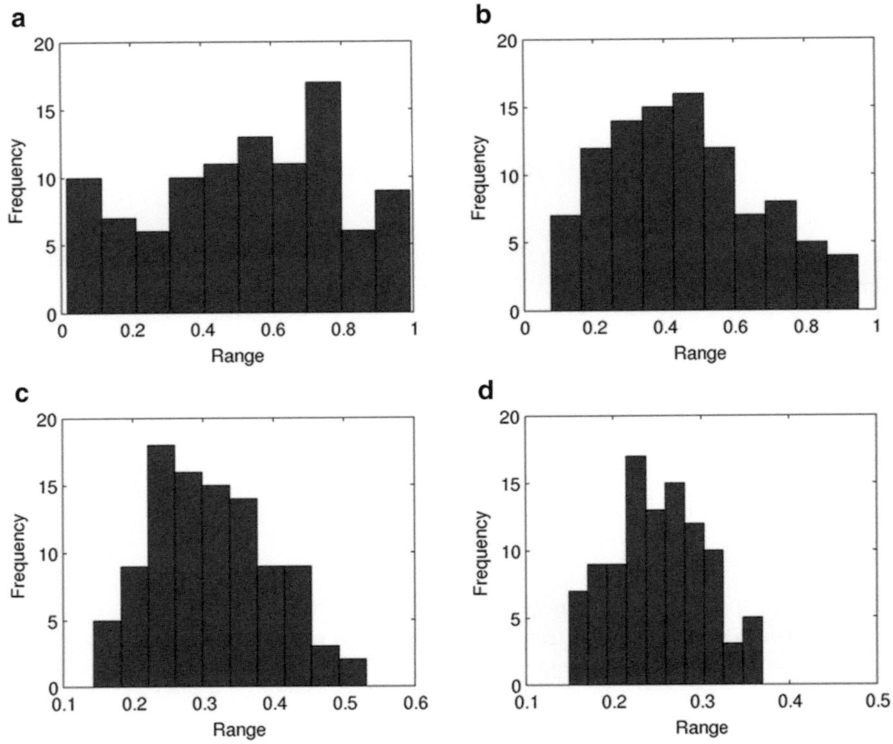

Fig. 12.2 Probabilistic distributions of the estimated rate constant C_7 over four iterations ($k = 2,3,4,5$) using ABC-SMC for inferring parameters in a stochastic model of a gene network [60]. The distribution of the first iteration ($k = 1$) is nearly a uniformly distribution

12.4.4 Criterion for Selecting Estimated Candidates

A major challenge in the inference of model parameters is that different estimates, whose value may vary over a wide range, all are able to faithfully realize experimental data. Thus additional criterion is needed to select the final estimate from candidates. Since robustness is a ubiquitously observed property of biological systems, this property has been widely used recently as an important measure to select the optimal network structure or model parameters from estimated candidates, including genetic regulatory networks and cell-signaling pathways [2, 7, 32]. Here robustness is defined as the ability of a system to function correctly in the presence of both internal and external uncertainty [3]. This theory has been extensively studied by Kitano and coworkers [23, 24]. A formal and abstract definition of the robustness property has been widely used in analyzing robustness properties of biological systems [3]. Recently more detailed definitions have been proposed to calculate the robustness property of biological systems [40].

According to the definition in Kitano [23], the robustness property of a mathematical model with respect to a set of perturbations P is defined as the average of an evaluation function $D^s_{a,P}$ of the system over all perturbations $p \in P$, weighted by the perturbation probabilities $prob(p)$, given by

$$R^s_{a,P} = \int_{p \in P} \text{prob}(p) D^s_{a,P} dp \tag{12.1}$$

We have also proposed to use the following measures to evaluate the average behavior

$$R^M_{a,P} = \sum_{i,j} \left[\int_{p \in P} \text{prob}(p) x_{ij}(p) dp \right] \tag{12.2}$$

which is the mean of simulated system dynamics that should be close to the simulated dynamics obtained from the unperturbed parameters [50]. In addition, the impact of perturbations on nominal behavior is defined by

$$R^N_{a,P} = \sum_{i,j} \left[\int_{p \in P} \text{prob}(p) \left(\overline{x_{ij}(p)} - x_{ij}(p) \right)^2 dp \right] \tag{12.3}$$

where $x_{ij}(p)$ and x_{ij} are the simulated dynamics x_i at time point t_j with perturbed and unperturbed parameters, respectively, and $\overline{x_{ij}(p)}$ is the mean of $x_{ij}(p)$ over all the perturbed parameters. For each parameter k_i, the perturbation is set to follow a uniform distribution or Gaussian distribution.

In addition, other additional criteria have been used to select the candidate estimates. Sloppy is a concept to measure sensitivity of model parameters to external perturbation. A model is termed as sloppy if its sensitivity eigenvalues were approximately equally spaced in their logarithm [20]. This study suggested that, even with precise datasets, many parameters are unknowable to the trajectory measurements. Using the epidermal growth factor (EGF) and neuronal growth factor (NGF) signaling pathways as the test systems, research results suggest optimism for the prospects for calibrating even large models. Success of parameter estimation is intimately linked to the experimental perturbations used. Thus experimental design methodology is important for parameter fitting of biological models and may determine the accuracy of estimation [1].

Another concept related to Bayesian inference is entropy, which is a measure in thermodynamics for the number of specific realizations or microstates that may realize a thermodynamic system in a defined state. Entropy has been used in ABC in a number of ways. For example, minimization of the estimated entropy of the posterior approximation is used as a heuristic for the selection of summary statistics [35]. Another related concept is the Shannon Information. The information theory has been combined with ABC to identify experiments that maximize the information contents of the resulting data [26].

12.5 ABC Software Packages

A number of computer software packages have been designed to implement ABC in different platforms using various computer languages. A software package, BioBayes, provides a framework for Bayesian parameter estimation and evidential model ranking over models of biochemical systems defined using ordinary differential equations. The package is extensible allowing additional modules to be included by developers [54]. Recently, a Python package, ABC-SysBio, implements parameter inference and model selection for dynamical systems in the ABC framework [27]. ABC-SysBio combines three algorithms: ABC rejection sampler, ABC-SMC for parameter inference, and ABC-SMC for model selection. It is designed to work with models written in Systems Biology Markup Language (SBML). Deterministic and stochastic models can be analyzed in ABC-SysBio. In addition, a computational tool SYSBIONS has been designed for model selection and parameter inference using nested sampling [22]. Using a data-based likelihood function, this package calculates the evidence of a model and the corresponding posterior parameter distribution. This is a C-based, GPU-accelerated implementation of nested sampling that is designed for biological applications.

A number of software packages have been designed in the R platform. Among them, *abc* implements ABC-rejection with many methods of regression post-processing; while easyABC implements a wide suite of ABC algorithms but not post-processing [36]. The abctools package has been designed to complement the existing software provision of ABC algorithms by focusing on tools for tuning them. It implements many previous unavailable methods from literature and makes them easy available to the research community [36]. In addition, there are two packages implemented as MATLAB toolbox, including EP-ABC for state space models and related models and ABC-SDE for inferring parameters in stochastic differential equations [38]. There are a number of other software packages that have been reviewed in [36]), including ABCreg, ABCtoolbox, Bayes SSC, DIY-ABC, and PopABC. Due to the limit of space, more information of these software packages can be found in Nunes and Prangle [36].

12.6 Applications

ABC has been applied to infer parameters in a wide range of regulatory networks in systems biology. This section only gives a few examples. For the mitogen-activated protein (MAP) kinase phosphorylation cascade, ABC allows to approximate the posterior distribution over parameters and shows how this can add insights into our understanding of the system dynamics [43]. This study highlights the added benefit of using the distribution of parameters rather than point estimates of parameter values when considering the notion of sloppy models in systems biology. In addition, a general statistical inference framework was designed based on ABC for constructing stochastic transcription–translation networks [37]. An exact inference algorithm and an efficient variational approximation allow scalable inference and learning of the model parameters.

The ABC-SMC method has been used to estimate the parameters in an ODE model of the glucose metabolism network using data from pH-controlled continuous culture experiments [48]. In addition, the profile likelihood estimation (PLE) was used to improve the calculation of confidence intervals for the same parameters. This study suggests that the approximate posterior is strongly non-Gaussian, and calculations of confidence intervals using the PLE method back this up [48]. In addition, ABC was used to infer parameters in a statistical thermodynamic model of gene regulation that combines the activity of a morphogen with the transcriptional network it controls [8]. A dynamical model was developed for the sonic hedgehog (Shh) patterning of the ventral neural tube for accurately predicting tissue patterning. This approach provides a unifying framework to explain the temporospatial pattern of morphogen-regulated gene expression.

Bayesian computation has also been applied to induce the network motifs through spike-timing dependent plasticity in combination with activity-dependent changes in the excitability of neurons. To deal with mixture of quantitative measures (i.e., likelihood and qualitative fitness) simultaneously, a Bayesian framework is formulated for hybrid fitness measures (HFM) [34]. The MCMC-HFM is applied to an apoptosis signal transduction network. The mixed use of quantitative and qualitative fitness measures narrowed down the acceptable range of parameters. In addition, for discrete chemical reactions systems, we developed two algorithms and applied them to infer rate constants in a model of genetic regulatory network [60].

12.7 Discussion

Recent decades have observed substantial progress in the development of Bayesian inference methods and in particular ABC algorithms. Various effective methods have been designed to improve the accuracy of inference results and reduce the huge amount of computing time of Monte Carlo methods. The open-access computer software packages accelerate the application of Bayesian inference methods to more practical problems. However, recent advances in mathematical modeling have raised complex and large-scale models that should be studied using computational statistical methods. The era of big data provides a large number of datasets with huge amount of information in biology, engineering, finance, and social sciences. Therefore Bayesian computation is still an exciting and productive research topic in the frontier of statistical studies.

This chapter only discusses the Bayesian computational methods, but the application of these methods has not been addressed in detail. In fact substantial challenges will immediately arise when applying these methods to develop mathematical models. For example, considering a gene network with a number of genes, a popular and important topic is how to infer the regulatory mechanisms using the microarray or RNA-seq time series data. The regulatory network should be sparse since a gene normally is regulated by a small number of genes. However, it is still a challenge to infer a sparse network structure using inference methods. There are a number of issues that are related to this challenge, including mathematical model to reflect biological regulation, inference methods to search global optimal model parameters, effective numerical methods with small simulation error, and reliable experimental data that is less influenced by noise. The development of Bayesian inference methods will be more or less influenced by the other issues.

Heterogeneous datasets in social sciences and single-cell biological studies have posed substantial need to develop stochastic and/or multi-scale mathematical models for describe random dynamics. However, parameter inference for stochastic dynamic models is harder and less developed than that for deterministic models because of the generally intractable analytical form of the likelihood. In addition, a large number of stochastic simulations are needed to derive reliable measure of the probability for transitional density. Although a number of inference methods have been designed for stochastic models [17, 21, 51, 57, 60], the huge computing time and reliability of estimates are still major questions. Generally the reliability of estimates can be improved when more stochastic simulations are used in inference.

The inference of network model requires a thorough understanding for the properties of network dynamics. Currently simulation error is the key measure to assess the quality of the generated parameter samples. In some cases the difference between the simulation errors of different samples is so small that it is difficult to distinguish which sample is really better than the others. In such cases, system property will help us to select optimal estimate from candidates that have similar simulation error. In recent year, robustness property, bistability, entropy, and Shannon information have been used as the additional criteria to infer model parameters. More work is needed in this area to find more selection criterion and develop effective methods to calculate these system properties.

12.8 Conclusions

This chapter gives a brief review for the statistic computational methods for inferring parameters in mathematical models. Focusing on Monte Carlo sampling methods, we discussed various approximate Bayesian computing algorithms and related issues in implementations. In addition, we discussed the property of network dynamics that is related to the inference of network parameters.

References

1. Apgar JF, et al. Sloppy models, parameter uncertainty, and the role of experimental design. Mol BioSyst. 2010;6(10):1890–900.
2. Apri M, et al. Efficient estimation of the robustness region of biological models with oscillatory behavior. PLoS One. 2010;5(4):e9865.
3. Bates DG, Cosentino C. Validation and invalidation of systems biology models using robustness analysis. IET Syst Biol. 2011;5(4):229–44.

4. Beaumont MA, et al. Adaptive approximate Bayesian computation. Biometrika. 2009;96(4):983–90.
5. Blum MGB, et al. A comparative review of dimension reduction methods in approximate Bayesian computation. Stat Sci. 2013;28(2):189–208.
6. Bruggeman FJ, Westerhoff HV. The nature of systems biology. Trends Microbiol. 2007;15(1):45–50.
7. Citri A, Yarden Y. EGF-ERBB signalling: towards the systems level. Nat Rev Mol Cell Biol. 2006;7(7):505–16.
8. Cohen M, et al. A theoretical framework for the regulation of Shh morphogen-controlled gene expression. Development. 2014;141(20):3868–78.
9. Cox J, Mann M. Quantitative, high-resolution proteomics for data-driven systems biology. Annu Rev Biochem. 2011;80:273–99.
10. Csillery K, et al. Approximate Bayesian Computation (ABC) in practice. Trends Ecol Evol. 2010;25(7):410–8.
11. Del Moral P, Doucet A, Jasra A. An adaptive sequential Monte Carlo method for approximate Bayesian computation. Stat Comput. 2012;22(5):1009–20.
12. Deng ZM, Tian TH. A continuous optimization approach for inferring parameters in mathematical models of regulatory networks. BMC Bioinf. 2014;15:256.
13. Drovandi CC, Pettitt AN. Estimation of parameters for macroparasite population evolution using approximate Bayesian computation. Biometrics. 2011;67(1):225–33.
14. Duffy DJ. Problems, challenges and promises: perspectives on precision medicine. Brief Bioinform. 2016;17(3):494–504.
15. Gardner TS, et al. Inferring genetic networks and identifying compound mode of action via expression profiling. Science. 2003;301(5629):102–5.
16. Goel G, Chou IC, Voit EO. System estimation from metabolic time-series data. Bioinformatics. 2008;24(21):2505–11.
17. Golightly A, Wilkinson DJ. Bayesian parameter inference for stochastic biochemical network models using particle Markov chain Monte Carlo. Interface Focus. 2011;1(6):807–20.
18. Goodacre R, et al. Metabolomics by numbers: acquiring and understanding global metabolite data. Trends Biotechnol. 2004;22(5):245–52.
19. Green PJ, et al. Bayesian computation: a summary of the current state, and samples backwards and forwards. Stat Comput. 2015;25(4):835–62.
20. Gutenkunst RN, et al. Universally sloppy parameter sensitivities in systems biology models. PLoS Comput Biol. 2007;3(10):e189.
21. Hurn AS, Jeisman JI, Lindsay KA. Seeing the wood for the trees: a critical evaluation of methods to estimate the parameters of stochastic differential equations. J Financ Econ. 2007;5(3):390–455.
22. Johnson R, Kirk P, Stumpf MPH. SYSBIONS: nested sampling for systems biology. Bioinformatics. 2015;31(4):604–5.
23. Kitano H. Towards a theory of biological robustness. Mol Syst Biol. 2007;3:137.
24. Kitano H. A robustness-based approach to systems-oriented drug design. Nat Rev Drug Discov. 2007;6(3):202–10.
25. Lenormand M, Jabot F, Deffuant G. Adaptive approximate Bayesian computation for complex models. Comput Stat. 2013;28(6):2777–96.
26. Liepe J, et al. Maximizing the information content of experiments in systems biology. PLoS Comput Biol. 2013;9(1):e1002888.
27. Liepe J, et al. A framework for parameter estimation and model selection from experimental data in systems biology using approximate Bayesian computation. Nat Protoc. 2014;9(2):439–56.
28. Lillacci G, Khammash M. Parameter estimation and model selection in computational biology. PLoS Comput Biol. 2010;6(3):e1000696.
29. Link H, et al. Real-time metabolome profiling of the metabolic switch between starvation and growth. Nat Methods. 2015;12(11):1091–7.
30. Maetschke SR, et al. Supervised, semi-supervised and unsupervised inference of gene regulatory networks. Brief Bioinform. 2014;15(2):195–211.

31. Marjoram P, et al. Markov chain Monte Carlo without likelihoods. Proc Natl Acad Sci U S A. 2003;100(26):15324–8.
32. Masel J, Siegal ML. Robustness: mechanisms and consequences. Trends Genet. 2009;25 (9):395–403.
33. Moles CG, Mendes P, Banga JR. Parameter estimation in biochemical pathways: a comparison of global optimization methods. Genome Res. 2003;13(11):2467–74.
34. Murakami Y, Takada S. Bayesian parameter inference by Markov chain Monte Carlo with hybrid fitness measures: theory and test in apoptosis signal transduction network. PLoS One. 2013;8(9):e74178.
35. Nunes MA, Balding DJ. On optimal selection of summary statistics for approximate Bayesian computation. Stat Appl Genet Mol Biol. 2010;9(1).
36. Nunes MA, Prangle D. abctools: an R package for tuning approximate Bayesian computation analyses. R J. 2015;7(2):189–205.
37. Ocone A, Millar AJ, Sanguinetti G. Hybrid regulatory models: a statistically tractable approach to model regulatory network dynamics. Bioinformatics. 2013;29(7):910–6.
38. Picchini U. Inference for SDE models via approximate Bayesian computation. J Comput Graph Stat. 2014;23(4):1080–100.
39. Quach M, Brunel N, d'Alche-Buc F. Estimating parameters and hidden variables in non-linear state-space models based on ODEs for biological networks inference. Bioinformatics. 2007;23 (23):3209–16.
40. Rizk A, et al. A general computational method for robustness analysis with applications to synthetic gene networks. Bioinformatics. 2009;25(12):i169–78.
41. Rung J, Brazma A. Reuse of public genome-wide gene expression data. Nat Rev Genet. 2013;14(2):89–99.
42. Sboner A, Elemento O. A primer on precision medicine informatics. Brief Bioinform. 2016;17 (1):145–53.
43. Secrier M, Toni T, Stumpf MP. The ABC of reverse engineering biological signalling systems. Mol BioSyst. 2009;5(12):1925–35.
44. Simon R. Microarray-based expression profiling and informatics. Curr Opin Biotechnol. 2008;19(1):26–9.
45. Sisson SA, Fan Y, Tanaka MM. Sequential Monte Carlo without likelihoods. Proc Natl Acad Sci U S A. 2007;104(6):1760–5.
46. Stumpf M, Balding DJ, Girolami M. Handbook of statistical systems biology. Chichester/ Hoboken: Wiley; 2011.
47. Sunnaker M, et al. Approximate Bayesian computation. PLoS Comput Biol. 2013;9(1): e1002803.
48. Thorn GJ, King JR. The metabolic network of Clostridium acetobutylicum: comparison of the approximate Bayesian computation via sequential Monte Carlo (ABC-SMC) and profile likelihood estimation (PLE) methods for determinability analysis. Math Biosci. 2016;271:62–79.
49. Tian T, Smith-Miles K. Mathematical modeling of GATA-switching for regulating the differentiation of hematopoietic stem cell. BMC Syst Biol. 2014;8 Suppl 1:S8.
50. Tian TH, Song JN. Mathematical modelling of the MAP kinase pathway using proteomic datasets. PLoS One. 2012;7(8):e42230.
51. Tian TH, et al. Simulated maximum likelihood method for estimating kinetic rates in gene expression. Bioinformatics. 2007;23(1):84–91.
52. Toni T, et al. Approximate Bayesian computation scheme for parameter inference and model selection in dynamical systems. J R Soc Interface. 2009;6(31):187–202.
53. Turner BM, Van Zandt T. A tutorial on approximate Bayesian computation. J Math Psychol. 2012;56(2):69–85.
54. Vyshemirsky V, Girolami M. BioBayes: a software package for Bayesian inference in systems biology. Bioinformatics. 2008;24(17):1933–4.
55. Wang J. Computational biology of genome expression and regulation – a review of microarray bioinformatics. J Environ Pathol Toxicol Oncol. 2008;27(3):157–79.

56. Wilkinson DJ. Bayesian methods in bioinformatics and computational systems biology. Brief Bioinform. 2007;8(2):109–16.
57. Wilkinson DJ. Stochastic modelling for quantitative description of heterogeneous biological systems. Nat Rev Genet. 2009;10(2):122–33.
58. Wolkenhauer O, et al. Enabling multiscale modeling in systems medicine. Genome Med. 2014;6(3):21.
59. Wu QQ, Smith-Miles K, Tian TH. Approximate Bayesian computation for estimating rate constants in biochemical reaction systems. In: IEEE international conference on bioinformatics and biomedicine (Bibm). 2013.
60. Wu QQ, Smith-Miles K, Tian TH. Approximate Bayesian computation schemes for parameter inference of discrete stochastic models using simulated likelihood density. BMC Bioinf. 2014;15:S3.

Chapter 13
Network-Based Biomedical Data Analysis

Yuxin Lin, Xuye Yuan, and Bairong Shen

Abstract Complex diseases are caused by disorders of both internal and external factors, and they account for a large proportion of human diseases. They are multigenetic and rarely a consequence of the dysfunction of single molecules. Systems biology views the living organism as an organic network. Compared with reductionism-based viewpoints, systems biology pays more attention to the interactions among molecules located at different omics levels. Based on this theory, the concepts of network biomarkers and network medicine have been proposed sequentially, which integrate clinical data with knowledge of network sciences, thereby promoting the investigation of disease pathogenesis in the era of biomedical informatics. The former aims to identify precise signals for disease diagnosis and prognosis, whereas the latter focuses on developing effective therapeutic strategies for specific patient cohorts. In this chapter, the basic concepts of systems biology and network theory are presented, and clinical applications of biomolecular networks, network biomarkers, and network medicine are then discussed.

Keywords Centrality • Cross-scale analysis • Network biomarker • Network medicine • Sequential network

13.1 Introduction

Complex diseases comprise a large class of common diseases, which originate from interactions among multiple factors such as gene mutations, environmental effects, and personal lifestyle choices [1]. The morbidity, mortality, and recurrence rates of these diseases are growing rapidly throughout the world at present. Due to the initiation and development of P4 (predictive, preventive, personalized, and participatory) medicine and precision medicine, medical paradigms are constantly shifting. Many traditional methods that focus only on single genes or proteins and

---

Y. Lin • X. Yuan • B. Shen (✉)
Center for Systems Biology, Soochow University, No. 1 Shizi Street, 206, 215006 Suzhou, Jiangsu, China
e-mail: bairong.shen@suda.edu.cn

B. Shen et al. (eds.), *Translational Biomedical Informatics*, Advances in Experimental Medicine and Biology 939, DOI 10.1007/978-981-10-1503-8_13

that view a living organism as simple systems cannot provide a clear understanding of the essential mechanisms of complex diseases such as cancer, diabetes, and cardiovascular and neuronal diseases [2]. Therefore, systematic theories and approaches need to be provided and translated into clinical practice.

Systems biology is one of the most effective and powerful weapons for fighting complex diseases [3], where it emphasizes the dynamic interactions among biological molecules at different omics levels as well as elucidating their in-depth behaviors or mechanisms from a systematic perspective. These interactions connect biological components to generate complex interacting modules or networks, which have great significance for case studies and clinical applications [4]. More importantly, most of these biological networks tend to have meaningful structural characteristics, which are of great value for discovering potential rules or patterns of occurrence and progression for complex diseases [5, 6].

Among the principles of systems biology, network analysis is now becoming the main approach for investigating biological processes and functions in the field of biomedical informatics [7]. Various holistic concepts such as network biomarkers [8] and network medicine [9], which break the shackles of reductionist viewpoints, offer new methods for exploring the complexities of human diseases as well as helping to address biomedical problems at the systems level. Due to the popularization of next-generation sequencing (NGS), increasing numbers of studies are combining static networks with large-scale dynamic expression data [10], thereby elucidating the changes in diseases at different time points. All of these innovations facilitate the diagnosis and treatment of complex diseases, as well as building a strong bridge between fundamental research and clinical sciences.

13.2 Networks and Graphs

A network is a description and abstraction of real things and their relationships, which can be represented as a graph model with two essential components: a set of vertexes (or nodes), $V = \{v_1, v_2, \ldots, v_N\}$, and a set of edges (or lines), $E = \{e_1, e_2, \ldots, e_M\}$, between pairs of vertexes. There are many instances of networks such as social networks, traffic networks, and financial networks, but we focus on biological networks.

13.2.1 Classification of Networks

13.2.1.1 Directed and Undirected Networks

Networks can be divided into two types according to the directivity that they indicate: directed networks and undirected networks. In a directed network, an edge (i, j) indicates that a relationship exists from vertex i to j, but not vice versa.

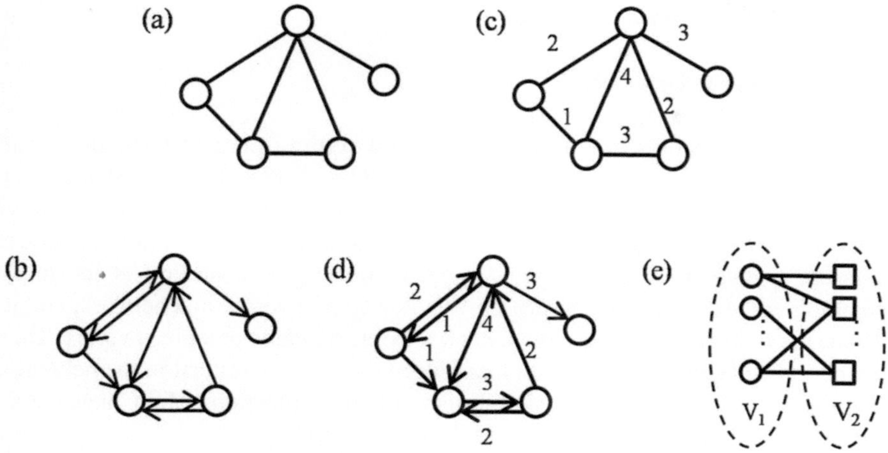

Fig. 13.1 Fundamental types of networks. (**a**) Unweighted undirected network. (**b**) Unweighted directed network. (**c**) Weighted undirected network. (**d**) Weighted directed network. (**e**) Bipartite network

Thus, *i* and *j* are known as the starting points and ending points, respectively. In an undirected network, two vertexes are bidirectionally reachable when their edges are linked. Among the different types of biological networks, a protein-protein interaction (PPI) network is typically an undirected network, whereas microRNA-mRNA regulatory networks are recognized more commonly as directed networks.

13.2.1.2 Weighted or Valued Networks

Network data often contain extra information regarding the extent or strength of each relationship. For example, in gene co-expression networks, correlation coefficients are usually calculated in order to quantify the extent of the interactions among genes [11]. This extra information is referred to as a weight or value in network science. Thus, the two basic network types mentioned above can be extended to four, as shown in Fig. 13.1a–d.

13.2.1.3 Bipartite Networks

Given a network $N = \{V, E\}$, if the vertex set V can be divided into two independent subsets V_1 and V_2 ($V_1 \cup V_2 = V$, $V_1 \cap V_2 = \Phi$), and all edges are between paired vertexes belonging to different subsets, then the network can be represented as a bipartite network (or bipartite graph; see Fig. 13.1e). Bipartite networks are used widely in biological research. For instance, a human gene-disease network is bipartite, where one set of vertexes are diseases and the other set are genes that are closely related to linked diseases.

13.2.2 *Centrality of Vertexes*

13.2.2.1 Degree Centrality

In undirected networks, the degree of a certain vertex i (termed $k(i)$) equals the number of edges that are incident on it. In directed networks, the vertex degree $k(i)$ is partitioned into two parts, the in-degree $k_I(i)$ and out-degree $k_O(i)$ (mathematically, $k(i) = k_I(i) + k_O(i)$), which are equivalent to the number of vertexes that are adjacent to and from the vertex i, respectively. Degree centrality is the most common property used to measure the importance of a vertex in a network, and it equals the ratio of the actual to theoretical maximum degree of a given vertex. This metric indicates that vertexes with larger degrees are more critical in the network. For example, old genes with significant biological functions often have large degrees, and they lie at the heart of a PPI network [12].

13.2.2.2 Closeness Centrality

This metric indicates how close the given vertex is to all of the other vertexes in the network. In general, the vertex with the highest closeness centrality is located at the optimum position for viewing the information flow. The vertex closeness centrality can be calculated for both nondirectional and directional relations. If we consider an undirected network as an example, the closeness centrality of vertex i ($CC(i)$) in an undirected network with N vertexes is

$$CC(i) = \frac{N}{\sum_{j=1}^{N} d(i,j)} \tag{13.1}$$

where $d(i, j)$ represents the distance from vertex i to j.

13.2.2.3 Betweenness Centrality

Interactions between two nonadjacent vertexes in a network can be affected by the actions of other vertexes, especially by those that lie between them. Some vertexes are important because all the shortest paths along which information flows from any vertex at one side to the other must pass through them, and thus the betweenness centrality is the metric used to describe the importance of a given vertex based on the number of shortest paths that it penetrates. Vertexes with higher betweenness centrality hint may have a greater capacity to control the flow of information. In an undirected network, the betweenness centrality of the given vertex i ($BC(i)$) is

$$BC(i) = \sum_{a \neq i \neq b} \frac{m^i_{ab}}{n_{ab}} \tag{13.2}$$

where n_{ab} is the number of shortest paths linking vertex a and b and m^i_{ab} is the number of shortest paths linking vertex a and b that contain vertex i.

13.2.3 *Topological Properties of Networks*

13.2.3.1 Degree Distribution

The degree distribution *P(k)* of a network is equivalent to the fraction of vertexes in the network with degree *k*. If the network is directional, the distribution should be refined as an in-degree distribution or out-degree distribution. Vertexes in different networks tend to follow different degree distributions, such as the normal distribution (or Gaussian distribution), binomial distribution, and long tail distribution (or scale-free distribution) (see Fig. 13.2a–c).

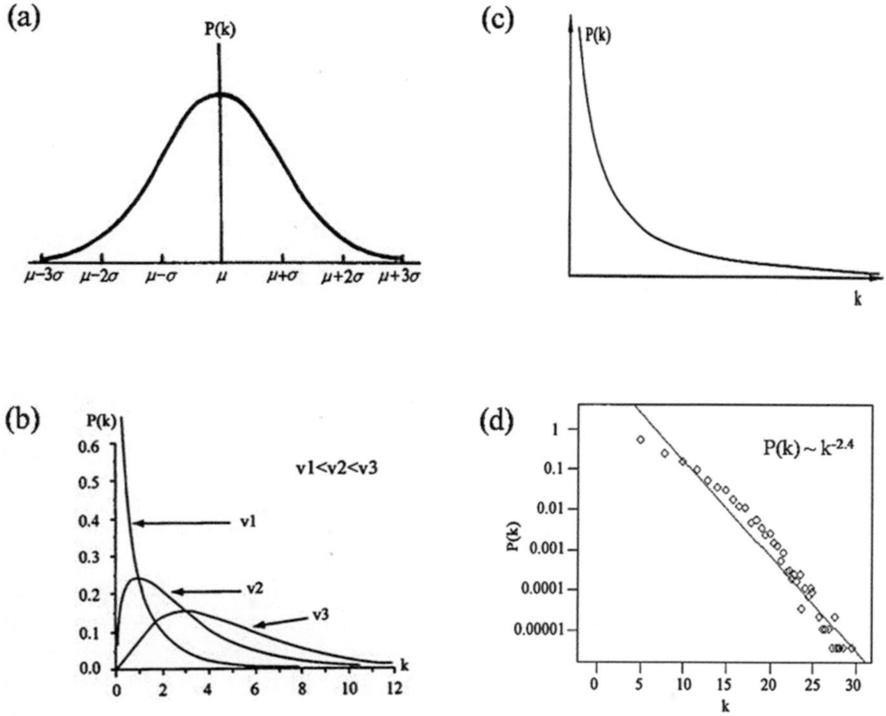

Fig. 13.2 Schematic diagrams of four common degree distributions. (**a**) Normal distribution. (**b**) Binomial distribution. (**c**) Long tail distribution. (**d**) Power-law distribution

13.2.3.2 Power-Law Distribution

In 1999, Barabasi and his colleagues showed that the in-degree and out-degree distribution of the World Wide Web has a power-law tendency [13]. In network science, networks with this property are often known as scale-free networks. Mathematically, $P(k) \sim k^{-\gamma}$, where γ is a parameter with a value that usually ranges among $2 < \gamma < 3$ (see Fig. 13.2d). In fact, many important biological networks are also scale-free. For example, the human PPI network has an approximately scale-free characteristic [14] with a degree exponent of 1.49 [12], which indicates that proteins (or genes) with large degrees (i.e., hubs) are few in number and they may affect the whole network greatly.

13.3 Biomolecular Networks and Their Clinical Applications

Biological molecules interact to promote the activity and evolution of living organisms. These interactions contribute to various types of biological networks, where they may influence the significance and complexity of biological processes in many ways.

13.3.1 Protein-Protein Interaction (PPI) Networks

Proteins are the direct products of functional genes, and they are large biological molecules that mediate the functions of living organisms. Accumulating evidence indicates that PPIs are closely associated with biological processes [15, 16], where they play pivotal roles in a large number of cellular behaviors and their abnormal activities may lead to the development of numerous diseases [17].

Protein interactions ("interactome") have generally been identified based on multiple biological experiments or computational approaches. However, due to the development of biological and computational techniques, the volume of PPI data increases year, and many publicly available databases have been created to store these data, thereby providing valuable information for interactome research.

Table 13.1 lists six manually curated PPI databases. The data in these databases have been verified by experiments or published studies. To fully exploit these data and analyze them at a higher level, Wu et al. integrated their interactions and constructed a PPI network analysis platform (called PINA) for investigating the underlying latent information [18]. The platform was then enhanced in version 2.0 by including interactome modules identified by the global PPI network for six model organisms [19].

Table 13.1 Protein-protein interaction databases

Name	Version	Link	Citation
BioGRID	3.4.132	http://www.thebiogrid.org/	Stark et al. [20]
DIP	2004 update	http://dip.doe-mbi.ucla.edu/	Salwinski et al. [21]
HPRD	Release 9	http://www.hprd.org/	Keshava Prasad et al. [22]
IntAct	4.2.3.1	http://www.ebi.ac.uk/intact/	Aranda et al. [23]
MINT	2012 update	http://mint.bio.uniroma2.it/mint/	Ceol et al. [24]
MIPS MPact	Not available	http://mips.gsf.de/genre/proj/mpact/	Guldener et al. [25]

It is widely acknowledged that the functions of biomolecules are affected by their structures [26, 27] and network structures with special characteristics can also indicate the possible mechanisms of interactions among given biomolecules [28]. Many studies have shown that PPI networks tend to be scale-free [29] and proteins/genes with large degrees or centralities (or hubs) may play important roles in relevant biological processes. In addition, PPI networks have been reported as modular [30], so some studies have addressed the substructural analysis of PPI networks. For example, Luo et al. [31] separated five modules related to the initiation of early-onset colorectal cancer using a PPI network based on gene expression data and cluster analysis and then screened five hub genes as key indicators or candidate therapeutic targets for this disease. Gene ontology and pathway enrichment analyses demonstrated the validity of their results. Zanzoni and Brun [32] designed a computational approach that considers both PPI network and stage-based proteomics profiles to identify dysregulated cellular functions during the progression of different cancers. They extracted several functional modules using the OCG algorithm [33] and annotated them based on gene ontology terms and pathway signals. Combined with actual proteomics datasets obtained at different stages of cancers, they selected modules with increasing, decreasing, or stage-specific importance during cancer progression. This study showed that protein modules are functional in different biological processes and that the interactions among them are usually as important as the proteins themselves. To some extent, PPI networks can provide a comprehensive understanding of molecular interactivity rather than single proteins, thereby presenting more opportunities for elucidating the potential mechanisms under different conditions. This could allow great breakthroughs in the diagnosis and treatment of complex diseases.

13.3.2 Gene Co-expression Networks

PPI networks represent the interactions among proteins/genes from a static perspective. However, these interactions might not be exactly the same in different conditions due to the specificity of samples or groups with different backgrounds. In recent decades, due to the rapid development of experimental technologies, the number of expression profile data identified by high-throughput screening has increased greatly, thereby providing unprecedented opportunities for integrative analyses of clinical diseases based on static networks and dynamic expression information.

Gene co-expression networks combine the similarity of expression among coordinated genes with the topological properties of networks, which can provide a systematic view of the dynamic changes of molecular activities and cellular functions during the evolution of biological processes. Using gene co-expression networks to analyze complex biological phenomena is simple and efficient [34]. Importantly, they are beneficial for building condition- or disease-specific networks, which are useful for elucidating the underlying mechanisms related to the progression of specific diseases [35].

Rotival and Petretto [36] reviewed some well-known computational methods for co-expression network analysis, which can be divided into two categories according to specific guiding principles. The first category comprises potential foundational factors, the influences of which may lead to changes in gene expression. These methods first select the principal factors and their induced genes from a pool of candidate factors based on principal components analysis or nonnegative matrix factorization algorithms [37], before extracting functional modules based on the factor-gene pairs. The other methods for co-expression network analysis are largely dependent on graph-based modeling, where vertexes or edges with similar features are clustered into the same modules. As shown in Fig. 13.3, co-expressed genes are usually measured by correlation analysis, such as Pearson's correlation coefficient (PCC), Spearman's rank correlation, or Kendall correlation tests, where the functional modules are finally inferred for further research.

In the era of biomedical informatics, co-expression network analysis greatly improves the speed and accuracy of disease-associated gene discovery. Zhang

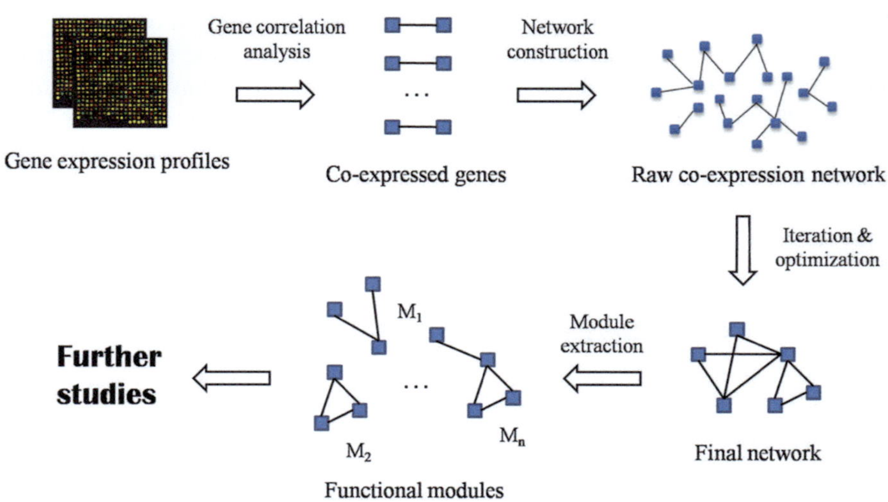

Fig. 13.3 Pipeline for graph-based gene co-expression network analysis

et al. [38] confirmed that five crucial genes can be used as prognostic markers for chronic lymphocytic leukemia, where they constructed a co-expression network using the CODENSE algorithm [39] and they focused mainly on modules containing the key gene *ZAP70*. Yang et al. [40] built gene co-expression networks for four different types of cancers and found that the features of prognostic genes did not lie at hub positions in cancer-specific co-expression networks, but instead they were often enriched in modules conserved among different cancer networks. This may be an important insight that could facilitate the identification of cancer prognostic genes in clinics.

13.3.3 MicroRNA-mRNA Regulatory Networks

MicroRNAs (miRNAs) are small noncoding RNAs that comprise approximately 22–24 nucleotides. miRNAs silence gene expression at the posttranscriptional level by base-paring with their target mRNAs [41]. According to previous studies, miRNAs are involved in a variety of important biological processes, such as cell proliferation, development, apoptosis, and immune responses [42, 43, 44]. In addition, the aberrant expressions of miRNAs may cause many serious diseases [45, 46, 47].

The relationships between miRNAs and their targets can be abstracted as a bipartite network (or bipartite graph), which is called a miRNA-mRNA regulatory network. The pairs in the network comprise miRNA-mRNA regulations, which can be determined using experimental and computational methods. Table 13.2 lists several useful databases that store miRNA-mRNA pairs.

Some well-known tools are also available for miRNA-target prediction. For example, TargetScan [55] infers miRNA targets by matching the seed region of

Table 13.2 miRNA-mRNA regulatory pair databases

Type	Name	Version	Link	Citation
Experimentally validated	miRTarBase	6.0	http://mirtarbase.mbc.nctu.edu.tw/	Chou et al. [48]
	TarBase	7.0	http://www.microrna.gr/tarbase/	Vlachos et al. [49]
	miRecords	4.0	http://miRecords.umn.edu/miRecords/	Xiao et al. [50]
	miR2Disease	Not available	http://www.miR2Disease.org/	Jiang et al. [51]
Computationally predicted	HOCTAR	2.0	http://hoctar.tigem.it/	Gennarino et al. [52]
	ExprTargetDB	Not available	http://www.scandb.org/apps/microrna/	Gamazon et al. [53]
	starBase	2.0	http://starbase.sysu.edu.cn/	Li et al. [54]

each input miRNA. miRanda [56] is an optimized method that relies only on sequence complementarity and user-specified rules to enhance the accuracy of predicted results. In general, the miRNA-mRNA pairs identified by low-throughput experiments such as real-time PCR are more convincing than those determined using high-throughput techniques such as microarrays or NGS, while the pairs predicted by computational algorithms often have a high false-positive rate. Thus, it is necessary to clean the data before constructing the final network.

miRNAs function in the development of many diseases, and many studies have attempted to discover disease-associated miRNAs based on miRNA-mRNA regulatory networks. One of the most popular approaches is based on the theory that miRNAs may be functionally synergistic so they can co-regulate the expression of their target genes. Bandyopadhyay et al. [57] found that the miRNAs included in a module may have a combinatorial effect on their targets, where those located next to the module appeared to have similar dysregulatory patterns. Based on this observation, several computational frameworks or programs have been developed to identify abnormal miRNAs or miRNA regulatory modules in human diseases [58, 59].

Instead of the synergistic functions of miRNAs, Zhang et al. [5] focused on the substructures of miRNA-mRNA regulatory networks and found evidence that miRNAs can regulate genes independently. They defined a novel bioinformatics model using the NOD (novel out-degree) parameter to quantify the independent regulatory power and employed it to detect key miRNAs in prostate cancer [5, 60], gastric cancer [61], and sepsis [62]. The model was expanded later by considering the biological functions of miRNA targets [6]. Unlike some machine learning-based methods that are highly reliant on the training data, the improved model identified crucial miRNAs without any prior knowledge, and its application to biomarker discovery for pediatric acute myeloid leukemia demonstrated its great predictive power.

Another typical application of miRNA-mRNA networks in clinical research is the approach proposed by Zhao et al. [63], who utilized a network as a bridge to infer cancer-related miRNAs from dysfunctional genes and their enriched pathways. The method is flexible because it can identify cancer-related miRNAs without requiring miRNA expression profiles. All of the studies mentioned above demonstrate the importance of miRNA-mRNA regulatory networks, especially in the field of disease-associated miRNA discovery.

13.3.4 Competing Endogenous RNA (CeRNA) Regulatory Networks

It has been widely reported that miRNAs may repress a large proportion of transcripts and they can act as oncogenes [64] or tumor suppressor genes [65] in

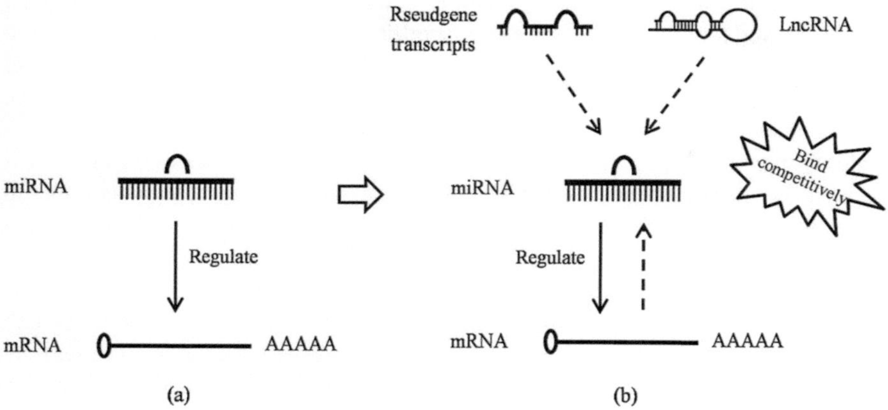

Fig. 13.4 Schematic diagram of two regulatory paradigms. (**a**) "miRNA→RNA" paradigm and miRNA regulatory network. (**b**) "RNA→miRNA→RNA" paradigm and ceRNA regulatory network

many diseases such as cancers. Recent studies have demonstrated that the transcriptome has a large number of components, including protein-coding RNAs (or mRNAs), pseudogene transcripts, and long noncoding RNAs (lncRNAs), which "talk" with each other using the "letter" miRNA response elements (MREs) by competitively binding the limited sites in common miRNAs to influence the regulatory effects of miRNAs on their targets [66]. Salmena et al. [67] formally proposed the ceRNA concept to represents the group of RNAs with these abilities.

The activities of competing endogenous RNA (ceRNAs) form a large-scale regulatory network at the posttranscriptional level, and thus the traditional paradigm of "miRNA→RNA" has gradually been replaced by "RNA→miRNA→RNA." As shown in Fig. 13.4, in this new model, miRNAs are often recognized as mediators, where different ceRNAs bind them competitively to promote changes in the expression of the target genes (or mRNAs) mediated by miRNAs.

In ceRNA regulatory networks, miRNAs can target a large number of co-expressed transcripts, and the expression of one targeted transcript can be affected by changes in the concentration of other transcripts [68]. Multiple RNA transcripts may share one miRNA via MREs in their 3' untranslated regions. Su et al. [69] found that overexpressed ceRNAs may increase the concentration of specific MREs to change the distribution of miRNAs, thereby leading to increases in the expression levels of their targets.

In recent years, studies have demonstrated that the initiation and progression of cancer are closely related to the dysregulation of ceRNA networks. Thus, Sumazin et al. [66] discovered a miRNA-mediated network with more than 248,000 interactions, and they showed that the network regulated various established genes and oncogenic pathways with close relationships to the initiation and development of glioblastoma. Tay et al. [70] confirmed that the ceRNA regulatory network was functional for protein-coding RNAs and tests based on the tumor suppressor gene *PTEN* showed that the expression patterns of protein-coding RNA transcripts were consistent with *PTEN*. Overall, it was concluded that ceRNAs and their networks may play crucial roles in disease development processes.

Understanding the competition mechanisms of ceRNAs may provide great insights into the pathogenesis of specific diseases. For instance, Zhou et al. [71] constructed a breast cancer-specific ceRNA regulatory network by combining miRNA-mRNA relationships with miRNA and mRNA expression datasets from patients with breast cancer, where they found that the network also tended to follow a power-law. Moreover, functional analysis indicated that the hub genes and dense clusters were strongly linked to cancer hallmarks, which proved valuable for risk assessments in breast cancer. Thus, ceRNA regulatory network-based analyses may inspire new approaches to both fundamental and clinical studies of complex diseases.

13.3.5 Others

Due to the complexity of disease progression, other biological networks such as drug-target interaction networks [72], metabolic networks [73], and epigenetic networks [74] may also have important functions during the occurrence and development of diseases. However, due to space limitations, please refer to the references cited for further details.

13.4 Network Biomarkers in Complex Diseases

Biological markers, also known as biomarkers, are unique molecules that can indicate changes or potential changes in biological conditions from normality to abnormality in living organism [75]. Clinically, biomarkers with high sensitivity and specificity could serve as powerful indicators for disease diagnosis and prognosis. Instead of using single biomarkers, network-based biomarkers are now becoming more popular because they can help to investigate the overall behaviors of biological molecules and they may reflect the system-level states of diseases.

13.4.1 Single Molecular Biomarkers and Network Biomarkers

Many studies have shown that single biological molecules can be effective biomarkers for both the diagnosis and prognosis of human diseases. For instance, the

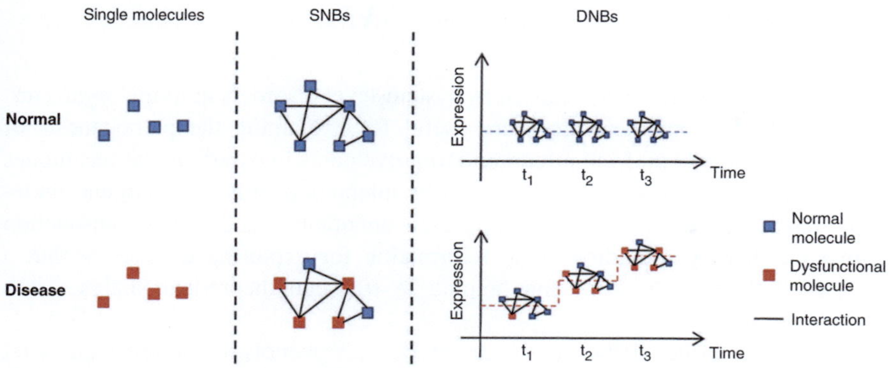

Fig. 13.5 Three different types of biomarkers: single molecular biomarkers, static network biomarkers (*SNBs*), and dynamic network biomarkers (*DNBs*)

protein prostate-specific antigen is widely used for the early detection of prostate cancer [76]. The *BRCA1* and *BRCA2* genes can also be useful markers for breast cancer [77]. In addition, some noncoding RNAs such as miRNAs may also have diagnostic or prognostic roles in many complex diseases [78, 79].

The traditional methods used to detect candidate biomarkers rely mainly on biological experiments. Most begin by identifying differentially expressed or deregulated molecules based on large-scale expression profiling data, before validating the selected candidates in low-throughput experiments [80]. Considering the limited availability of samples and time-consuming pipelines, several computational approaches have been developed to improve the efficiency of biomarker signature discovery [81].

Single molecules may be dysfunctional in many cellular processes, but they are still not sufficiently powerful to explore the underlying mechanisms of certain diseases due to the diversity and complexity of disease development. In fact, complex diseases are usually due to interactions among multiple factors rather than the breakdown of single molecules. Moreover, single biomarkers identified in samples from patients with similar diseases by different methods tend to exhibit high heterogeneity [82]. Complex diseases should be considered more as disorders in a system; therefore, the concept of network biomarkers has been proposed, and novel strategies have been developed for explaining genetic or epigenetic changes across diseases.

There are two main types of network biomarkers: static network biomarkers (SNBs) and dynamic network biomarkers (DNBs). As shown in Fig. 13.5, the former integrates the interactions, annotations, and pathway signals of molecules by focusing only on the static nature of networks, whereas the latter pays considers the states of a disease at different time points, which is useful for monitoring the progression of diseases [83].

13.4.2 Static Network Biomarkers (SNBs)

Complex diseases are always caused by system-level disorders in living organisms. Thus, network biomarkers are more useful for explaining the pathogenesis of diseases than single molecular markers. Improvements in experimental techniques and theories of informatics mean that more interactions among biological molecules have been elucidated as well as their annotations and signal transduction pathways, thereby providing static information for exploring diseases within a systems biology framework and helping to translate theoretical analyses into clinical research.

As a solid bridge between the genotype and phenotype, proteins are vital biological molecules with significant roles in the occurrence and evolution of diseases. Thus, many studies have focused on protein-based network biomarkers, and they are valuable for validating mechanistic hypotheses related to the progression of diseases. The main pipeline is shown in Fig. 13.6. First, disease-associated proteins/genes are selected by analyzing experimental data or other publications, which are then mapped onto the reference PPI network where the knowledge-based PPIs are integrated. Thus, a disease-specific PPI network is constructed. Second, subnetworks of candidate biomarkers are scored and identified from the disease-

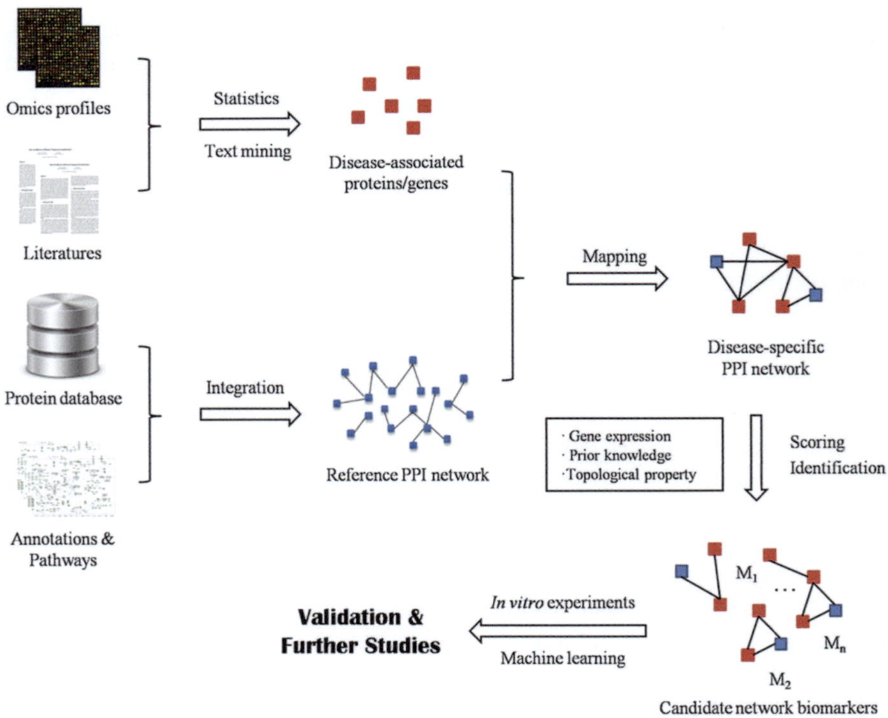

Fig. 13.6 Pipeline for protein-based network biomarker discovery

specific network according to their actual expression levels, existing knowledge, or the topological properties of the network. Finally, in vitro experiments or machine learning methods such as support vector machines (SVMs) [84] or artificial neural networks [85] can be used to validate the results and to perform further research.

To highlight the carcinogenic mechanisms of lung cancer, Wang and Chen [86] constructed a biomarker network based on microarray data and PPIs. They identified 40 proteins that had significant associations with lung carcinogenesis using the network, and they found that three-quarters of the total (30/40) had annotations related to cell growth. The biomarker network had the potential to diagnose smokers with signs of lung cancer, which could be an effective therapeutic target to fight cancer.

In addition to disease diagnosis, biomarker networks are capable of distinguishing metastatic and non-metastatic tumors. Chuang et al. [87] combined breast cancer metastatic and non-metastatic data with a PPI network using the "subnetwork activity matrix" and greedy algorithm to prioritize high-ranked subnetworks as candidate biomarkers. They found that genes in these biomarker networks were enriched for the hallmarks of cancer, and the results of SVM classification showed that these network biomarkers were highly accurate in separating metastatic and non-metastatic breast tumors, which may have significant utility for tumor progression investigations.

In addition to protein-based biomarker networks, noncoding RNAs are essential during the disease development process. It is obvious that interactions among these RNAs and their targets or regulators can form functional or even biomarker networks. Lu et al. [88] built miRNA biomarker networks containing miRNA targets and relevant transcription factors and applied them to the diagnosis of gastric cancer. Cui et al. [89] identified three lncRNA co-expression modules connected with prostate cancer, one of which may be recognized as a module biomarker for prostate cancer diagnosis.

13.4.3 Dynamic Network Biomarkers (DNBs)

Traditional molecular biomarkers and network biomarkers can only distinguish between diseases in two stable states. This static information limits their capacity to detect certain pre-disease states. However, pre-disease states may reflect crucial signs of disease progression, and they could be key indicators for early diagnosis and the prevention of diseases.

The novel DNB concept was proposed to overcome these limitations and to elucidate more dynamic changes in diseases. Based on complex network theory and nonlinear dynamical theory, DNBs can evaluate the stages of diseases at different time points and represent molecules and their relations in a three-dimensional image, as well as facilitating the discovery of stage-specific or personalized biomarkers in the era of biomedical informatics [83].

Chen et al. [90, 91] partitioned the process of disease development into three stages: normal, pre-disease, and disease. The normal stage is stable, and it represents the state of health or early disease. In this stage, changes are usually gradual. The pre-disease stage indicates the state immediately before critical changes have been reached. Molecules in living systems undergo dramatic transitions during this stage until another stable stage (the state of disease or advanced disease) occurs. The pre-disease stage is crucial because it may provide latent signals of disease progression, which could be pivotal markers for the early diagnosis of disease.

To quantify signals and detect DNBs during system-level transitions, a composite index (*CI*) is defined as follows [90]:

$$CI = \frac{\mathrm{SD}_d \times PCC_d}{PCC_o} \tag{13.3}$$

where SD_d is the average standard deviation (SD) of the DNB molecules (molecules in DNB), PCC_d represents the average PCC among DNB molecules as absolute values, and PCC_o represents the average PCC among DNB molecules and other molecules as absolute values. In fact, the DNB comprises a group of molecules in the system, which can provide significant information about the changes at critical points of the pre-disease stage. These molecules are functional compared with other non-DNB molecules in the same system. The expression of these molecules is identified mainly using experimental data, especially those obtained from high-throughput omic experiments.

The theory of DNB has been employed to detect early-warning signs for both type 1 and type 2 diabetes, especially recognizing the key points at which the state reverses. In a study of type 1 diabetes [92], two DNBs were built to predict sudden changes during the progressive disease deterioration. Previous studies and functional analyses demonstrated that these two DNBs are highly relevant to type 1 diabetes and they may be useful for its early diagnosis. Based on this study, tissue-specific DNBs were constructed for type 2 diabetes mellitus, and two significant states were identified that had strong associations with severe inflammation and insulin resistance [90]. The genes in the DNBs were shown to be dysfunctional at the point of disease deterioration according to a cross-tissue analysis. Importantly, they were mostly located upstream of the signaling pathways, and they acted as leaders during the transcriptional processes. These results demonstrate that DNB can be predictors of the occurrence of disease, as well as transducers that may facilitate a better understanding of the molecular mechanisms of disease development.

13.4.4 Evolution of Network Biomarkers During Disease Progression

Network biomarkers are system-level molecular modules that are helpful for investigating the evolutionary mechanism of disease progression. Wong et al. [93] constructed two PPI-based network biomarkers for the early and late stages of bladder cancer. First, they downloaded microarray data for the two stages of bladder cancer and for normal samples from the Gene Expression Omnibus repository, before constructing two different networks for the two stages of bladder cancers using statistical methods. Second, proteins/genes were extracted with significant carcinogenesis relevance values together with their network structures. The activities of these proteins tended to exhibit remarkable changes in normal and disease samples, and these changes may be essential in bladder cancer carcinogenesis. The results obtained by pathway enrichment analysis showed that proteins in the biomarker network for early-stage bladder cancer were significantly more enriched in pathways related to ordinary cancer mechanisms such as the cell cycle, pathways in cancer, and Wnt signaling pathway, and these proteins may also be functional in other cancers such as prostate cancer, chronic myeloid leukemia, and small cell lung cancer. By contrast, the ribosome and spliceosome pathway were the top two pathways targeted by the biomarker network for late-stage bladder cancer. Obviously, during the evolution of bladder cancer, proteins and their interactions change gradually, but ultimately there is a shift in the enriched pathways from universal to specific types.

Meaningful evolutionary patterns were also discovered in a study of hepatocellular carcinoma (HCC) [94], where Wong et al. analyzed the evolution of network biomarkers from the early to late stages of HCC using a framework analogous to that employed for bladder cancer research. However, NGS datasets were used in this study. They found that the common pathways enriched for network biomarkers in both the early and late stages of HCC were associated with the ordinary mechanisms of cancers, where the spliceosome pathway was prominent in the late stages of both hepatocellular and bladder cancer.

Both of these studies provide new insights into disease-targeted therapies at different stages or time points, and they merit further clinical research.

13.5 Network Medicine in Clinical Practice

The Human Genome Project shifted genome-wide studies from isolated genes or proteins to the networks of interconnections among them. The traditional methods for disease diagnosis and drug discovery are symptom-based or molecule-based. However, the occurrence of diseases is rarely a consequence of the disorder of single molecules, and different diseases are likely to share similar symptoms. Thus, the concept of network medicine, which emphasizes treating disease progression at the systems level, may provide new directions for disease analysis and therapeutics.

13.5.1 Paradigm of Network Medicine

The pattern for disease classification and drug discovery has changed greatly due to the continuous deepening of biomedical ideas and techniques. During the early period, diseases were often simply classified based on knowledge of clinical symptoms. However, this method is inaccurate, and it may miss opportunities for disease prevention due to its low sensitivity and specificity. Clearly, symptoms may be totally absent during the early stage of a disease, and most ordinary symptoms are not specific to a certain disease [95].

The emergence and development of genomic research has provided various types of molecular data, which facilitate investigations of the underlying mechanisms of disease progression. Therefore, the disease analysis paradigm has gradually shifted from studies of outward manifestations to internal mechanisms. For example, complex diseases can be caused by multiple changes in biomolecules, such as DNA methylation, single nucleotide polymorphisms, and DNA copy number variations. Similar disease symptoms may be apparent, but the treatments will be quite different according to the differences in pathogenesis. Therefore, molecule-based methods are more beneficial for the personalized and precise treatment of diseases [96].

Recently, many analyses have shown that complex diseases are multigenic, resulting from the synergistic actions of genetic and environmental factors. Simple molecule-based methods focus only on biological molecules that act as key players in the system. However, these single components are not sufficient to create system-level disruption. Instead, network medicine treats disease diagnosis and therapeutics from a global perspective by linking the potential factors that are relevant to disease occurrence and development to form an organic network, thereby identifying reasonable therapeutic strategies at specific time points according to both the static and dynamic properties of the network. The pathogenic behavior of complex interactions among molecules can be uncovered at various omics levels using this systemic approach, and effective drugs may be obtained to reach the goal of precision medicine [97].

13.5.2 Foundations and Resources

Network medicine is based on a series of hypotheses that are widely acknowledged by researchers. However, the theory continues to improve due to the development of systems biology and network science. The main focus is on linking network structures and disease occurrence. Thus, the topological structures of biological networks might potentially reflect the roles of specific molecules during disease initiation and progression. In particular, evidence has shown that essential proteins/genes often lie at the heart of a PPI network, whereas nonessential disease proteins are not found in these central locations. This is quite similar to a social network where important people or leaders are usually hub nodes who can control the information flow. In addition, proteins appear to cooperate with each other, especially those involved in the same diseases. Many studies have shown that proteins often participate in biological processes in the form of modules, which highlights the existence of synergistic mechanisms. Moreover, cells that exist in a microenvironment of diseases with similar phenotypes tend to have common disease-associated components. This may help to explain why comorbidities usually occur. Finally, the causal molecular pathways are parsimonious, and they often form the shortest paths between known components of diseases [95].

The essential resources for network medicine study are suitable data or datasets. It is obvious that sufficient data can drive research to become more precise and specific due to differences between omics levels, disease stages, and even individuals or groups. Chen and Butte [96] summarized eight publicly available data sources for network medicine, which offer great opportunities for disease analysis and drug discovery. Furthermore, databases such as HMDD [98] and DriverDBv2 [99] aim to represent the relationships between biomolecules and diseases, thereby providing great insights into the pathogenic nature of diseases. Details of these databases are listed in Table 13.3.

Bioinformatics approaches perform well at mining functional molecules or molecular modules for disease diagnosis and treatment. The most remarkable

Table 13.3 Publicly available data sources for network medicine study

Name	Description	Link
CCLE	Cancer Cell Line Encyclopedia: genetic and pharmacological characterization of cancer models	http://www.broadinstitute.org/ccle/
CMAP	Connectivity Map: a collection of genome-wide transcriptional expression profiles	http://www.broadinstitute.org/cmap/
ChEMBL	Biological activities for drug-like molecules	https://www.ebi.ac.uk/chembl/
DriverDBv2	Relationships between driver genes/mutations and cancers	http://driverdb.tms.cmu.edu.tw/driverdbv2/
ENCODE	Encyclopedia of DNA Elements: comprehensive database of genome-wide functional elements	http://genome.ucsc.edu/ENCODE/
GEO	Gene Expression Omnibus: a functional genomics data repository	http://www.ncbi.nlm.nih.gov/geo/
HMDD	Human microRNA Disease Database: a collection of human microRNAs and their related diseases	http://cmbi.bjmu.edu.cn/hmdd/
ICGC	International Cancer Genome Consortium: a comprehensive description of changes at different omics levels in different cancers	https://icgc.org/
ImmPort	Immunology Database and Analysis Portal: data and advanced techniques in immunology	https://immport.niaid.nih.gov/
LINCS	Library of Integrated Cellular Signatures: signatures of different cellular states and development tools for data analysis	http://www.lincscloud.org/
PubChem	Connects PubChem substance, compound, and bioassay data	http://pubchem.ncbi.nlm.nih.gov/
TCGA	The Cancer Genome Atlas: a platform for searching, downloading, and analyzing cancer-related data	http://cancergenome.nih.gov/

achievement is the discovery and application of network biomarkers. As described in Sect. 11.4, network biomarkers indicate dysfunctional modules during disease progression, and they facilitate the development of macroscopic explanations of disease initiation. It is clear that they are indispensable components of network medicine because they can provide important signals, which are sensitive and specific for both disease research and drug design.

13.5.3 Research Significance and Practical Challenges

Network medicine combines systematic thinking with clinical sciences, and by utilizing network theory as a mediator, it is poised to promote the understanding of disease pathogenesis and to forecast disease development trajectories or tendencies. It focuses on predicting the key players in disease progression, with the aim of providing better therapeutic strategies for patients [100].

The development of ideas is always accompanied by opportunities and challenges, and network medicine is not an exception. The volume of data available for network medicine study is huge, but that with practical value may be limited. Furthermore, the structure of the data is inconsistent, especially clinical data, which is rooted in different schemas and ontologies [96]. Thus, necessary criteria should be established for data representation, or the process of data integration and further analysis may be hindered. Due to the complexity of biological mechanisms, networks should be more specific. It has been reported that response networks for the same drug tend to exhibit distinct heterogeneity in different cell lines. The components and activities of real living organisms are more complex than those in computational models because networks or functions are generally not condition-specific and they fail to consider the effects of the external environment. Thus, effective methods and tools should be developed for constructing models across different omics levels in the era of big data, as well as to aid discovery in personalized therapeutics for different populations with different diseases under the guidance of precision medicine.

13.6 Conclusions

Analyzing complex biological problems within a network framework facilitates deeper investigations of the behaviors of biomolecules and their interconnections. The application of network biomarkers and network medicine may accelerate the understanding of disease pathogenesis, as well as promoting the transformation from fundamental research to clinical practice in the era of biomedical informatics. In particular, the human system is far more complex than simply emulating networks, where even the size or shape of cells may affect their biological functions. Thus, cross-scale analyses and dynamic simulations are urgently needed in the future.

References

1. Sudlow C, et al. UK biobank: an open access resource for identifying the causes of a wide range of complex diseases of middle and old age. PLoS Med. 2015;12:e1001779.
2. Chen L, Wu J. Systems biology for complex diseases. J Mol Cell Biol. 2012;4:125–6.
3. Bjorkegren J, Tegner J. Systems biology makes detailed understanding of complex diseases possible. Arteriosclerosis is an example. Lakartidningen. 2007;104:3042–5.
4. Liu R, et al. Identifying critical transitions and their leading biomolecular networks in complex diseases. Sci Rep. 2012;2:813.
5. Zhang W, et al. Identification of candidate miRNA biomarkers from miRNA regulatory network with application to prostate cancer. J Transl Med. 2014;12:66.
6. Yan W, et al. MicroRNA biomarker identification for pediatric acute myeloid leukemia based on a novel bioinformatics model. Oncotarget. 2015;6:26424–36.
7. Cho DY, Kim YA, Przytycka TM. Chapter 5: network biology approach to complex diseases. PLoS Comput Biol. 2012;8:e1002820
8. Zhao XM, Chen L. Network-based biomarkers for complex diseases. J Theor Biol. 2014;362:1–2.
9. Silverman EK, Loscalzo J. Network medicine approaches to the genetics of complex diseases. Discov Med. 2012;14:143–52.
10. Xin J, et al. Identifying network biomarkers based on protein-protein interactions and expression data. BMC Med Genet. 2015;8 Suppl 2:S11.
11. Lu YY, et al. Transcriptional profiling and co-expression network analysis identifies potential biomarkers to differentiate chronic hepatitis B and the caused cirrhosis. Mol BioSyst. 2014;10:1117–25.
12. Zhang W, et al. New genes drive the evolution of gene interaction networks in the human and mouse genomes. Genome Biol. 2015;16:202.
13. Albert R, Jeong H, Barabasi AL. Internet – diameter of the world-wide web. Nature. 1999;401:130–1.
14. Barabasi AL. Scale-free networks: a decade and beyond. Science. 2009;325:412–3.
15. Cho S, et al. Protein-protein interaction networks: from interactions to networks. J Biochem Mol Biol. 2004;37:45–52.
16. Thakur S, et al. A review on protein-protein interaction network of APE1/Ref-1 and its associated biological functions. Cell Biochem Funct. 2015;33:101–12.
17. Hu Z. Analysis strategy of protein-protein interaction networks. Methods Mol Biol. 2013;939:141–81.
18. Wu J, et al. Integrated network analysis platform for protein-protein interactions. Nat Methods. 2009;6:75–7.
19. Cowley MJ, et al. PINA v2.0: mining interactome modules. Nucleic Acids Res. 2012;40: D862–5.
20. Stark C, et al. The BioGRID interaction database: 2011 update. Nucleic Acids Res. 2011;39: D698–704.
21. Salwinski L, et al. The database of interacting proteins: 2004 update. Nucleic Acids Res. 2004;32:D449–51.
22. Keshava Prasad TS, et al. Human protein reference database-2009 update. Nucleic Acids Res. 2009;37:D767–72.
23. Aranda B, et al. The IntAct molecular interaction database in 2010. Nucleic Acids Res. 2010;38:D525–31.
24. Ceol A, et al. MINT, the molecular interaction database: 2009 update. Nucleic Acids Res. 2010;38:D532–9.
25. Guldener U, et al. MPact: the MIPS protein interaction resource on yeast. Nucleic Acids Res. 2006;34:D436–41.
26. Dall E, Brandstetter H. Structure and function of legumain in health and disease. Biochimie. 2015;122:126–50.

27. Zanoli L, et al. Arterial structure and function in inflammatory bowel disease. World J Gastroenterol. 2015;21:11304–11.
28. Selbig J, SteinfathM, Repsilber D. Network structure and biological function: reconstruction, modeling, and statistical approaches. EURASIP J Bioinform Syst Biol. 2009;2009:714985.
29. Maslov S, Sneppen K. Specificity and stability in topology of protein networks. Science. 2002;296:910–3.
30. Rives AW, Galitski T. Modular organization of cellular networks. Proc Natl Acad Sci U S A. 2003;100:1128–33.
31. Luo T, et al. Network cluster analysis of protein-protein interaction network identified biomarker for early onset colorectal cancer. Mol Biol Rep. 2013;40:6561–8.
32. Zanzoni A, Brun C. Integration of quantitative proteomics data and interaction networks: identification of dysregulated cellular functions during cancer progression. Methods. 2015;93:103–9.
33. Becker E, et al. Multifunctional proteins revealed by overlapping clustering in protein interaction network. Bioinformatics. 2012;28:84–90.
34. Ma S, et al. Incorporating gene co-expression network in identification of cancer prognosis markers. BMC Bioinf. 2010;11:271.
35. Zhao W, et al. Weighted gene coexpression network analysis: state of the art. J Biopharm Stat. 2010;20:281–300.
36. Rotival M, Petretto E. Leveraging gene co-expression networks to pinpoint the regulation of complex traits and disease, with a focus on cardiovascular traits. Brief Funct Genomics. 2014;13:66–78.
37. Lee DD, Seung HS. Learning the parts of objects by non-negative matrix factorization. Nature. 1999;401:788–91.
38. Zhang J, et al. Using gene co-expression network analysis to predict biomarkers for chronic lymphocytic leukemia. BMC Bioinf. 2010;11 Suppl 9:S5.
39. Hu H, et al. Mining coherent dense subgraphs across massive biological networks for functional discovery. Bioinformatics. 2005;21 Suppl 1:i213–21.
40. Yang Y, et al. Gene co-expression network analysis reveals common system-level properties of prognostic genes across cancer types. Nat Commun. 2014;5:3231.
41. Bartel DP. MicroRNAs: genomics, biogenesis, mechanism, and function. Cell. 2004;116:281–97.
42. Dar AA, et al. miRNA-205 suppresses melanoma cell proliferation and induces senescence via regulation of E2F1 protein. J Biol Chem. 2011;286:16606–14.
43. Shivdasani RA. MicroRNAs: regulators of gene expression and cell differentiation. Blood. 2006;108:3646–53.
44. Lindsay MA. microRNAs and the immune response. Trends Immunol. 2008;29:343–51.
45. Jian B et al. Downregulation of microRNA-193-3p inhibits tumor proliferation migration and chemoresistance in human gastric cancer by regulating PTEN gene. Tumour Biol. 2016.
46. Yan W, et al. Comparison of prognostic microRNA biomarkers in blood and tissues for gastric cancer. J Cancer. 2016;7:95–106.
47. Kong XM, et al. MicroRNA-140-3p inhibits proliferation, migration and invasion of lung cancer cells by targeting ATP6AP2. Int J Clin Exp Pathol. 2015;8:12845–52.
48. Chou CH, et al. miRTarBase 2016: updates to the experimentally validated miRNA-target interactions database. Nucleic Acids Res. 2016;44:D239–47.
49. Vlachos IS, et al. DIANA-TarBase v7.0: indexing more than half a million experimentally supported miRNA:mRNA interactions. Nucleic Acids Res. 2015;43:D153–9.
50. Xiao F, et al. miRecords: an integrated resource for microRNA-target interactions. Nucleic Acids Res. 2009;37:D105–10.
51. Jiang Q, et al. miR2Disease: a manually curated database for microRNA deregulation in human disease. Nucleic Acids Res. 2009;37:D98–104.
52. Gennarino VA, et al. HOCTAR database: a unique resource for microRNA target prediction. Gene. 2011;480:51–8.

53. Gamazon ER, et al. Exprtarget: an integrative approach to predicting human microRNA targets. PLoS One. 2010;5:e13534.
54. Li JH, et al. starBase v2.0: decoding miRNA-ceRNA, miRNA-ncRNA and protein-RNA interaction networks from large-scale CLIP-Seq data. Nucleic Acids Res. 2014;42:D92–7.
55. Lewis BP, Burge CB, Bartel DP. Conserved seed pairing, often flanked by adenosines, indicates that thousands of human genes are microRNA targets. Cell. 2005;120:15–20.
56. John B, et al. Human microRNA targets. PLoS Biol. 2004;2:e363.
57. Bandyopadhyay S, et al. Development of the human cancer microRNA network. Silence. 2010;1:6.
58. Xu J, et al. Prioritizing candidate disease miRNAs by topological features in the miRNA target-dysregulated network: case study of prostate cancer. Mol Cancer Ther. 2011;10:1857–66.
59. Zhang S, et al. A novel computational framework for simultaneous integration of multiple types of genomic data to identify microRNA-gene regulatory modules. Bioinformatics. 2011;27:i401–9.
60. Zhu J, et al. Screening key microRNAs for castration-resistant prostate cancer based on miRNA/mRNA functional synergistic network. Oncotarget. 2015;6:43819–30.
61. Yan W, et al. Identification of microRNAs as potential biomarker for gastric cancer by system biological analysis. BioMed Res Int. 2014;2014:901428.
62. Huang J, et al. Identification of microRNA as sepsis biomarker based on miRNAs regulatory network analysis. BioMed Res Int. 2014;2014:594350.
63. Zhao XM, et al. Identifying cancer-related microRNAs based on gene expression data. Bioinformatics. 2015;31:1226–34.
64. Chen B, et al. MicroRNA-346 functions as an oncogene in cutaneous squamous cell carcinoma. Tumour Biol. 2015;37(2):2765–71.
65. Song N, et al. microRNA-107 functions as a candidate tumor suppressor gene in renal clear cell carcinoma involving multiple genes. Urol Oncol. 2015;33(205):e201–11.
66. Sumazin P, et al. An extensive microRNA-mediated network of RNA-RNA interactions regulates established oncogenic pathways in glioblastoma. Cell. 2011;147:370–81.
67. Salmena L, et al. A ceRNA hypothesis: the Rosetta Stone of a hidden RNA language? Cell. 2011;146:353–8.
68. Marques AC, Tan J, Ponting CP. Wrangling for microRNAs provokes much crosstalk. Genome Biol. 2011;12:132.
69. Su X, et al. microRNAs and ceRNAs: RNA networks in pathogenesis of cancer. Chin J Cancer Res. 2013;25:235–9.
70. Tay Y, et al. Coding-independent regulation of the tumor suppressor PTEN by competing endogenous mRNAs. Cell. 2011;147:344–57.
71. Zhou X, Liu J, Wang W. Construction and investigation of breast-cancer-specific ceRNA network based on the mRNA and miRNA expression data. IET Syst Biol. 2014;8:96–103.
72. Li ZC et al. Identification of drug-target interaction from interactome network with 'guilt-by-association' principle and topology features. Bioinformatics. 2015.
73. Duarte NC, et al. Global reconstruction of the human metabolic network based on genomic and bibliomic data. Proc Natl Acad Sci U S A. 2007;104:1777–82.
74. Cheung N, et al. Targeting aberrant epigenetic networks mediated by PRMT1 and KDM4C in acute myeloid leukemia. Cancer Cell. 2016;29:32–48.
75. Chen J, Sun M, Shen B. Deciphering oncogenic drivers: from single genes to integrated pathways. Brief Bioinform. 2015;16:413–28.
76. Barry MJ. Clinical practice. Prostate-specific-antigen testing for early diagnosis of prostate cancer. N Engl J Med. 2001;344:1373–7.
77. Ford D, et al. Genetic heterogeneity and penetrance analysis of the BRCA1 and BRCA2 genes in breast cancer families. The breast cancer linkage consortium. Am J Hum Genet. 1998;62:676–89.

78. Tsujiura M, et al. Circulating miR-18a in plasma contributes to cancer detection and monitoring in patients with gastric cancer. Gastric Cancer. 2015;18:271–9.
79. Ge W, et al. Expression of serum miR-16, let-7f, and miR-21 in patients with hepatocellular carcinoma and their clinical significances. Clin Lab. 2014;60:427–34.
80. Kojima S, et al. [MiRNA profiling in prostate cancer], Nihon rinsho. Japanese J Clin Med. 2011;69 Suppl 5:92–5.
81. Cun Y, Frohlich H. netClass: an R-package for network based, integrative biomarker signature discovery. Bioinformatics. 2014;30:1325–6.
82. Ein-Dor L, Zuk O, Domany E. Thousands of samples are needed to generate a robust gene list for predicting outcome in cancer. Proc Natl Acad Sci U S A. 2006;103:5923–8.
83. Chen H, et al. Pathway mapping and development of disease-specific biomarkers: protein-based network biomarkers. J Cell Mol Med. 2015;19:297–314.
84. Vangala RK, et al. Novel network biomarkers profile based coronary artery disease risk stratification in Asian Indians. Adv Biomed Res. 2013;2:59.
85. Chowdhury SA, et al. Subnetwork state functions define dysregulated subnetworks in cancer. J Comput Biol. 2011;18:263–81.
86. Wang YC, Chen BS. A network-based biomarker approach for molecular investigation and diagnosis of lung cancer. BMC Med Genet. 2011;4:2.
87. Chuang HY, et al. Network-based classification of breast cancer metastasis. Mol Syst Biol. 2007;3:140.
88. Lu L, Li Y, Li S. Computational identification of potential microRNA network biomarkers for the progression stages of gastric cancer. Int J Data Min Bioinform. 2011;5:519–31.
89. Cui W, et al. Discovery and characterization of long intergenic non-coding RNAs (lincRNA) module biomarkers in prostate cancer: an integrative analysis of RNA-Seq data. BMC Genomics. 2015;16 Suppl 7:S3.
90. Li M, et al. Detecting tissue-specific early warning signals for complex diseases based on dynamical network biomarkers: study of type 2 diabetes by cross-tissue analysis. Brief Bioinform. 2014;15:229–43.
91. Liu R, et al. Early diagnosis of complex diseases by molecular biomarkers, network biomarkers, and dynamical network biomarkers. Med Res Rev. 2014;34:455–78.
92. Liu X, et al. Detecting early-warning signals of type 1 diabetes and its leading biomolecular networks by dynamical network biomarkers. BMC Med Genet. 2013;6 Suppl 2:S8.
93. Wong YH, Li CW, Chen BS. Evolution of network biomarkers from early to late stage bladder cancer samples. BioMed Res Int. 2014;2014:159078.
94. Wong YH, et al. Applying NGS data to find evolutionary network biomarkers from the early and late stages of hepatocellular carcinoma. BioMed Res Int. 2015;2015:391475.
95. Barabasi AL, Gulbahce N, Loscalzo J. Network medicine: a network-based approach to human disease. Nat Rev Genet. 2011;12:56–68.
96. Chen B, Butte AJ. Network medicine in disease analysis and therapeutics. Clin Pharmacol Ther. 2013;94:627–9.
97. Chan SY, Loscalzo J. The emerging paradigm of network medicine in the study of human disease. Circ Res. 2012;111:359–74.
98. Li Y, et al. HMDD v2.0: a database for experimentally supported human microRNA and disease associations. Nucleic Acids Res. 2014;42:D1070–4.
99. Chung IF, et al. DriverDBv2: a database for human cancer driver gene research. Nucleic Acids Res. 2016;44:D975–9.
100. Baffy G. The impact of network medicine in gastroenterology and hepatology. Clin Gastroenterol Hepatol. 2013;11:1240–4.

MIX
Papier aus verantwortungsvollen Quellen
Paper from responsible sources
FSC® C105338

If you have any concerns about our products,
you can contact us on
ProductSafety@springernature.com

In case Publisher is established outside the EU,
the EU authorized representative is:
Springer Nature Customer Service Center GmbH
Europaplatz 3, 69115 Heidelberg, Germany

Printed by Libri Plureos GmbH
in Hamburg, Germany